Advances in Operational Researches

Advances in Operational Researches

Edited by:

Jovan Pehcevski

ARCLER
PRESS

www.arclerpress.com

Advances in Operational Researches

Jovan Pehcevski

Arcler Press
224 Shoreacres Road
Burlington, ON L7L 2H2
Canada
www.arclerpress.com
Email: orders@arclereducation.com

ISBN: 978-1-77469-175-5 (Hardcover)

DECLARATION

Some content or chapters in this book are open access copyright free published research work, which is published under Creative Commons License and are indicated with the citation. We are thankful to the publishers and authors of the content and chapters as without them this book wouldn't have been possible.

ABOUT THE EDITOR

Jovan obtained his PhD in Computer Science from RMIT University in Melbourne, Australia in 2007. His research interests include big data, business intelligence and predictive analytics, data and information science, information retrieval, XML, web services and service-oriented architectures, and relational and NoSQL database systems. He has published over 30 journal and conference papers and he also serves as a journal and conference reviewer. He is currently working as a Dean and Associate Professor at European University in Skopje, Macedonia.

TABLE OF CONTENTS

Section 3: Operational Research Applications in Logistics

Section 4: Operational Research Applications in Logistics

LIST OF CONTRIBUTORS

Farhana Ahmed Simi
Department of Mathematics, Dhaka University, Dhaka, Bangladesh

Md. Shahjalal Talukder
Department of Natural Sciences, Daffodil International University, Dhaka, Bangladesh

Mahboobeh Honarvar
Department of Industrial Engineering, Yazd University, Yazd, Iran

Majid Alimohammadi Ardakani
Department of Industrial Engineering, Ardakan University, Yazd, Iran

Mohammad Modarres
Department of Industrial Engineering, Sharif University of Technology, Tehran, Iran

Robert Rich
Industrial & Systems Engineering, Liberty University, Lynchburg, VA, USA

Alireza Noroziroshan
IMSE Department, The University of Texas at Arlington, Arlington, USA

H. W. Corley
IMSE Department, The University of Texas at Arlington, Arlington, USA

Jay M. Rosenberger
IMSE Department, The University of Texas at Arlington, Arlington, USA

Hochung Liu
College of Intelligence Engineering and Mathematics, Liaoning Technical University, Fuxin, China

Peizhuang Wang
College of Intelligence Engineering and Mathematics, Liaoning Technical University, Fuxin, China

Zhengang Shu
School of Business and Administration, South China University of Technology, Guangzhou, China

Xiaoli Wu
School of Business and Administration, South China University of Technology, Guangzhou, China

Ming Meng
Department of Economics and Management, North China Electric Power University, Baoding, Hebei 071003, China

Yun Ma
Department of Economics and Management, North China Electric Power University, Baoding, Hebei 071003, China

Deo Chimba
Department of Civil Engineering, Tennessee State University, Nashville, USA

Daniel Emaasit
Department of Civil Engineering, Tennessee State University, Nashville, USA

Boniphace Kutela
Department of Civil Engineering, Tennessee State University, Nashville, USA

Jing He
Southwest Jiaotong University, GIS Engineering Center, Chengdu, China

Yongshu Li
Southwest Jiaotong University, GIS Engineering Center, Chengdu, China

Keke Zhang
Engineering Design Co., Ltd., CNPC, Chengdu, China

M. C. Agarana
Department of Mathematics, College of Science and Technology, Covenant University, Ota, Nigeria

T. A. Anake
Department of Mathematics, College of Science and Technology, Covenant University, Ota, Nigeria

H. I. Okagbue
Department of Mathematics, College of Science and Technology, Covenant University, Ota, Nigeria

Farahanim Misni
Department of Mathematics, Faculty of Science, Universiti Putra Malaysia, Serdang, Selangor, Malaysia.
Faculty of Sciences & Technology, Universiti Malaysia Pahang, Gambang, Kuantan, Malaysia.

Lai Soon Lee
Department of Mathematics, Faculty of Science, Universiti Putra Malaysia, Serdang, Selangor, Malaysia.
Laboratory of Computational Statistics and Operations Research, Institute for Mathematical Research, Universiti Putra Malaysia, Serdang, Selangor, Malaysia.

Jiuyi An
Beijing Wuzi University, Beijing, China

Lei Wang
Beijing Benz Automotive Co. Ltd., Beijing, China

Xiaohua Lv
Beijing Wuzi University, Beijing, China

Otieno Robert Kennedy
Transportation Management College, Dalian Maritime University, Dalian, China

Khin Lin
Transportation Management College, Dalian Maritime University, Dalian, China

Hualong Yang
Management College, Dalian Maritime University, Dalian, China

Banomyong Ruth
Department of International Business, Logistics & Transport, Thammasat University, Bangkok, Thailand

Bashkim Çerkini
Faculty of Engineering and Informatics, University of Applied Sciences in Ferizaj, Ferizaj, Kosovo

Fakije Zejnullahu
Faculty of Engineering and Informatics, University of Applied Sciences in Ferizaj, Ferizaj, Kosovo

Dhuratë Hyseni
Faculty of Engineering and Informatics, University of Applied Sciences in Ferizaj, Ferizaj, Kosovo

Morteza Seddighin
Mathematics Department, Indiana University East, Richmond, IN, USA

Enrico Del Re
University of Florence, Florence, Italy
CNIT, Florence Research Unit, Florence, Italy

Renato Pucci
University of Florence, Florence, Italy
CNIT, Florence Research Unit, Florence, Italy

Luca Simone Ronga
CNIT, Florence Research Unit, Florence, Italy

Ahmad Nahhas
Center on Stochastic Modeling, Optimization, and Statistics, The University of Texas at Arlington, Arlington, TX, USA

H. W. Corley
Center on Stochastic Modeling, Optimization, and Statistics, The University of Texas at Arlington, Arlington, TX, USA

Chuma R. Nwozo
Department of Mathematics, University of Ibadan, Ibadan, Nigeria

Charles I. Nkeki
Department of Mathematics, University of Benin, Benin City, Nigeria

LIST OF ABBREVIATIONS

ACA	Ant Colony Algorithm
ADT	Annual Daily Traffic
AVE	Average Variance Extracted
BPL	Bilevel programming
BPR	Bureau of Public Roads
COST	Constraint optimal selection technique
CPACA	Coarse-Grain Parallel Ant Colony Algorithm
DO	Dynamic optimization
EEIWF	Energy Efficient Iterative Water Filling
ELV	End-of-life vehicle
EMA	Enterprise du Metro d'Alger's
EA	Evolutionary algorithm
ENR	Egyptian National Railway
EVRPTW	Electric vehicle routing problem with time windows
EVs	Electric vehicles
GAMS	Generalized algebraic modeling system
GAP	Good agricultural practices
GSCND	Green supply chain network design
GVRP	Green vehicle routing problem
GSD	Ground sampled distance
IFI	Incremental Fit Index
ISO	International Standard Organization
LCA	Life cycle assessment
LFP	Linear fractional programming
LIPS	Linear programming solver
LOS	Level of Service
LP	Linear program
LR	Likelihood Ratio

MILP	Mixed-integer linear programming model
MIP	Mixed integer programming
MST	Minimum spanning tree
NE	Nash equilibrium
NFI	Normed Fit Index
NNLP	Nonnegative linear program
O-D	Origin-destination
OEM	Original equipment manufacturer
PCA	Principal component analysis
PIGA	Priority-based encoding genetic algorithm
PPP	Public Private Partnership
PSO	Particle Swarm Optimization
RAG	Resource allocation games
RMSE	Root-Mean-Square Error Approximation
SA	Simulated Annealing
SCND	Supply chain network design
TLI	Tucker-Lewis Index
SINR	Signal to noise plus interference ratio
SRMR	Standardized Root Mean Residual
SUE	Stochastic User Equilibrium
UAV	Unmanned aerial vehicle
UE	User equilibrium
VNW	Von Neumann-Morgenstern
VRP	Vehicle routing problem
VRPTW	Vehicle routing problem with time windows
WEEE	Waste of electrical and electronic equipment

PREFACE

Operational research is a set of quantitative and other scientific methods used to determine optimal economic and technical solutions to complex problems. Operational researches are mathematical disciplines, but at the same time they are also basic disciplines in management. They are named after the research of operations in organizational systems with the purpose of their optimization. They were first developed for military purposes, and only later was their usefulness in managing business systems noticed.

Known areas of operational research include: linear programming, nonlinear programming, integer programming, transport problems, network optimization, multicriteria optimization, etc. As a rule, when modeling specific problems, the research operates with many quantities (variables) upon which appropriate restrictions are imposed. The set of those values of given quantities – that satisfy the set system of constraints – is called the set of admissible solutions.

The set of admissible solutions can have many elements, even an infinite number of admissible solutions. To choose an admissible solution, we must know the criterion on the basis of which it can be concluded, so we can decide whether one admissible solution is better than the other.

The problems that are most often solved by applying linear programming are generally problems of allocating limited resources between competing activities in the best (optimal) way. The word programming in the name LP does not refer to computer programming (as in writing programs), but is a synonym for planning. This means that linear programming involves planning activities to get the optimal result, i.e., the best result from possible solutions according to a set goal, or a so called goal function.

This edition covers different topics from advanced operational researches, including: optimization and linear programming, graph-based techniques and route planning, operations research in logistics problems, and methods for resource allocation.

Section 1 focuses on optimization and linear programming, describing a new approach for solving linear fractional programming problems with duality concept, a particle swarm optimization algorithm for solving pricing and lead time quotation in a dual-channel supply chain with multiple customer classes, inverting the multiple-assisting tool network problem to solve for optimality, a dynamic active-set method for linear programming, and the sliding gradient algorithm for linear programming.

Section 2 focuses on graph-based techniques and route planning, describing route optimization of electric vehicle considering soft time windows and two ways of power replenishment, the route planning on campus bus in a University, integrating origin-destination survey and stochastic user equilibrium as a case study for route relocation, and research of UAV flight planning parameters.

Section 3 focuses on operations research in logistics problems, describing optimization of urban rail transportation in emerging countries using operational research techniques, a review on strategic, tactical and operational decision planning in reverse logistics of green supply chain network design, research on agri-food cold chain logistics management system, a use case of sea-port operational efficiency by an evaluation of five Asian ports using stochastic frontier production function model, and the design of urban traffic in Ferizaj through operational research.

Section 4 focuses on methods for resource allocation, describing application of the two non-zero component lemma in resource allocation, energy efficient non-cooperative methods for resource allocation in cognitive radio networks, an alternative interpretation of mixed strategies in n-person normal form games via resource allocation, and a dynamic optimization technique for resource allocation problems in a production company.

Section 1: Optimization and Linear Programming

CHAPTER 1

A New Approach for Solving Linear Fractional Programming Problems with Duality Concept

Farhana Ahmed Simi[1], Md. Shahjalal Talukder[2]

[1]Department of Mathematics, Dhaka University, Dhaka, Bangladesh.

[2]Department of Natural Sciences, Daffodil International University, Dhaka, Bangladesh.

ABSTRACT

Most of the current methods for solving linear fractional programming (LFP) problems depend on the simplex type method. In this paper, we present a new approach for solving linear fractional programming problem in which the objective function is a linear fractional function, while constraint functions are in the form of linear inequalities. This approach does not depend on the simplex type method. Here first we transform this LFP problem into linear

Citation: Ahmed Simi, F. and Talukder, M. (2017), A New Approach for Solving Linear Fractional Programming Problems with Duality Concept. Open Journal of Optimization, 6, 1-10. doi: 10.4236/ojop.2017.61001.

programming (LP) problem and hence solve this problem algebraically using the concept of duality. Two simple examples to illustrate our algorithm are given. And also we compare this approach with other available methods for solving LFP problems.

Keywords:- Linear Fractional Programming, Linear Programming, Duality

INTRODUCTION

The linear fractional programming (LFP) problem has attracted the interest of many researches due to its application in many important fields such as production planning, financial and corporate planning, health care and hospital planning.

Several methods were suggested for solving LFP problem such as the variable transformation method introduced by Charnes and Cooper [1] and the updated objective function method introduced by Bitran and Novaes [2] . The first method transforms the LFP problem into an equivalent linear programming problem and uses the variable transformation $y = tx, t \geq 0$ in such a way that $dt + \beta = \gamma$ where $\gamma \neq 0$ is a specified number and transform LFP to an LP problem. And the second method solves a sequence of linear programming pro- blems depending on updating the local gradient of the fractional objective function at successive points. But to solve this sequence of problems, sometimes may need much iteration. Also some aspects concerning duality and sensitivity analysis in linear fractional program were discussed by Bitran and Magnant [3] and Singh [4] , in his paper made a useful study about the optimality condition in fractional programming. Assuming the positivity of denominator of the objective function of LFP over the feasible region, Swarup [5] extended the well- known simplex method to solve the LFP. This process cannot continue infinitely, since there is only a finite number of basis and in non-degenerate case, no basis can ever be repeated, since F is increased at every step and the same basis cannot yield two different values of F. While at the same time the maximum value of the objective function occurs at of the basic feasible solution. Recently, Tantawy [6] has suggested a feasible direction approach and the main idea behind this method for solving LFP problems is to move through the feasible region via a sequence of points in the direction that improves the objective function. Tantawy [7] also proposed a duality approach to solve a linear fractional programming problem. Tantawy [8] develops another technique for solving LFP which can be used for sensitivity analysis. Effati and Pakdaman [9] propose a method for solving interval-valued linear fractional programming problem. A method for solving multi objective linear plus linear fractional programming problem based on Taylor series approximation is proposed by Pramanik et al. [10] . Tantawy and Sallam [11] also propose a new method for solving linear programming problems.

In this paper, our main intent is to develop an approach for solving linear fractional programming problem which does not depend on the simplex type method because method based on vertex information may have difficulties as the problem size increases; this method may prove to be less sensitive to problem size. In this paper, first of all, a linear fractional programming problem is transformed into linear programming problem by choosing an initial feasible point and hence solves this problem algebraically using the concept of duality.

DEFINITION AND METHOD OF SOLVING LFP

A linear fractional programming problem occurs when a linear fractional function is to be maximized and the problem can be formulated mathematically as follows:

Maximize $F(x)=\frac{c^T x+\gamma}{d^T x+\beta}$.

Subject to,

$$x \in X=\{x : Ax \le b\}, x \ge 0 \tag{1}$$

where c, d and $x \in \mathbb{R}^n$, A is an $(m+n)n$ matrix, $b \in \mathbb{R}^{m+n}$ and γ and β are scalars.

We point out that the nonnegative conditions are included in the set of constraints and that $d^T x+\beta > 0$ has to be satisfied over the compact set X.

To transform the LFP problem into LP problem, we choose a feasible point x^* of the compact set X. Then

$$F^* = F(x^*) = \frac{c^T x^* + \gamma}{d^T x^* + \beta} \tag{2}$$

is a given constant vector computed at a given feasible point x^*. Thus the level curve of objective function for (1) can be written as

$$(c^T - F^* d^T)x = \beta F^* - \gamma$$

Hence the linear programming problem is as follows:

Maximize $\varphi(x)=\left(c^{\mathrm{T}}-F^{*}d^{\mathrm{T}}\right)x$

Subject to,

$$x\in X=\{x:Ax\leq b\},x\geq 0 \tag{3}$$

Proposition

If x^{*} solves the LFP problem (1) with objective function values F^{*} then x^{*} solves the LP problem defined by (3) with objective function value $\varphi^{*}=\beta F^{*}-\gamma$.

Now rewrite the LP problem (3) in the form

Maximize $H(x)=C^{\mathrm{T}}x$

Subject to,

$$x\in X=\{x:Ax\leq b\} \tag{4}$$

where, C^{T} is a matrix whose row is represented by $\left(c^{\mathrm{T}}-F^{*}d^{\mathrm{T}}\right)$ and $C,x\in\mathbb{R}^{n}$, A is a $(m+n)\times n$ matrix, $b\in\mathbb{R}^{m+n}$ we point out that the nonnegative conditions are included in the set of constraints.

Now consider the dual problem for the linear program (4) in the form

Minimize $w=u^{\mathrm{T}}b$

Subject to,

$$u^{\mathrm{T}}A=C^{\mathrm{T}},u\geq 0 \tag{5}$$

Since the set of constraints of this dual problem is written in the matrix form hence we can multiply both side by a matrix $T=(T_1|T_2)$, where $T_1=C\left(C^{\mathrm{T}}C\right)^{-1}$ and the columns of the matrix T_2 constitute the bases of $\{x:C^{\mathrm{T}}x=0\}$.

Thus this implies

$$u^{\mathrm{T}}AT_1=1,\ u^{\mathrm{T}}AT_2=0 \text{ and } u\geq 0. \tag{6}$$

If we define $l\times(m+n)$ matrix P of nonnegative entries such that $PAT_2=0$, then (6) can be written as

$v^{\mathrm{T}}G=1,v\geq 0$ (7)

where $G = PAT_1$ and $v^T P = u^T$, Equation (7) will play an important role for finding the optimal solution of the LP problem (4). Using the Equation (7) the equivalent LP problem of (5) can be written as

Minimize $w = v^T g$

Subject to,

$$v^T G = 1,\ v \geq 0 \tag{8}$$

With $G = PAT_1, g = Pb, v^T P = u^T$, the linear programming (8) has the dual programming problem in just one unknown Z in the form.

Maximize Z

Subject to,

$$GZ \leq g,\ Z \geq 0 \tag{9}$$

Note: The set of constraints of the above linear programming problem will give the maximum value z^* and also will define only one active constraint for this optimal value. We have to note that from the complementary slackness theorem the corresponding dual variable will be positive and the remaining dual variables will be zeros for the corresponding non active constraints.

ALGORITHM FOR SOLVING LFP PROBLEMS

The method for solving LFP problems summarize as follows:

- Step 1: Select a feasible point x^* and using Equation (2) to compute F^*.
- Step 2: Find the level curve of objective function

$$\left(c^T - F^* d^T\right)x = \beta F^* - \gamma$$

Hence find the LP problem (2) which can be rewritten as (3).

- Step 3: Compute $T_1 = C\left(C^T C\right)^{-1}$, and the matrix T_2 as the bases of $\{x : C^T x = 0\}$.
- Step 4: Find the matrix P of nonnegative entries such that $PAT_2 = 0$ and hence compute $G = PAT_1,\ g = Pb$.
- Step 5: Find the LP problem (8) and dual of this LP (9). Use the LP (9) to find the optimal value z^* and also determine the corresponding active constraints and use the constraint of (8) to compute v^T.

- Step 6: Find the dual variables $u^{\mathrm{T}} = v^{\mathrm{T}}P$, for each positive variable $u_i, i = 1, 2, \cdots, m$ find the corresponding active set of constraint of the matrix A.
- Step 7: Solve a n×n system of linear equations for these set of active constraints (a subset from a m+n constraints) to get the optimal solution of LP problem (4) and hence for the LFP problem (1).

COMPUTATIONAL PROCESS

Choose x^* in such a way that

$$x^* \in X = \{x^* : Ax^* \le b\}$$

$$F^* \leftarrow F(x^*) \leftarrow \frac{c^{\mathrm{T}}x^* + \gamma}{d^{\mathrm{T}}x^* + \beta}$$

$$d^{\mathrm{T}}x^* + \beta > 0.$$

The level curve is $(c^{\mathrm{T}} - F^*d^{\mathrm{T}})x = \beta F^* - \gamma$.

Then $\varphi(x) \leftarrow (c^{\mathrm{T}} - F^*d^{\mathrm{T}})x$ or $H(x) \leftarrow C^{\mathrm{T}}x$; $C^{\mathrm{T}} \leftarrow (c^{\mathrm{T}} - F^*d^{\mathrm{T}})$;

$T_1 \leftarrow C(C^{\mathrm{T}}C)^{-1}$; $T_2 \leftarrow \{x : C^{\mathrm{T}}x = 0\}$;

Find P such that $PAT_2 = 0$..

Compute $G \leftarrow PAT_1, g \leftarrow Pb$;

Formulate, Maximize Z

Subject to, $GZ \le g,\ Z \ge 0.$

Find Z^* and corresponding active constraint and compute v^{T} for $v^{\mathrm{T}}G = 1$;

Then $u^{\mathrm{T}} \leftarrow v^{\mathrm{T}}P$; hence find x^{T} from corresponding n×n active constraints satisfied by positive u^{T};

Compute H^* and F^*.

NUMERICAL EXAMPLES

Here we illustrate two examples to demonstrate our method.

Example 1: Consider the linear fractional programming (LFP) problem

Maximize $F(x) = \dfrac{x_2 + 1}{x_1 + 3}$

Subject to,

$$-x_1 + x_2 \le 1$$

$$x_2 \le 2$$

$$x_1 + 2x_2 \le 1$$

$$x_1 \le 5$$

$$x_1, x_2 \ge 0.$$

Solution:

- Step 1: Let $x^* = \begin{pmatrix} 1 \\ 1 \end{pmatrix}$, then $F^* = \frac{1+1}{1+3} = \frac{1}{2}$ and hence we have

$$\left(c^{\mathrm{T}} - F^* d^{\mathrm{T}}\right)x = \left[(0 \quad 1) - \frac{1}{2}(1 \quad 0)\right]\begin{pmatrix} x_1 \\ x_2 \end{pmatrix} = -\frac{1}{2}x_1 + x_2$$

- Step 2: Therefore we have the following LP problem

Maximize $H(x) = -\frac{1}{2}x_1 + x_2$

Subject to,

$$-x_1 + x_2 \le 1$$

$$x_2 \le 2$$

$$x_1 + 2x_2 \le 1$$

$$x_1 \le 5$$

$$-x_1 \le 0$$

$$-x_2 \le 0$$

Dual problem for this LP problem is

Minimize $w(x) = u_1 + 2u_2 + 7u_3 + 5u_4$

Subject to,

$$-u_1 + u_3 + u_4 - u_5 = \frac{1}{2}$$

$$u_1 + u_2 + 2u_3 - u_6 = 1$$

$$u_1, u_2, u_3, u_4, u_5, u_6 \geq 0$$

- Step 3: Compute $T_1 = \begin{pmatrix} -\frac{1}{2} \\ 1 \end{pmatrix} \left[\begin{pmatrix} -\frac{1}{2} & 1 \end{pmatrix} \begin{pmatrix} -\frac{1}{2} \\ 1 \end{pmatrix} \right]^{-1} = \frac{4}{5} \begin{pmatrix} -\frac{1}{2} \\ 1 \end{pmatrix} = \begin{pmatrix} -\frac{2}{5} \\ \frac{4}{5} \end{pmatrix}.$.

And the matrix $T_2 = \begin{pmatrix} 2 \\ 1 \end{pmatrix}$.

- Step 4: Compute nonnegative matrix P such that $PAT_2 = 0$,

$$P = \begin{pmatrix} 1 & 1 & 0 & 0 & 0 & 0 \\ 1 & 0 & 0 & 1 & 0 & 1 \\ 0 & 1 & 0 & 0 & 0 & 1 \\ 0 & 0 & 1 & 0 & 2 & 0 \\ 0 & 0 & 0 & 1 & 1 & 0 \\ 0 & 0 & 1 & 0 & 1 & 2 \end{pmatrix}.$$

Also compute

$$G = PAT_1 = \begin{pmatrix} 1 & 1 & 0 & 0 & 0 & 0 \\ 1 & 0 & 0 & 1 & 0 & 1 \\ 0 & 1 & 0 & 0 & 0 & 1 \\ 0 & 0 & 1 & 0 & 2 & 0 \\ 0 & 0 & 0 & 1 & 1 & 0 \\ 0 & 0 & 1 & 0 & 1 & 2 \end{pmatrix} \begin{pmatrix} -1 & 1 \\ 0 & 1 \\ 1 & 2 \\ 1 & 0 \\ -1 & 0 \\ 0 & -1 \end{pmatrix} \begin{pmatrix} -\frac{1}{2} \\ 1 \end{pmatrix} = \begin{pmatrix} 2 \\ 0 \\ 0 \\ 2 \\ 0 \\ 0 \end{pmatrix},$$

$$g = Pb = \begin{pmatrix} 1 & 1 & 0 & 0 & 0 & 0 \\ 1 & 0 & 0 & 1 & 0 & 1 \\ 0 & 1 & 0 & 0 & 0 & 1 \\ 0 & 0 & 1 & 0 & 2 & 0 \\ 0 & 0 & 0 & 1 & 1 & 0 \\ 0 & 0 & 1 & 0 & 1 & 2 \end{pmatrix} \begin{pmatrix} 1 \\ 2 \\ 1 \\ 5 \\ 0 \\ 0 \end{pmatrix} = \begin{pmatrix} 3 \\ 6 \\ 2 \\ 7 \\ 5 \\ 7 \end{pmatrix}$$

- Step 5: We get the LP problem of the form

Maximize Z

Subject to,

$2Z \le 3$

$0Z \le 6$

$0Z \le 2$

$2Z \le 7$

$0Z \le 5$

$0Z \le 7$

For this LP problem we get that the first constraint is the only active constraint and this active constraint shows that the maximum optimal value is

$Z^* = \frac{3}{2}$. Corresponding this active constraint of (8), we get the dual variables $v^T = \left(\frac{1}{2}, 0, 0, 0, 0, 0\right)$

Step 6: Compute $u^T = v^T P = \left(\frac{1}{2}, \frac{1}{2}, 0, 0, 0, 0\right)$ with objective value $w^* = \frac{3}{2}$. This indicates that in the original set of constraints the first and the second constraints are the only active constraints.

- Step 7: Solve the system of linear equations

$-x_1 + x_2 = 1$

$x_2 = 2$

We get the optimal solution $x^* = \begin{pmatrix} 1 \\ 2 \end{pmatrix}$ of the LP problem with objective value $H^* = \frac{3}{2}$.

Finally we get our desired optimal solution of the given LFP problem is $x^* = \begin{pmatrix} 1 \\ 2 \end{pmatrix}$ with the optimal value $F^* = \frac{3}{4}$.

Example 2: Consider the linear fractional programming (LFP) problem

Maximize $F(x) = \dfrac{5x_1 + 3x_2}{5x_1 + 2x_2 + 1}$

Subject to,

$3x_1 + 5x_2 \le 15$

$5x_1 + 2x_2 \le 10$

$x_1, x_2 \ge 0$

Solution:

- Step 1: Let $x^* = \begin{pmatrix} 1 \\ 1 \end{pmatrix}$, then $F^* = \frac{5+3}{5+2+1} = 1$ and hence we have

$$\left(c^{\mathrm{T}} - F^* d^{\mathrm{T}}\right)x = \left[(5 \quad 3) - 1*(5 \quad 2)\right]\begin{pmatrix} x_1 \\ x_2 \end{pmatrix} = x_2$$

- Step 2: Therefore we have the following LP problem

Maximize $H(x) = x_2$

Subject to,

$$3x_1 + 5x_2 \le 15$$

$$5x_1 + 2x_2 \le 10$$

$$-x_1 \le 0$$

$$-x_2 \le 0$$

Dual problem for this LP problem is

Minimize $w(x) = 15u_1 + 10u_2$

Subject to,

$$3u_1 + 5u_2 - u_3 = 0$$

$$5u_1 + 2u_2 - u_4 = 1$$

$$u_1, u_2, u_3, u_4 \ge 0$$

- Step 3: Compute $T_1 = \begin{pmatrix} 0 \\ 1 \end{pmatrix}\left[(0 \quad 1)\begin{pmatrix} 0 \\ 1 \end{pmatrix}\right]^{-1} = \begin{pmatrix} 0 \\ 1 \end{pmatrix}$.

And the matrix $T_2 = \begin{pmatrix} 1 \\ 0 \end{pmatrix}$.

- Step 4: Compute nonnegative matrix P such that $PAT_2 = 0$,

$$P = \begin{pmatrix} 1 & 0 & 3 & 0 \\ 0 & 1 & 5 & 0 \\ 0 & 1 & 6 & 1 \\ 0 & 0 & 1 & 1 \end{pmatrix}.$$

.

Also compute

$$G = PAT_1 = \begin{pmatrix} 1 & 0 & 3 & 0 \\ 0 & 1 & 5 & 0 \\ 0 & 1 & 6 & 1 \\ 0 & 0 & 1 & 1 \end{pmatrix}\begin{pmatrix} 3 & 5 \\ 5 & 2 \\ -1 & 0 \\ 0 & -1 \end{pmatrix}\begin{pmatrix} 0 \\ 1 \end{pmatrix} = \begin{pmatrix} 5 \\ 2 \\ 1 \\ -1 \end{pmatrix},$$

$$g = Pb = \begin{pmatrix} 1 & 0 & 3 & 0 \\ 0 & 1 & 5 & 0 \\ 0 & 1 & 6 & 1 \\ 0 & 0 & 1 & 1 \end{pmatrix}\begin{pmatrix} 15 \\ 10 \\ 0 \\ 0 \end{pmatrix} = \begin{pmatrix} 15 \\ 10 \\ 10 \\ 0 \end{pmatrix}$$

- Step 5: We get the LP problem of the form

Maximize Z

Subject to,

$5Z \leq 15$

$2Z \leq 10$

$Z \leq 10$

$-Z \leq 0$

For this LP problem we get that the first constraint is the only active constraint and this active constraint shows that the maximum optimal value is $Z^* = 3$. Corresponding to this active constraint of (8), we get the dual variables

$v^T = \left(\frac{1}{5}, 0, 0, 0, 0, 0\right).$

- Step 6: Compute $u^T = v^T P = \left(\frac{1}{5}, 0, \frac{3}{5}, 0, 0, 0\right)$ with objective value $w^* = 3$.

This indicates that in the original set of constraints the first and the third constraints are the only active constraints.

- Step 7: Solve the system of linear equations

$-3x_1 + 5x_2 = 15$

$x_1 = 0$

We get the optimal solution $x^* = \begin{pmatrix} 0 \\ 3 \end{pmatrix}$ of the LP problem with objective value $H^* = 3$.

Finally we get our desired optimal solution of the given LFP problem is $x^* = \begin{pmatrix} 0 \\ 3 \end{pmatrix}$ with the optimal value $F^* = \frac{9}{7}$.

Table 1. Results of existing and our methods for Example 1 and Example 2.

	Bitran and Novea	Swarup	Tantawy	Our Method
Example 1	3 iterations with lots of calculations	3 iterations with clumsy calculations	2 iterations	1 iterations with simple calculations
Example 2	3 iterations	3 iterations	2 iterations	1 iterations

Now different methods can be compared with our method and all the methods give the same results for Example 1 and Example 2. Table 1 shows the results of number of iterations that are required for our method and the existing methods for these Examples.

COMPARISON

In this Section, we find that our method is better than any other available method. The reason can be given as follows:

§ Any type of LFP problem can be solved by this method.

§ The LFP problem can be transformed into LP problem easily with initial guess.

§ In this method, problems are solved by algebraically with duality concept. So that it's computational steps are so easy from other methods.

§ The final result converges quickly in this method.

§ In some cases of numerator and denominator, other existing methods are failed but our method is able to solve any kind of problem easily.

CONCLUSION

In this paper, we give an approach for solving linear fractional programming problems. The proposed method differs from the earlier methods as it is based upon solving the problem algebraically using the concept of duality. This method does not depend on the simplex type method which searches along the boundary from one feasible vertex to an adjacent vertex until the optimal solution is found. In some certain problems, the number of vertices is quite large, hence the simplex method would be prohibitively expensive in computer time if any substantial fraction of the vertices had to be evaluated. But our proposed method appears simple to solve any linear fractional programming problem of any size.

CONFLICTS OF INTEREST

The authors declare no conflicts of interest.

REFERENCES

1. Charnes, A. and Cooper, W.W. (1962) Programming with Fractional Functions. Naval Research Logistic Quarterly, 9, 181-186. https://doi.org/10.1002/nav.3800090303
2. Bitran, G.R. and Novaes, A.G. (1973) Linear Programming with a Fractional Objective Functions. Operations Research, 21, 22-29. https://doi.org/10.1287/opre.21.1.22
3. Bitran, G.R. and Magnanti, T.L. (1976) Duality and Sensitivity Analysis with Fractional Objective Function. Journal of Operation Research, 24, 675-699. https://doi.org/10.1287/opre.24.4.675
4. Sing, H.C. (1981) Optimality Condition in Fractional Programming. Journal of Optimization Theory and Applications, 33, 287-294. https://doi.org/10.1007/BF00935552
5. Swarup, K. (1964) Linear Fractional Programming. Operation Research, 13, 1029- 1036. https://doi.org/10.1287/opre.13.6.1029
6. Tantawy, S.F. (2008) A New Procedure for Solving Linear Fractional Programming Problems. Mathematical and Computer Modelling, 48, 969-973. https://doi.org/10.1016/j.mcm.2007.12.007
7. Tantawy, S.F. (2014) A New Concept of Duality for Linear Fractional Programming Problems. International Journal of Engineering and Innovative Technology, 3, 147- 149.
8. Effati, S. and Pakdaman, M. (2012) Solving the Interval-Valued Linear Fractional Programming Problem. American Journal of Computational Mathematics, 5, 51-55. https://doi.org/10.4236/ajcm.2012.21006
9. Pramanik, S., Dey, P.P. and Giri, B.C. (2011) Multi-Objective Linear Plus Linear Fractional Programming Problem Based on Taylor Series Approximation. International Journal of Computer Applications, 32, 61-68.
10. Tantawy, S.F. (2008) An Iterative Method for Solving Linear Fractional Programming (LFP) Problem with Sensitivity Analysis. Mathematical and Computational Mathematics, 13, 147-151.
11. Tantawy, S.F. and Sallam, R.H. (2013) A New Method for Solving Linear Fractional Programming Problems. International Journal of Recent Scientific Research, 4, 623- 625.

CHAPTER 2

A Particle Swarm Optimization Algorithm for Solving Pricing and Lead Time Quotation in a Dual-Channel Supply Chain with Multiple Customer Classes

Mahboobeh Honarvar,[1] Majid Alimohammadi Ardakani,[2] and Mohammad Modarres[3]

[1]Department of Industrial Engineering, Yazd University, Yazd, Iran

[2]Department of Industrial Engineering, Ardakan University, Yazd, Iran

[3]Department of Industrial Engineering, Sharif University of Technology, Tehran, Iran

ABSTRACT

The combination of traditional retail channel with direct channel adds a new dimension of competition to manufacturers' distribution system. In this paper, we consider a make-to-order manufacturer with two channels of sale,

Citation: MahboobehHonarvar, Majid AlimohammadiArdakani, Mohammad Modarres, "A Particle Swarm Optimization Algorithm for Solving Pricing and Lead Time Quotation in a Dual-Channel Supply Chain with Multiple Customer Classes", Advances in Operations Research,vol. 2020,Article ID 5917126, 21 pages, 2020. https://doi.org/10.1155/2020/5917126.

sale through retailers and online direct sale. The customers are classified into different classes, based on their sensitivity to price and due date. The orders of traditional retail channel customers are fulfilled in the same period of ordering. However, price and due date are quoted to the online customers based on the available capacity as well as the other orders in the pipeline. We develop two different structures of the supply chain: centralized and decentralized dual-channel supply chain which are formulated as bilevel binary nonlinear models. The Particle Swarm Optimization algorithm is also developed to obtain a satisfactory near-optimal solution and compared to a genetic algorithm. Through various numerical analyses, we investigate the effects of the customers' preference of a direct channel on the model's variables.

INTRODUCTION

The rapidly expanding Internet provides an opportunity for organizations to distribute their products via both direct channel and traditional retail channel. Some personal computer manufacturers (like Dell company), apparel retailers, and automotive industries (like General Motors) are examples of companies that use hybrid of both direct and retailer channels.

In the direct channel, the firm interacts with consumers directly through Internet. There are a number of benefits from direct channel distribution such as controlling the distribution and pricing directly, providing a broader product selection, and improving firms' visibility [1].

Despite hybrid channel's benefits such as capturing a larger share of the market, combining the retail distribution channel with direct channel may pose some challenges, including pricing policies, distribution strategies, and conflicting demands placed on internal company resources such as capital, personal, products, and technology by multiple channels [2].

To overcome these challenges, supply chain's members negotiate the retail as well as wholesale price to cooperate with each other. In addition to the product price, there are other factors such as product availability and service quality that contribute to consumer preference of the direct channel [3].

Delivery lead time is one of the important factors that can affect customers' demand [3]. This is the reason many e-retailers, such as Amazon.com, BestBuy.com, Walmart.com, and FYE.com, try to offer competitively quoted lead times [4].

This paper is focused on competitive pricing strategy as well lead time decisions in a supply chain for a manufacturer that sells the products through two channels. One channel is the traditional one in which the firm uses an intermediary to reach final consumers, while the other is a direct channel in which the customer places direct orders through the Internet. It is assumed that there are multiple classes of customers in the market. The demand of each class depends on the price and lead time. A finite planning horizon is considered, where the production capacity in each period is finite but varies. The manufacturer has to respond to the customers quickly, based on the available capacity and the orders in pipeline. Due to the finite production capacity and the demand sensitivity to price and lead time changes, the manufacturer needs to decide on the selling price and the lead time of direct customers, the contract terms with the retailer, and the wholesale price as well as the production schedule to maximize his own profit.

We consider two different cases, centralized and decentralized dual-channel supply chain. We propose a model to determine suitable lead time and price, simultaneously for each case. In a decentralized dual-channel supply chain, the two members interact within the framework of Stackelberg game. Under this framework, the manufacturer, as the leader, determines the wholesale price for retailer and also a price and lead time for direct sale in each period. The intermediary then reacts by choosing a retail price to maximize its own profit. On the other hand, for a centralized dual-channel supply chain, the direct sale price, the traditional retail price, and the quoted lead time in the direct channel are determined by a vertically integrated manufacturer.

The resulting models happen to be binary nonlinear programs. These kinds of problems with large dimensions are not usually solved by exact algorithms. Therefore, an alternative method for solving the models is to use metaheuristic algorithms to obtain a near-optimal solution with reasonable computational time. Thus, an algorithm of Particle Swarm Optimization (PSO) is developed to solve the models. The results will be compared, in terms of solution quality and computational time, with genetic algorithm (GA).

The remainder of this paper is structured as follows. Section 2 reviews the more relevant literature. We formulate the models in Section 3. The principles of the PSO metaheuristic and our algorithm are explained in Section 4. Section 5 presents some numerical examples of our model to examine the effects of customer preference of a direct channel on the pricing strategies and lead time decisions. Our conclusions are summarized in Section 6.

LITERATURE REVIEW

This paper focuses on competition in a dual-channel supply chain. A comprehensive review of multichannel models can be found in [1, 5].

Several researchers and practitioners have focused on dual-channel supply chains during the last decade. Different aspects of this chain are investigated. Most of the papers that formulated the dual-channel supply chain focused on the competition context in the supply chain and pricing optimization issues in each channel [6–9]. The researchers combined the pricing issues with other aspects of supply chain. In this regard, we can refer to sales effort determining and service management in each channel [10–13], contract optimization [6, 14], disruption management [15–17], product variety in supply chain [18, 19], and multiperiod model [20].

Besides parameters such as price, services, and quality that can affect the demand process, the lead time (or due date) is also an important parameter. Some researchers have considered that the quoted lead times (or due dates) also affect customers' decisions on placing an order (e.g., Duenyas and Hopp [21]; So and Song [22]; Easton and Moodie [23]; Keskinocak et al. [24]; Webster [25]; Watanapa and Techanitisawad [26]; Charnsirisakskul et al. [27]; Mustafa et al. [28]; Chaharsooghi et al. [29]; and Chaharsooghi et al. [30]). However, the above papers have not addressed the dual-channel distribution issue. The papers by Chen et al. [31]; Hua et al. [3]; Xu et al. [32]; Batarfi et al. [33]; Yang et al. [4]; and Modak and Kelle [34] are some examples that investigated lead time optimization in the dual-channel supply chain.

Xu et al. [32] considered the unit delivery cost (m/t) with $m > 0$ if the product is to be delivered with lead time t. Batarfi et al. [33] considered the combination of two production approaches: make-to-order and make-to-stock in direct online channel and indirect offline channel, respectively. The demand depends on the prices, the quoted delivery lead time, and the product differentiation.

Yang et al. [4] modeled delivery lead time optimization for perishable products in a dual-channel supply chain. Modak and Kelle [34] considered that the demand not only is dependent on price and delivery time but also is stochastic.

The lead times in the above papers are determined exogenously (determined by the sales department, without knowing the actual production schedule). In addition, the production capacity is unlimited. Table 1 illustrates the major literature review with our paper included.

Table 1. Related literature.

Research paper	Production type		Pricing decision	Due date quotation		Capacity constraint	Production planning	Customer classes	Competition approach
	Make-to-stock	Make-to-order		Internally	Exogenously				
Tsay and Agrawal [13]	*		*						Nash
Chiang et al. [7]	*		*						Nash
Yao and Liu [9]	*		*						Bertrand and Stackelberg
Cai et al. [6]	*		*						Stackelberg, Nash
Dan et al. [10]	*		*						Stackelberg
Huang et al. [15]	*		*						Stackelberg
Soleimani et al. [17]	*		*						Nash
Roy et al. [12]	*		*						Stackelbarg
Hua et al. [3]	*		*				*		Nash
Xu et al. [32]	*		*		*				Nash
Batarfi et al. [33]	*	*	*		*				-
Yang et al. [4]	*		*		*				Nash

Rofin and Mahanty [18]	*		*		*				Nash
Pi et al. [16]	*		*						Stackel-berg
Modak and Kelle [34]	*		*		*				Stackel-berg
This paper		*	*	*		*	*	*	Stackel-berg

Based on the above literature, this paper investigates the joint decision on production, pricing, and lead time in a dual-channel distribution system where the lead times are assigned internally by the scheduling model. Aside from considering the above literature, we address a new model in which the production capacity in each period is limited. We also consider multiple customer classes that differ in their arrival (commitment) times, quantities demanded, and sensitivity to price and lead time.

PROBLEM DESCRIPTION

We consider a dual-channel supply chain in which a manufacturer sells to retailers as well as directly to end customers. The planning horizon is assumed to be finite and divided into periods of equal length. The capacity of the manufacturer is limited but may vary in different periods. Customers can be classified with respect to their sensitivity to price and lead time. Furthermore, the attributes of customers such as arrival (commitment) times, quantities demanded, unit production, and holding costs are different.

In each period, the manufacturer must set a wholesale price for the traditional retail channel as well as setting both price and lead time (or due date) for the customers of direct channel. For the customers of traditional retail channel, the production must be scheduled at the same period, at which the order arrives. However, the manufacturer can quote a lead time (due date) to the customers of direct channel. The production for these customers is scheduled within any period between the arrival time of order and the quoted due date. A holding cost occurs for any order of direct channel which is completed before the quoted due date.

As mentioned before, in order to evaluate the effects of the delivery lead time and customer's preference of the direct channel on the pricing decisions of the manufacturer and retailer, we consider two different dual-channel supply chains, centralized and decentralized systems.

Notation

We use the following notation.

Sets

$\Psi = \{1, \ldots, N\}$: set of customer classes, based on sensitivity to price and due date

$T = \{1, \ldots, T\}$: set of planning periods

Parameters

e(i) : arrival time of customer of class i Ψ

Ch_i: third party holding cost per time period per unit of customer of class i $\in \Psi$

$Cp^1_{i,t}$: production cost of customer of class i $\in \Psi$ in the direct channel in period t $\in$ T

$Cp^2_{i,t}$: production cost of customer of class i $\in \Psi$ in the retail channel in period t $\in$ T

Cr_i: operational cost in the retail channel for customer of class i $\in \Psi$

Ce_i: operational cost in the direct channel for customer of class i $\in \Psi$

K_t: production capacity available in period t $\in$ T

$D^r_{i,j}$: demand (in terms of production capacity units) for customer of class i $\in \Psi$ in the retail channel, quoted due date j $\in$ [e(i), . . . , T]

$D^s_{i,j}$: demand (in terms of production capacity units) for customer of class i $\in \Psi$ in the direct channel, quoted due date j $\in$ [e(i), . . . , T]

Decision Variables

$Z_{i,j}$: 1 if the due date j, j $\in$ [e(i), . . . , T], is selected (quoted) for customer of class i $\in \Psi$ in direct channel; 0 otherwise

P^r_i : price in the retail channel for customer of class i $\in \Psi$

P^s_i: price in the direct channel for customer of class i $\in \Psi$

W_i: wholesale price charged for customer of class i $\in \Psi$ in the retail channel

$Y_{i,t}$: total production (in units of capacity) for customer of class i ∈ Ψ in the direct channel in period t ∈ T

H_i: total inventory of customer of class i ∈ Ψ

Demand Function

In line with Kurata et al. [35], Cai et al. [6], and Hua et al. [3], we assume that the demand is a linear function of the self- and cross-price and lead time, as follows:

$$D_i^s = a_i \cdot \theta_i - b_i^s \cdot P_i^s + \alpha_i^r \cdot P_i^r - c_i^s \cdot L_i^s, \quad (1)$$

$$D_i^r = a_i \cdot (1 - \theta_i) - b_i^r \cdot P_i^r + \alpha_i^s \cdot P_i^s + c_i^r L_i^s, \quad (2)$$

where D_i^s and D_i^r denote the consumer demand to the manufacturer and the consumer demand to the retailer, respectively, a_i denotes the base level of industry demand or the demand rate for customer of class i, and $\theta_i (0 < \theta_i < 1)$ represents the initial portion of customers of class i that prefer the direct channel if the price and lead time are zero. Thus, $1 - \theta_i$ represents the portion of customers of class i that prefer purchasing from the retailer. The coefficients b_i^s and b_i^r are the coefficients of the price elasticity in the direct channel and retail channel demand functions, respectively. The cross-price sensitivities α_i^r and α_i^s reflect the substitution's degree of the two channels, and c_i^s is the lead time sensitivity of the demand in the direct channel. If the lead time L_i^s increases by one unit, c_i^r units of demand will transfer to the retail channel and c_i^s units of direct channel's demand will be decreased. The total demand of the two channels should have a negative slope with respect to the retailer's price, the direct sale price, and the quoted lead time. Thus, we have $\alpha_i^r < b_i^r$, $\alpha_i^s < b_i^s$, and $c_i^s > c_i^r$.

Following Hua et al. [3], it is assumed that the manufacturer uses dual channels to sell their goods, and the base level of industry demand or demand rate in both channels is very large; i.e., θ_i should not be unreasonably small or large.

Model of the Decentralized Dual-Channel Supply Chain

In this section, we study a decentralized dual-channel supply chain. In this model both the manufacturer and the retailer make their own decisions separately to maximize their profits. The manufacturer, as the Stackelberg leader, first determines the wholesale price W_i, the direct sale price P_i^s, and the direct sale quoted lead time L_i^s. Then, the retailer chooses his own optimal retail price P_i^r based on the manufacturer's decisions.

Manufacturer's Best Response

The goal of the manufacturer as a leader is to maximize his profit, considering the capacity constraints and demand constraints in the systems and constraints determined by the retailer optimization problem. The problem is formulated within the framework of bilevel programming (BLP), first level (the manufacturer model called the leader) and second level (the retailer model called a follower). In BLP model, each decision maker tries to optimize its own objective function without considering the objective of the other party. However, the decision of each party affects the objective value of the other one as well as the decision space.

First-Level Model: Manufacturer Model

$$Max\ \Pi^s = \left(\sum_{\Psi} \sum_{t} \left(b_i^t \times D_i^{rt} \times S^{rt} \right) - \sum_{\Psi} \sum_{t} \left(\lambda^{rt} \times \left(Cb_i^{rt} - C^{st} \right) \right) + \sum_{\Psi} \sum_{t} \left(M^t - Cb_j^{r,e(i)} \right) \left(D_i^{rt} \times S^{rt} \right) - \sum_{\Psi} C^{ht} \times H^t \right). \quad (3)$$

S.T.

$$W_i \leq P_i^s, \quad \forall i \in \Psi, \quad (4)$$

$$\sum_{j=e(i)}^{T} Z_{i,j} = 1, \quad \forall i \in \Psi, \quad (5)$$

$$\sum_{t=e(i)}^{T} Y_{i,t} = \sum_{j=e(i)}^{T} D_{i,j}^s Z_{i,j}, \quad \forall i \in \Psi, \quad (6)$$

$$\sum_{i \in \Psi | e(i) \leq t} Y_{i,t} + \sum_{i \in \Psi | e(i)=t} \sum_{j=e(i)}^{T} D_{i,j}^r Z_{i,j} \leq K_t, \ t = 1, \ldots, T, \quad (7)$$

$$\sum_{t=j}^{T} Y_{i,t} \leq M\left(1 - Z_{i,j}\right), \quad \forall i \in \Psi, \ j = e(i), \ldots, T, \quad (8)$$

$$H_i \geq \sum_{t=e(i)}^{j} (j-t)\left(Y_{i,t}\right) + M\left(Z_{i,j} - 1\right), \quad \forall i \in \Psi, \ j = e(i), \ldots, T, \quad (9)$$

$$D_{i,j}^s = a_i \cdot \theta_i - b_i^s \leq P_i^s + \alpha_i^r \cdot P_i^r - c_i^s \cdot (j - e(i) + 1), \forall i \in \Psi, \ j = e(i), \ldots \infty, T. \quad (10)$$

Second-Level Model: Retailer Model

$$\underset{P_i^r}{\text{Max}} \quad \Pi_r = \sum_{i \in \psi} (P_i^r - W_i - Cr_i) \cdot \left(a_i \cdot (1-\theta_i) - b_i^r \cdot P_i^r + \alpha_i^s \cdot P_i^s + c_i^r \sum_{j=e(i)}^{T} (j - e(i) + 1) \cdot Z_{i,j} \right), \tag{11}$$

S.T.

$$\sum_{i|t=e(i)} a_i \cdot (1-\theta_i) - b_i^r \cdot P_i^r + \alpha_i^s \cdot P_i^s + c_i^r \sum_{j=e(i)}^{T} (j - e(i) + 1) \cdot Z_{i,j} \le K_t, \ t = 1, \ldots, T, \tag{12}$$

$$a_i \cdot (1-\theta_i) - b_i^r \cdot P_i^r + \alpha_i^s \cdot P_i^s + c_i^r \sum_{j=e(i)}^{T} (j - e(i) + 1) \cdot Z_{i,j} \ge 0, \ \forall i \in \Psi, \tag{13}$$

$$Y_{i,t}, H_i, D_{i,j}^s, P_i^r, P_i^s, W_i \ge 0, \ \forall i \in \Psi, \ t = e(i), \ldots, T, \ j = e(i), \ldots, T,$$
$$Z_{i,j} \in \{0, 1\}, \ \forall i \in \Psi, \ j = e(i), \ldots, T. \tag{14}$$

The first and second terms in the first-level objective represent the total revenue and production cost of direct sales. The third term represents the total profit obtained by sales to the retailer and the fourth one is the carrying cost.

Constraint (4) indicates that the wholesale price cannot be higher than the direct channel price; otherwise, the retailer may purchase from the direct channel at a lower price. Constraints (5) ensure that only one due date is chosen for each direct channel customer order. Constraint (6) ensures that if due date j is selected for customer of class i in the direct channel, then exact $D_{i,j}^s$ units must be produced and delivered, where $D_{i,j}^s$ depends on the selected price and due date. Constraint (7) is a capacity constraint that ensures that the production capacity in each period is not exceeded. An order of a retail channel customer is produced in the customer's arrival time period and delivered instantaneously in the same period, whereas an order of a direct channel customer can be produced in any period between its arrival period and the quoted due date. The first term in the left hand side of constraint (7) indicates the total production for direct channel customers and the second term is the total production for retail channel customers. Constraint (8) indicates that an order of a direct channel customer can be produced in any period between its arrival period and the quoted due date.

The required inventory for orders is scheduled in any period prior to its commitment, and the negotiated due date is controlled by constraint (9), where M is a sufficiently large number. Constraint (10) is demand functions for customer orders in the direct channel. The term $j - e(i) + 1$ is a time interval between the arrival time of an order and the quoted due date, which is called the lead time. The objective function of second-level optimization problem for the retailer channel is represented by (11). Constraints (12) and

(13) control feasibility as well as demand restrictions.

Without considering constraints (12)–(14) for retailer, its best response to the wholesale price can be defined in Proposition 1.

Proposition 1. *The retailer's best response to the wholesale price W_i, the direct sale price P_i^s, and the direct quoted lead time L_i^s set by the manufacturer is as follows:*

$$P_i^r = \frac{a_i \cdot (1-\theta_i) + W_i \cdot b_i^r + \alpha_i^s P_i^s + c_i^r L_i^s + Cr_i b_i^r}{2b_i^r}, \tag{15}$$

The BLP model (4)–(14) can be formulated as a single level mixed binary problem. This is achieved by replacing the lower level problem (12)–(14) with its Kuhn–Tucker conditions, which we name Model I, as follows.

Model I

$$\text{Max} \quad \Pi_s = \left(\sum_{i=1}^{N} \sum_{j=e(i)}^{T} \left(p_i^s \times D_{i,j}^s \times Z_{i,j}\right) - \sum_{i=1}^{N} \sum_{t=e(i)}^{T} \left(Y_{i,t} \times Cp_{i,t}^1\right) + \sum_{i=1}^{N} \sum_{j=e(i)}^{T} \left(W_i - Cp_{i,e(i)}^1\right)\left(D_{i,j}^r \times Z_{i,j}\right) - \sum_{i=1}^{N} Ch_i \times H_i \right), \tag{16}$$

S.T.

$$W_i \le P_i^s, \tag{17}$$

$$\sum_{t=e(i)}^{T} Y_{i,t} = \sum_{j=e(i)}^{T} D_{i,j}^s Z_{i,j}, \forall i \in \Psi, \tag{18}$$

$$\sum_{j=e(i)}^{T} Z_{i,j} = 1, \forall i \in \Psi, \tag{19}$$

$$\sum_{i \in \Psi | e(i) \le t} Y_{i,t} + \sum_{i \in \Psi | e(i) = t} \sum_{j=e(i)}^{T} D_{i,j}^r Z_{i,j} \le K_t, \ t = 1, \ldots, T, \tag{20}$$

$$\sum_{t=j+1}^{T} Y_{i,t} \le M\left(1 - Z_{i,j}\right), \forall i \in \Psi, \ j = e(i), \ldots, T-1, \tag{21}$$

$$H_i \ge \sum_{t=e(i)}^{j} (j-t)\left(Y_{i,t}\right) + M\left(Z_{i,j} - 1\right), \forall i \in \Psi, \ j = 1, \ldots, T, \tag{22}$$

$$D_{i,j}^s = a_i \cdot \theta_i - b_i^s \cdot P_i^s + \alpha_i^r \cdot P_i^r - c_i^s \cdot (j - e(i) + 1), \forall i \in \Psi, \ j = e(i), \ldots, T, \tag{23}$$

$$a_i(1-\theta_i) - 2P_i^r \cdot b_i^r + \alpha_i^s P_i^s + c_i^s \sum_{j=e(i)}^{T} (j - e(i) + 1) \cdot Z_{i,j} + W_i \cdot b_i^r + Cr_i \cdot b_i^r + \lambda_{e(i)} \cdot b_i^r - \lambda_i^\prime \cdot b_i^r + \nu_i = 0, \forall i \in \Psi, \tag{24}$$

$$\sum_{i|e(i)=t} \left(a_i \cdot (1-\theta_i) - b_i^r \cdot P_i^r + \alpha_i^s \cdot P_i^s + c_i^r \sum_{j=e(i)}^{T} (j - e(i) + 1) \cdot Z_{i,j} \right) + S_t = K_t, \ t = 1. \tag{25}$$

$$a_i . (1-\theta_i) - b_i^r \cdot P_i^r + \alpha_i^s \cdot P_i^s + c_i^r \sum_{j=e(i)}^{T} (j - e(i) + 1) \cdot Z_{i,j} - S_i^\prime = 0, \tag{26}$$

$$\lambda_t \cdot S_t + \lambda'_i \cdot S'_i + \nu_i \cdot P^r_i = 0, \tag{27}$$

$$Y_{i,t}, H_i, D^s_{i,j}, P^r_i, P^s_i, W_i, \lambda_i, \lambda'_i, S'_i, S_i, \nu_i \geq 0,\ \forall i \in \Psi,\ t = e(i), \ldots, T,\ j = e(i), \ldots, T,$$
$$Z_{i,j} \in \{0, 1\},\ \forall i \in \Psi,\ j = e(i), \ldots, T. \tag{28}$$

Constraints (24)–(27) indicate the optimality condition for the retailer to convert the bilevel problem into a single level one.

Proposition 2. *For any given due date j, the wholesale price W_i, and the direct sale price P^s_i for customer class i, the optimal retail price P^r_i is always positive.*

As a result of Proposition 2, it can be concluded that v_i is always equal to zero, because $\nu_i \cdot P^r_i = 0$ in Model I.

Proposition 3. *Consider an unlimited production capacity in each period for a decentralized dual-channel supply chain. If $8b^r_i \cdot b^s_i - (\alpha^s_i)^2 - (\alpha^r_i)^2 - 6\alpha^s_i \cdot \alpha^r_i > 0,$, the dual channel Π_s is strictly jointly concave in W_i and P^s_i.*

From Proposition 3, if $\alpha^s_i = \alpha^r_i$, then $8b^r_i \cdot b^s_i - (\alpha^s_i)^2 - (\alpha^r_i)^2 - 6\alpha^s_i \cdot \alpha^r_i$ is always greater than zero. Thus, in this case, the manufacturer profit Π_s is always strictly jointly concave in W_i and P^s_i. Thus, for the sake of analytical tractability, we will assume that the cross-price effects are symmetric and $\alpha^s_i = \alpha^r_i$.

Proposition 4. *Let the production capacity be unlimited in each period for a decentralized dual-channel supply chain and suppose there is a $L^0_i \in [L^l_i, L^u_i]$ with*

$$L^0_i = \frac{-a_i\alpha_i(1-\theta_i) - a_i\theta_i(b^r_i - \alpha_i) + a_i b^s_i(1-\theta_i) - Cr_i\left(b^r_i b^s_i - (\alpha_i)^2\right)}{c^r_i(b^s_i - \alpha_i) + c^s_i(b^r_i - \alpha_i)}. \tag{29}$$

Then, the optimal lead time $L^s_i \in [L^l_i, L^u_i]$ can be found from L^l_i, L^u_i, *and* L^0_i by comparing their Π_s values; the one with the largest Π_s value is the optimal lead time.

Model of the Centralized Dual-Channel Supply Chain

In a centralized dual-channel supply chain, a vertically integrated manufacturer controls the wholesale price W_i, the direct sale price P^s_i, the direct sale quoted lead time L^s_i, and the retailer sale price P^r_i. The model, called Model II, is as follows.

Model II

$$\text{Max} \quad \Pi_c = \Pi_s + \Pi_r = \left(\sum_{i=1}^{N} \sum_{j=e(i)}^{T} \left(p^s_i \times D^s_{i,j} \times Z_{i,j}\right) - \sum_{i=1}^{N} \sum_{t=e(i)}^{T} \left(Y_{i,t} \times Cp^1_{i,t}\right) + \sum_{i=1}^{N} \sum_{j=e(i)}^{T} \left(p^r_i - Cp^2_{i,e(i)} - Cr_i\right)\left(D^r_{i,j} \times Z_{i,j}\right) - \sum_{i=1}^{N} Ch_i \times H_i \right), \tag{30}$$

S.T.

$$\sum_{j=e(i)}^{T} Z_{i,j} = 1, \quad \forall i \in \Psi, \tag{31}$$

$$\sum_{t=e(i)}^{T} Y_{i,t} = \sum_{j=e(i)}^{T} D_{i,j}^{s} Z_{i,j}, \quad \forall i \in \Psi, \tag{32}$$

$$\sum_{i \in \Psi | e(i) \le t} Y_{i,t} + \sum_{i \in \Psi | e(i) \le t} \sum_{j=e(i)}^{T} D_{i,j}^{r} Z_{i,j} \le K_t, \; t = 1, \ldots, T, \tag{33}$$

$$\sum_{t=j+1}^{T} Y_{i,t} \le M\left(1 - Z_{i,j}\right), \quad \forall i \in \Psi, \; j = e(i), \ldots, T-1, \tag{34}$$

$$H_i \ge \sum_{t=e(i)}^{j} (j-t)\left(Y_{i,t}\right) + M\left(Z_{i,j} - 1\right), \quad \forall i \in \Psi, \; j = e(i), \ldots, T, \tag{35}$$

$$D_{i,j}^{s} = a_i \cdot \theta_i - b_i^{s} \cdot P_i^{s} + \alpha_i^{r} \cdot P_i^{r} - c_i^{s} \cdot (j - e(i) + 1), \quad \forall i \in \Psi, \; j = e(i), \ldots, T, \tag{36}$$

$$D_{i,j}^{r} = a_i \cdot (1 - \theta_i) - b_i^{r} \cdot P_i^{r} + \alpha_i^{s} \cdot P_i^{s} + c_i^{r} \cdot (j - e(i) + 1), \quad \forall i \in \Psi, \; j = e(i), \ldots, T, \tag{37}$$

$$Y_{i,t}, H_i, D_{i,j}^{s}, P_i^{r}, P_i^{s} \ge 0, \quad \forall i \in \Psi, \; t = e(i), \ldots, T, \; j = e(i), \ldots, T, \quad Z_{i,j} \in \{0,1\}, \; \forall i \in \Psi, \; j = e(i), \ldots, T. \tag{38}$$

The objective function (30) maximizes the total profit of the centralized supply chain, which is the sum of manufacturer's profit and retailer's profit. The remaining constraints are the same as those in Model I.

Proposition 5. *Consider an unlimited production capacity in each period for a centralized dual-channel supply chain. If* $4b_i^{r} \cdot b_i^{s} - (\alpha_i^{s})^2 - (\alpha_i^{r})^2 - 2\alpha_i^{s} \cdot \alpha_i^{r} > 0,$ *, then the optimal lead time* $L_i^{s} \in [L_i^{l}, L_i^{u}]$ *can be found from* L_i^{l} and L_i^{u} *by comparing their* Π_c *values; the largest* Π_c *value is the optimal lead time.*

SOLUTION APPROACH

Models I and II are 0-1 nonlinear programs with continuous price and production decision variables and binary variables for due date selection. It is well known that the problem of nonlinear 0-1 programming is NP-hard [36]. Thus, these problems cannot be solved easily, and exact algorithms can be time consuming when the number of 0-1 variables is large. An alternative for solving these models is to use approximate algorithms or heuristic methods with reasonable computational times.

In this paper, PSO metaheuristic is used and compared to the GA for modeling and optimization of the pricing and due date setting problems in a dual-channel supply chain.

PSO and GA can handle the whole MINLP easily and naturally, and it is easy to apply it to various problems for comparison with conventional methods.

If we fix the set of binary variables *Z* in the models, then both models are converted to quadratic programs that can be solved by the nonlinear programming solver LINGO.

Particle Swarm Optimization (PSO)

The PSO algorithm was first proposed by Kennedy and Eberhart [37]. This method is based on the information shared among members of a species and then used for evolution.

Simple structure, ease of implementation, speed of acquiring solutions, and robustness are the advantages of the PSO that persuade us to use it in solving the models.

Recently, PSO algorithms were successfully applied to a wide range of applications. A comprehensive survey of PSO algorithms and applications can be found in the paper by Kennedy et al. [38].

In the proposed PSO algorithm, particle *k* is represented as $L_k = \{L_{k,1}, L_{k,2}, \ldots, L_{k,N}\}$, which denotes the due dates quoted to *N* customers. For each $L_{k,i}$ in N-dimensional space, the $Z_{iL_{k,i}}$ is set to one in Models I and II, and the other variables $Z_{i,j}$, $j = e(i), \ldots T$, and $j \neq L_{k,i}$, are fixed to zero for customer order $i \in \Psi$.

This section will present the application of the PSO algorithm for Models I and II. The PSO algorithm for solving these models is illustrated in Figure 1.

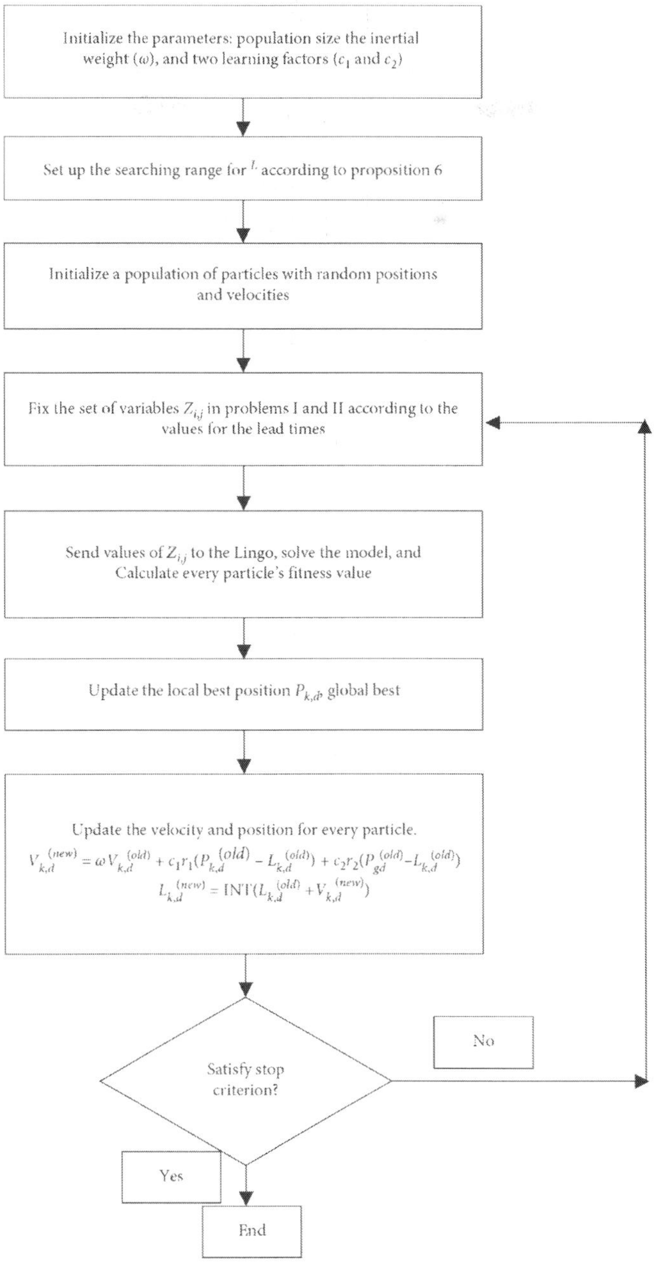

Figure 1. PSO flowchart for pricing and due date quotation.

The key point in a constrained optimization process is to deal with the constraints. We must modify the original PSO method for constrained optimization.

Many methods were proposed for handling constraints, such as methods based on preserving the feasibility of solutions, methods based on penalty functions, methods that make a clear distinction between feasible and infeasible solutions, and other hybrid methods [39].

The most straightforward method is the one based on preserving the feasibility of solutions. In this method, each particle searches the whole space but only keeps track of the feasible solutions.

To accelerate the process, we derive an upper bound on the search range by finding the largest lead times such that both demand in retail and direct channel will be nonnegative; i.e., $D^s_{i,j} \geq 0\, D^r_{i,j} \geq 0.$. The upper bounds on lead times are given in the following proposition.

Proposition 6. *The best due date for each customer i ∈ Ψ is restricted to a range* $L^l_i \leq L_i \leq L^u_i$. *The lower bound of the range is the arrival time of the customer, i.e.,* $e(i)$, *and the upper bound of the range is*

$$L^u_j = \frac{a_i(\theta_i \cdot b^r_i - \theta_i \cdot \alpha^r_i + \alpha^r_i)}{c^s_i \cdot b^r_i - c^r_i \cdot \alpha^r_i} + e(i). \qquad (39)$$

Genetic Algorithm (GA)

GA is a well-known metaheuristic optimization technique based on Darwin's theory of the "survival of the fittest," proposed by Holland [40]. The GA starts with a group of individuals created randomly.

The individuals in the population are then evaluated using a fitness value. Two individuals are then selected based on their fitness. These individuals "reproduce" to create one or more offspring by crossover and mutation operators.

The one-point crossover and swap mutation are used in this article. According to the description provided in PSO algorithm, each solution can be represented by an integer string with length *N* (number of customers). Each gene represents the due date quoted to each customer.

The GA algorithm for solving these models is illustrated in Figure 2:

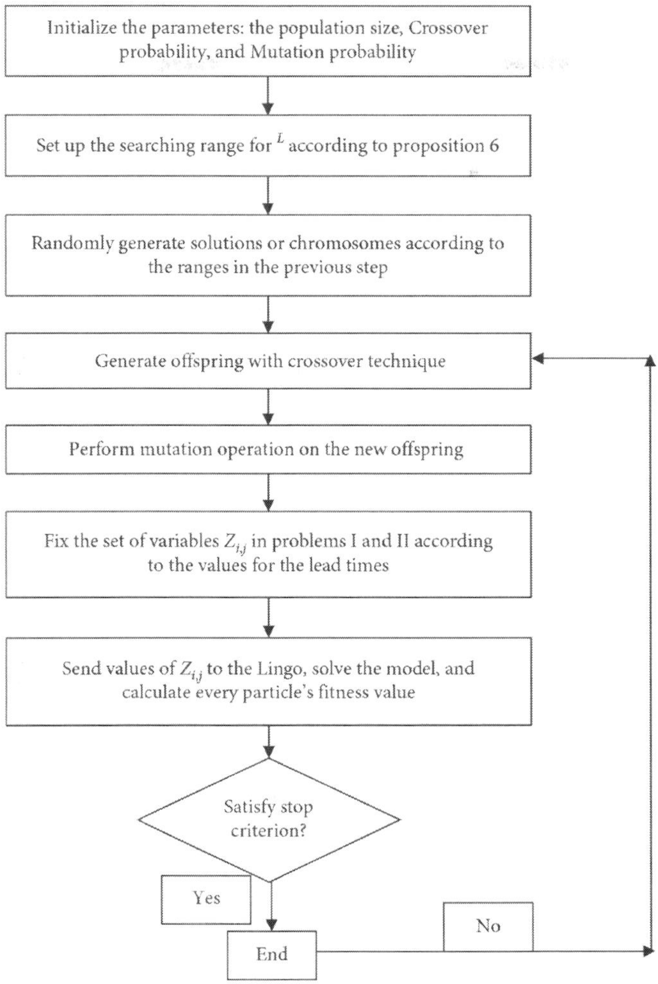

Figure 2. GA flowchart for pricing and due date quotation.

NUMERICAL STUDIES

In this section, we investigate the relation between the wholesale price, the direct sale price, and the retail price in the decentralized and centralized supply chain. The pricing strategies and quoted lead time decisions are compared for the two models using numerical experiments. The parameters of the proposed PSO algorithm are selected based on parameter tuning. With testing different values for PSO' parameters, the ones selected are those that gave the best results. The best values of the computational experiments are as follows: (1) the values of a population of 30 individuals are used for both GA and PSO, (2) the initial inertia weight is set to 0.9, and (3) the values of the acceleration constants c1 and c2 are fixed to 0.9. The maximum velocity is set as the difference between the upper and the lower bounds, which ensures that the particles are able to fly across the problem-specific constraints region. The other parameters used in GA are crossover rate of 0.80 and mutation rate of 0.3. Each instance was run for 50 iterations, and 30 replications were conducted for each instance. The performance of PSO algorithm was compared with MINLP solver GAMS and GA algorithm by testing on a set of 8 small size instances in the decentralized (Model I) and centralized (Model II) supply chain. Table 2 reports the comparison between the GAMS solutions and the presented PSO algorithm. The sum of manufacturer's and retailer's profit is shown in Table 2. All the results show that PSO algorithm can achieve better results than those obtained by GAMS solver.

Table 2. A comparison between the GAMS solutions, the presented PSO algorithm, and the GA algorithm.

Problem number	*N*	*T*	Model I			Model II		
			PSO	GAMS solver	GA	PSO	GAMS solver	GA
Test1	6	12	104202.90	99659.64	103960.30	131059	126182.57	131190.62
Test2	6	12	191024.60	179887.90	186743.10	204614.50	192186.88	203760.16
Test3	12	12	700107.10	656490.40	687800.42	786710.60	745207.09	784619.42
Test4	12	12	125705.30	116692.20	118314.32	135097.20	125816.00	135266.22
Test5	12	12	160890.30	139234.50	161420.60	185068.60	164681.83	186343.64
Test6	12	20	123685.50	107878.50	119659.70	129199.40	119983.29	129934.9
Test7	20	20	138419.10	No feasible	139520.10	146737.70	82849.83	147780.33
Test8	30	20	149965.6	No feasible	148832.40	156788.81	No feasible	157613.04

N: number of customer classes; *T*: planning periods.

Comparing the results of the GA and PSO methods, we can see that there is no conclusive winner. In some cases, the GA method results in a better solution and in some cases the PSO method does. Due to the simplicity of the PSO structure, we use it for our sensitivity analysis.

To analyze the effects of models' parameters on prices and profits, we considered 18 different problem groups to carry out this numerical study. For each group of problems, the production capacity of each period (k) is categorized as high and low capacity. The ratio of operational costs in direct channel and retailer (Ce/Cr) is categorized as high, medium, and low. We can have single class customer problems where the price and lead time sensitivities are the same across the customers and multiclass customer problems based on the variability of these sensitivities. With this assumption, the additional categories are based on price sensitivity variability (PV) and lead time sensitivity variability (LV). Table 3 presents the characteristics of different groups of problems.

Table 3. Specifications of different groups of problems.

Problem number	K	(Ce/Cr)	PV	LV
1	M	M	No	No
2	H	M	No	No
3	M	M	No	Yes
4	H	M	No	Yes
5	M	M	Yes	No
6	H	M	Yes	No
7	M	H	No	No
8	H	H	No	No
9	M	H	Yes	No
10	H	H	Yes	No
11	M	H	No	Yes
12	H	H	No	Yes

13	M	L	No	No
14	H	L	No	No
15	M	L	Yes	No
16	H	L	Yes	No
17	M	L	No	Yes
18	H	L	No	Yes

The other parameters are considered as follows:

The holding costs per unit of time for each customer of class i are Ch_i =5, and the production cost in each period t for each customer of class i is $Cp_i^t = 10$. The demand rate for customer of class i, a_i, is generated randomly from $a_i \in [500, 3000]$.

We considered a planning horizon with 12 time periods and 6 customer classes arriving during the planning horizon for all problems. Two types of arrival time distributions are considered: the arrival times near the beginning of planning horizon (B) $(e(i) \sim$ (uniform [1, 4]) and the uniform distribution along planning horizon (U) (($e(i) \sim$ uniform [1, 12]) [29, 33].

In all instances, we investigated the effect of customer preference of the direct channel, , on pricing and lead time decisions. The results of some carried instances are summarized in Figures 3–9 and Tables 4 and 5.

Tables 4 and 5 show the differences between average values of retailer prices ($\overline{P}_{II}^r - \overline{P}_I^r$) and direct prices ($\overline{P}_{II}^s - \overline{P}_I^s$) obtained for customer classes and the profits $(\Pi_{II} - \Pi_I)$ under both decentralized (Model I) and centralized (Model II) supply chains. In the decentralized supply chain, as illustrated in Figures 3 and 4, in some examples, the manufacturer's profit increases as the customer preference of the direct channel increases, whereas the profit decreases for others. Almost all the examples with decreasing profit functions are consequence of high operational cost in the direct channel. The retailer's profit is decreasing with customer preference of the direct channel.

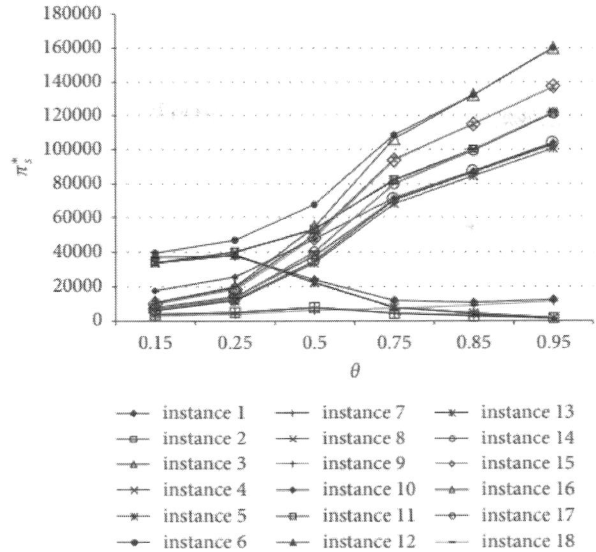

Figure 3. Manufacturer' profit in the decentralized dual-channel supply chain for different instances.

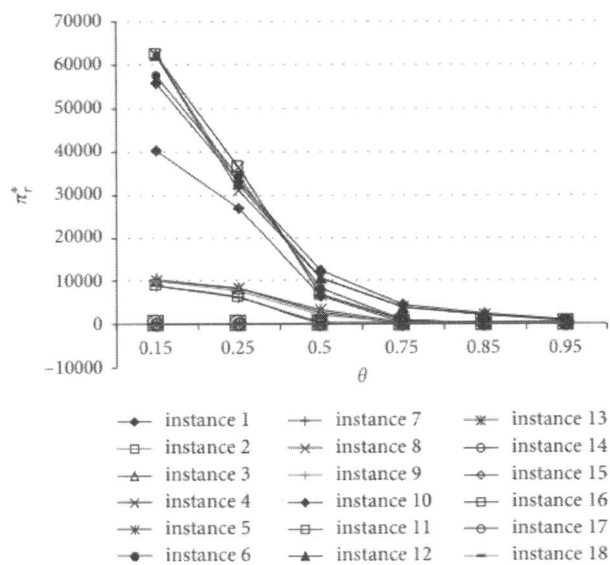

Figure 4. Retailer' profit in the decentralized dual-channel supply chain for different instances.

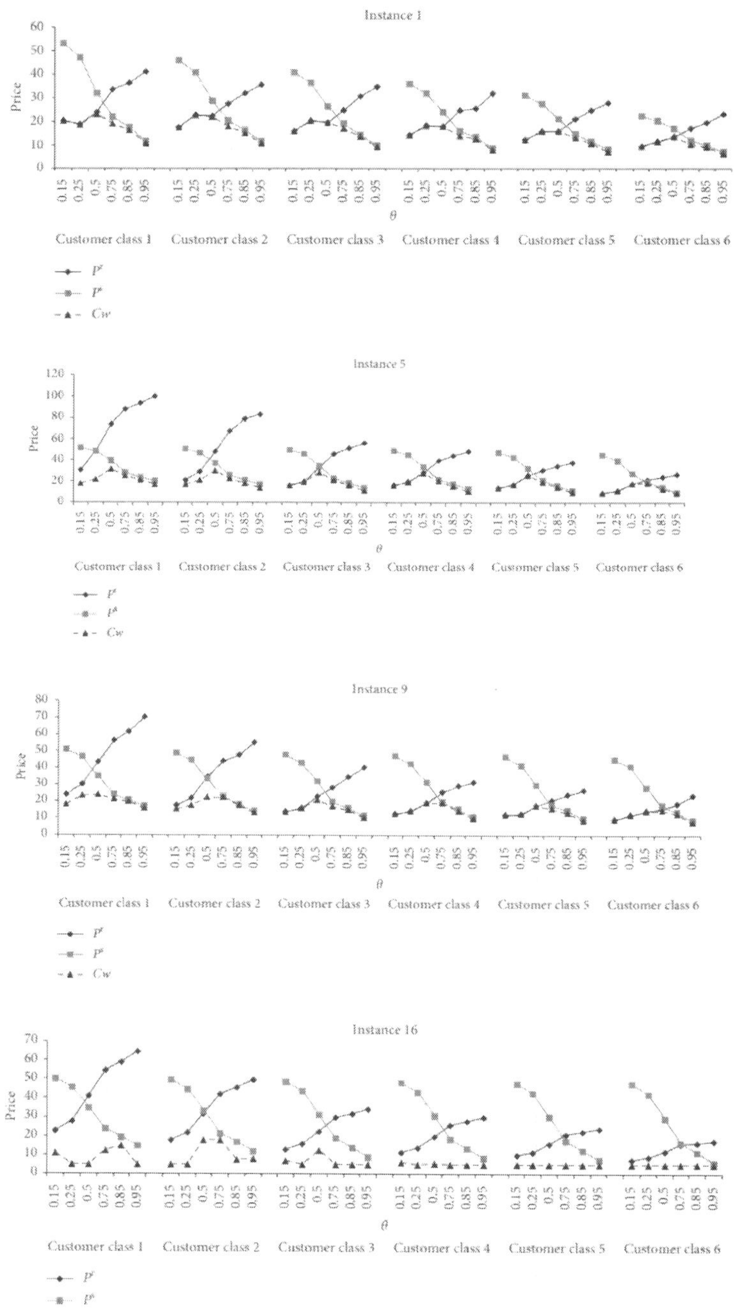

Figure 5. Comparison of the retail price, direct sale price, and wholesale price in decentralized dual-channel supply chain for different instances.

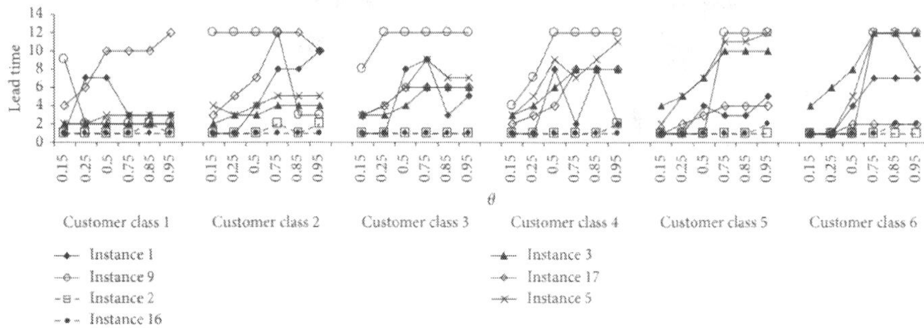

Figure 6. Comparison of the optimal quoted lead time in decentralized dual-channel supply chain for different instances.

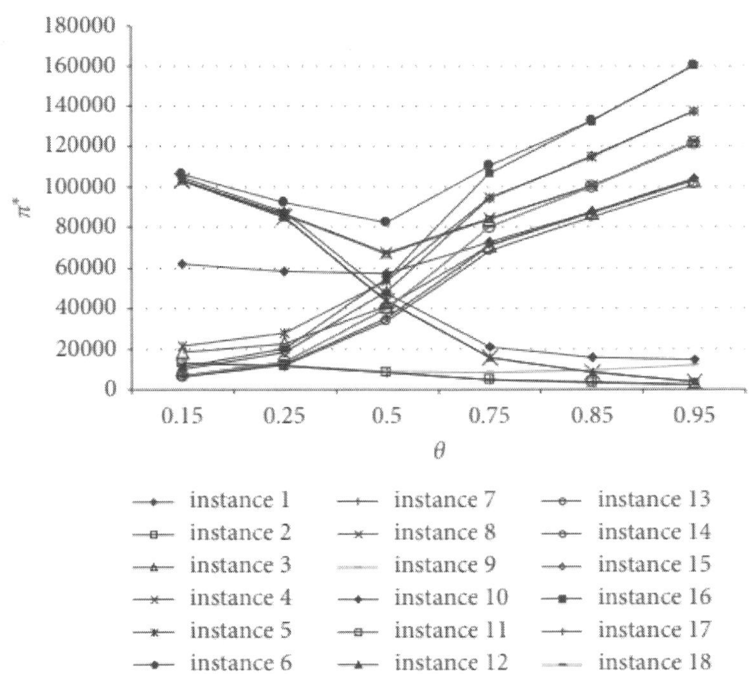

Figure 7. Supply chain profit in the centralized dual-channel supply chain for different instances.

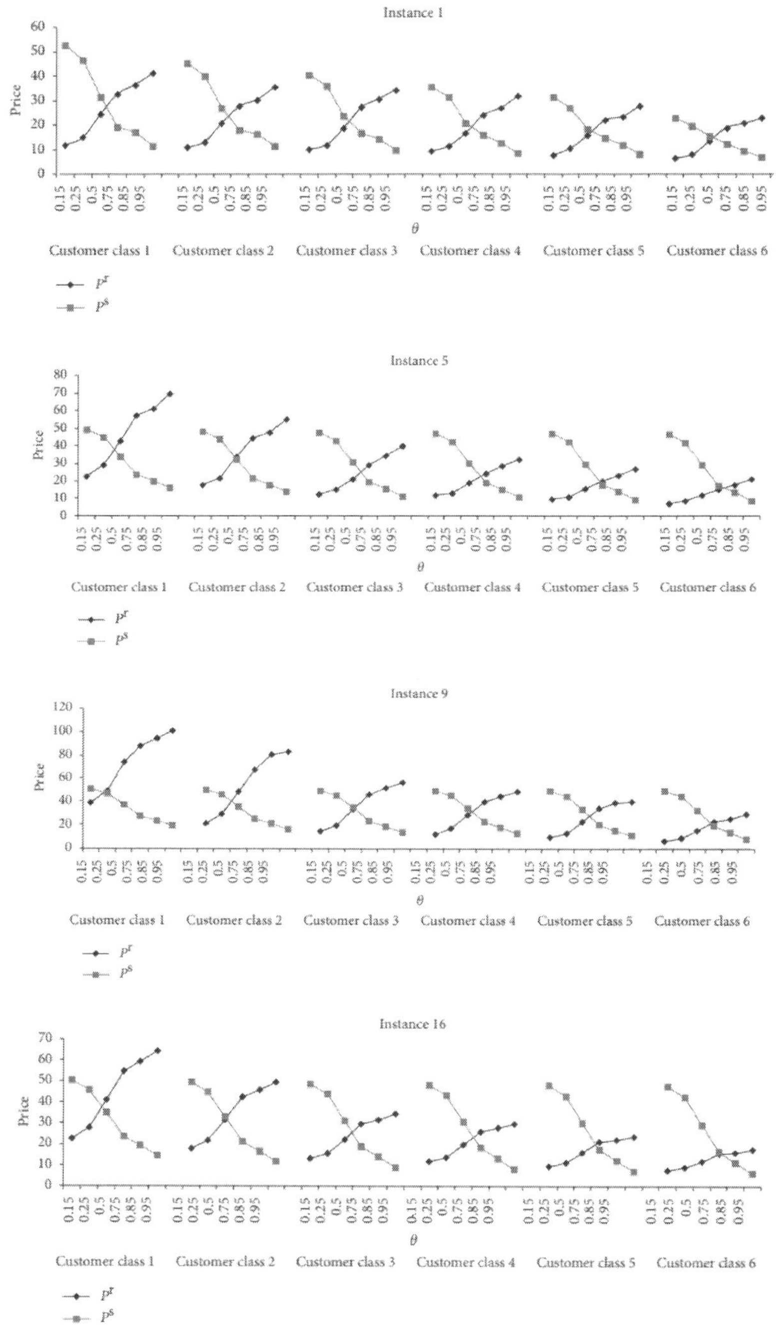

Figure 8. Comparison of the retail price and direct sale price in centralized dual-channel supply chain for different instances.

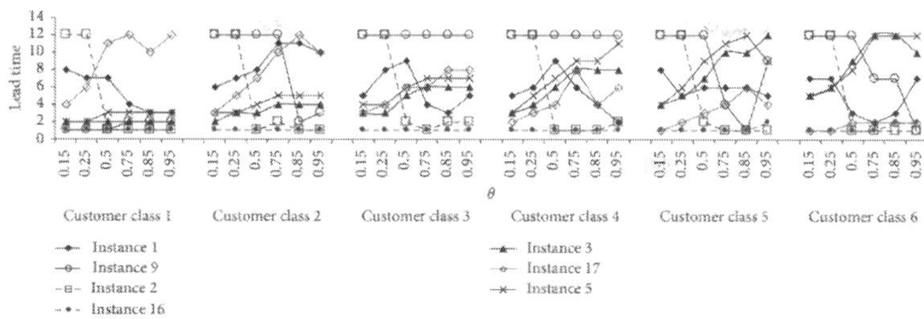

Figure 9. Comparison of the optimal quoted lead time in centralized dual-channel supply chain for different instances.

Table 4. Comparison of retailer and direct prices under both decentralized and centralized supply chain.

Problem number	$\overline{P}^r_{II} - \overline{P}^r_{I}$						$\overline{P}^s_{II} - \overline{P}^s_{I}$					
	$\theta = 0.15$	$\theta = 0.25$	$\theta = 0.5$	$\theta = 0.75$	$\theta = 0.85$	$\theta = 0.95$	$\theta = 0.15$	$\theta = 0.25$	$\theta = 0.5$	$\theta = 0.75$	$\theta = 0.85$	$\theta = 0.95$
1	−5.20	−5.99	−0.23	0.88	0.15	0.00	0.06	−0.36	−1.94	−0.93	−0.06	0.00
2	−5.24	−5.74	0.15	0.65	0.37	0.02	−0.19	−4.24	−5.18	−1.39	−0.36	−0.01
3	−0.45	−0.44	−0.97	0.05	0.00	−0.05	−0.05	−0.08	−0.12	−0.05	0.00	−0.01
4	−5.64	−4.52	0.19	−0.02	0.37	0.01	−0.26	−4.64	−4.92	−2.16	−0.36	−0.01
5	−0.82	−1.11	−0.86	0.44	0.01	−3.49	0.03	−0.02	−0.07	−0.14	0.00	−0.62
6	−4.47	−3.76	−0.34	−0.01	0.27	−3.81	−1.53	−5.06	−5.06	−2.18	−0.61	−0.69
7	−4.22	−4.22	−1.53	0.00	0.00	0.00	1.28	1.28	0.47	0.00	0.00	0.00
8	−5.24	−6.21	−1.32	−0.79	−0.58	−0.37	−0.19	−5.31	−6.99	−4.20	−3.08	−1.97
9	−0.78	−1.42	−1.07	0.68	1.29	−5.67	0.67	0.55	0.46	−0.28	−0.29	−1.25
10	−5.31	−4.60	−1.82	−0.74	−1.42	−6.64	−1.74	−5.44	−7.01	−4.46	−3.20	−2.89
11	−4.99	−4.99	−2.63	0.00	0.00	−0.76	1.14	1.14	0.29	0.00	0.00	−0.11
12	−5.24	−5.96	−1.32	−0.79	−0.58	−1.41	−0.19	−5.34	−6.99	−4.20	−3.08	−1.94
13	0.00	0.00	0.00	0.00	0.00	0.45	0.00	0.00	0.00	0.00	0.00	−0.10
14	0.00	0.00	0.00	0.00	0.00	0.02	0.00	0.00	0.00	0.00	0.00	−0.01
15	0.00	0.00	0.07	0.09	−0.01	−3.96	0.00	0.00	0.01	0.02	0.00	−0.69
16	0.00	0.00	0.00	0.00	0.00	−3.81	0.00	0.00	0.00	0.00	0.00	−0.69
17	0.00	0.00	−0.01	0.05	0.00	0.28	0.00	0.00	0.03	0.01	0.00	−0.14
18	0.00	0.00	0.00	0.00	0.00	0.01	0.00	0.00	0.00	0.00	0.00	−0.01

Table 5. Comparison of obtained profits under both decentralized and centralized supply chain.

	$\Pi_{II} - \Pi_{I}$					
Problem number	$\theta = 0.15$	$\theta = 0.25$	$\theta = 0.5$	$\theta = 0.75$	$\theta = 0.85$	$\theta = 0.95$
1	3884.36	5657.56	3091.33	136.57	91.42	16.27
2	6589.72	9055.78	6495.73	839.46	75.91	0.00
3	75.40	98.32	336.52	8.66	0.00	−167.17
4	6713.26	9965.70	7062.69	908.76	75.91	0.00
5	440.87	582.24	502.25	272.85	−438.03	609.73
6	8812.77	10719.97	5635.36	965.10	123.80	0.00
7	385.00	385.00	140.00	0.00	0.00	67.57
8	6589.72	15270.77	10452.74	3773.87	2034.84	971.90
9	566.32	612.44	912.49	101.92	96.21	108.02
10	11481.49	16019.67	10816.05	4407.86	2449.25	1022.60
11	354.53	354.53	81.87	3.20	3.20	54.06
12	6603.00	13601.09	10455.94	3777.07	2038.04	1062.36
13	0.00	0.00	0.00	0.00	0.00	0.00
14	0.00	0.00	0.00	0.00	0.00	0.00
15	0.00	0.00	111.02	130.35	147.26	−185.62
16	0.00	0.00	0.00	0.00	0.00	0.00
17	0.00	0.00	−47.52	156.25	−10.91	166.67
18	0.00	0.00	0.00	0.00	0.00	0.00

As we can see from Figures 5 and 8, in both the centralized and decentralized dual-channel supply chain, when customer preference of the direct channel θ is below a certain level, the retail price is higher than the direct sale price; conversely, when θ exceeds that level, the direct price becomes higher than the retail sale price. This result shows that if the base level of demand or demand rate in one channel is relatively higher than a certain threshold, the sale price in that channel should be set higher than the one in the other channel.

From Figure 5, in some examples, when θ is relatively low, the direct sale price should be set equal to the wholesale price, and when θ is relatively high, the direct sale price should be set higher than the wholesale price. In other words, when θ is lower than a certain threshold, the base level of demand in the retail channel is high. Thus, the retail price can also be set to be high, but according to the low base demand in the direct channel, the direct sale price is set to be low.

The wholesale price must be less than or equal to the direct sale price. Therefore, the direct sale price is equal to the wholesale price, and the retail price is higher than the wholesale price; conversely, when θ exceeds the threshold, the base level of demand in the retail channel is low, and the retail price is generally also set to be low.

We know that the wholesale price must be lower than the retail price. Therefore, with increasing θ, the retail price and consequently the wholesale price decrease.

As we can see from Figures 5 and 8, for some values of θ, both the retail price and the direct price in the decentralized dual-channel supply chain should be set higher than those in the centralized one, while the total profit in the decentralized dual-channel supply chain is lower than that in the centralized one.

The negative value of gap in price differences in Table 4 and the positive value of gap for profits in Table 5 show this fact. This shows that when each partner in the decentralized dual-channel supply chain maximizes his own profit, this leads to double marginalization. Double marginalization means that the retailer and the manufacturer independently set their price to maximize their profit margins; as a result, the price is higher and the sales volume and profits are lower than those of a vertically integrated channel [3].

In instances with low (Ce/Cr) or with high Cr, the direct sale price in the centralized setting tends to equal the direct sale price in decentralized settings. This trend is also true for retail prices in the two settings. Moreover, when Cr is very large, the retail channel has a small impact on the manufacturer's profits, and in both settings, the manufacturer has a main role in determining prices.

Therefore, the prices defined in the two settings become close. As illustrated in Figure 7, in the centralized dual-channel supply chain, when customer preference of the direct channel θ is lower than a certain threshold, the total profit decreases with increasing θ; conversely, when θ exceeds the threshold, the total profit increases with increasing θ. This threshold differs from one instance to another. In examples with high direct channel operational cost, the threshold is near 1, meaning the function seems as decreasing function. To compare the profitability changes of the firm according to different direct and retail channel operational costs, we carried some more examples by varying these two parameters. Figure 10 shows the results. As we can see from Figure 10, under high value of direct channel operational cost, the profit decreases with increasing θ. Therefore, it is more profitable to encourage customers buying from retail channel.

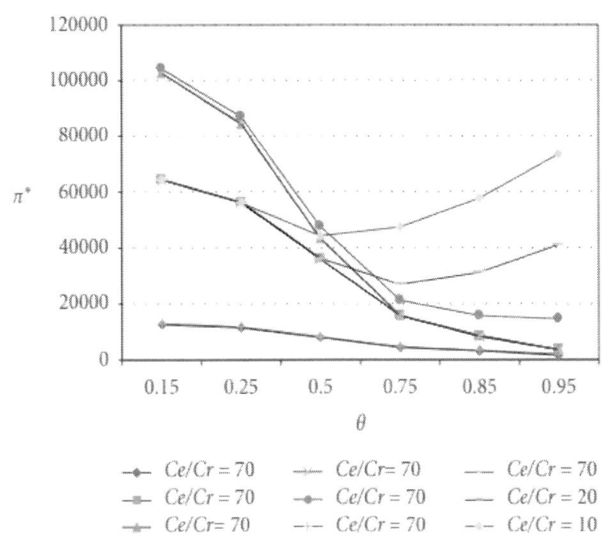

(a)

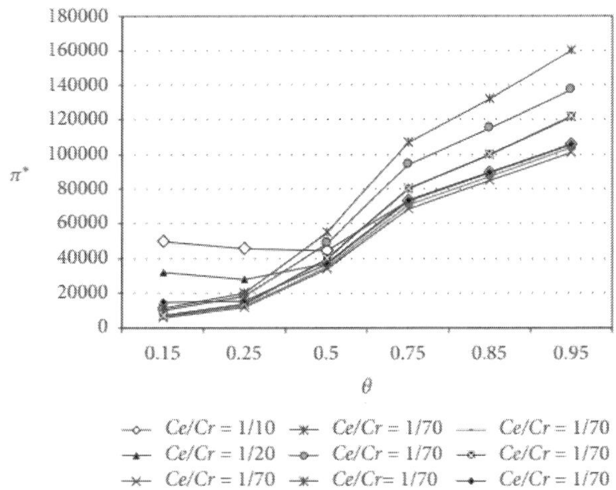

(b)

Figure 10. Comparison of the profitability of the firm according to different direct and retail channel operational costs.

As we can see from Figures 6 and 9, the high lead time changes depending on in two centralized and decentralized setting show that the delivery lead time has a strong effect on the manufacturer's and the retailer's pricing strategies and profits. We can see the results of Propositions 4 and 5 in instances 2 and 16, where the unlimited capacity is considered. The quoted lead times for the customers are equal to the lower bound or the upper bound of the lead time.

We perform a numerical study to compare how the lead time flexibility affects the profitability of the firm.

We consider two policies of lead time flexibility: P1 (lead time flexibility), P2 (no lead time flexibility). With lead time flexibility, we quote different lead times to different customers. When there is no lead time flexibility, a single (fixed) lead time is quoted to all customers.

We have generated the problems in three situations: the customers' sensitivity to price in direct channel is 1- greater, 2- equal, and 3- lower than customers' sensitivity to price in retail channel. We have considered the same price sensitivities for all customers. The categorization for production capacity and variability of customers' lead time sensitivity is shown in Table 3. The percentage of profit increases over P2 policy with no lead time flexibility under each policy and category is represented in Table 6. The main conclusions one can draw from Table 6 are as follows:

(1)The effect of production capacity: comparing instances 2 and 4 (instance 4 with high production capacity), we can see that more capacity makes the manufacturer able to charge equal lead times to all customers, and manufacturers usually determine a lead time equal to that of the first period for all customers.

(2)The effect of variability of customers' sensitivity to lead time: comparing instances 1 and 2 (instance 1 with high variability in customers' sensitivity), we can see that if there is high variability in customers' lead time sensitivities, the manufacturer can obtain more profit from lead time flexibility. The manufacturer can charge the high lead time for customer with lower sensitivity to lead time and reserve the capacity of the first period for customers with higher sensitivity to lead time.

Table 6. The percentage of profit increases over P2 policy with no lead time flexibility.

Problem number	Capacity (K)	Lead time sensitivity variability (LV)	Gap = $((P1 - P2)/P2) * 100$		
			$b_i^s > b_i^r$	$b_i^s = b_i^r$	$b_i^s < b_i^r$
1	M	Yes	19.74	19.81	22.86
2	M	No	17.20	15.98	18.34
3	H	Yes	1.08	1.64	0.11
4	H	No	0	0.80	0

CONCLUSION

In this paper, we presented a pricing and due date setting model for a manufacturer with a dual sale channel, an online direct channel, and a traditional retail channel. We see that a large number of e-retailers, such as Amazon.com, BestBuy.com, and Walmart.com, attempt to offer a proper delivery lead time (Yang et al. [4]). Also the model can be developed for 's auto industry which is the second largest industry in Iran after oil and gas. Currently, Iran's auto industry has an online sale and the proposed list price is sometimes based on delivery time.

We assumed a finite planning horizon divided into periods of equal length. We considered the limit production capacity in each period and multiple classes of customers arriving during these periods. We developed a Stackelberg game to model the decentralized dual-channel supply chain. Under this game framework, the supplier, as the leader, announces the wholesale price to an intermediary in addition to the direct channel sales price and lead time. The intermediary then reacts by choosing a retail price to maximize its own profit. Bilevel programming was developed to model this situation. For a centralized dual-channel supply chain, the traditional retail price, the direct sale price, and the quoted lead time in the direct channel are determined in an integrated manner. Considering the lead time selection for each customer order, mixed binary integer nonlinear programming models for the design of the centralized and decentralized dual-channel supply chains were used.

Because exact algorithms for solving the proposed models can be expensive and time consuming for instances with large numbers of 0-1 variables, the PSO algorithm solving model is adopted to get a satisfactory near-optimal solution efficiently.

Through numerical analyses, we have examined the effects of the customer preference of a direct channel on the manufacturers' and retailers' pricing behaviors. We also compared the optimal lead times, prices, and profits in the two settings of a centralized and decentralized dual-channel supply chain.

Numerical examples have shown that the retail price is decreasing with increasing customer preference of the direct channel and that the direct sale price is increasing with increasing customer preference of the direct channel. Therefore, when the customer preference of the direct channel is low, the retail price is higher than the direct sale price; conversely, when the customer preference of the direct channel is high, the retail price is lower than the direct sale price. Comparing the two settings of dual-channel supply chains, we found that when θ is relatively low and, thus, the base level demand in the retail channel is high, both the retail price and the direct price in the decentralized dual-channel supply chain should be set to be higher than those in the centralized supply chain, while the total profit in the decentralized dual-channel supply chain is lower than that in the centralized supply chain. This shows double marginalization in the decentralized dual-channel supply chain which each member maximizes his own profit.

The high changes in the lead time as a function of θ in the centralized and decentralized settings show that the delivery lead time strongly influences the manufacturers' and the retailers' pricing strategies and profits. Therefore, considering a constant value for the lead time in all scenarios leads to a decrease in profits.

This research can be extended in several directions in future work. First, our model is deterministic, and we can study a model with stochastic demand. Second, we can consider competition among several manufacturers and retailers. Third, it is worth investigating the coordination of the dual-channel supply chain by contracts when the price and lead time are both considered in the models.

DATA AVAILABILITY

The data used to support the findings of this study are available from the corresponding author upon request.

CONFLICTS OF INTEREST

The authors declare that they have no conflicts of interest.

REFERENCES

1. A.A. Tsay and N. Agrawal, "Modeling conflict and coordination in multi-channel distribution systems: a review," in *Handbook of Quantitative Supply Chain Analysis: Modeling in the eBusiness Era 557–606)*, S. D. Simchi-Levi Wu and Z. M. Shen, Eds., Kluwer Academic Publishers, Norwell, MA, USA, 2004.View at: Google Scholar
2. K. L. Webb and J. E. Hogan, "Hybrid channel conflict: causes and effects on channel performance," *Journal of Business & Industrial Marketing*, vol. 17, no. 5, pp. 338–356, 2002.View at: Publisher Site | Google Scholar
3. G. Hua, S. Wang, and T. C. E. Cheng, "Price and lead time decisions in dual-channel supply chains," *European Journal of Operational Research*, vol. 205, no. 1, pp. 113–126, 2010.View at: Publisher Site | Google Scholar
4. J. Q. Yang, X. M. Zhang, H. Y. Fu, and C. Liu, "Inventory competition in a dual-channel supply chain with delivery lead time consideration," *Applied Mathematical Modelling*, vol. 42, pp. 675–692, 2017.View at: Publisher Site | Google Scholar
5. K. Cattani, W. G. Gilland, and J. M. Swaminathan, "Coordinating traditional and Internet supply chains," in *Handbook of Quantitative Supply Chain Analysis: Modeling in the eBusiness Era*, D. Simchi-Levi, & M. Wu, and Shen, Eds., pp. 643–680, Kluwer Academic Publishers, Norwell, MA, USA, 2004.View at: Google Scholar
6. G. Cai, Z. G. Zhang, and M. Zhang, "Game theoretical perspectives on dual-channel supply chain competition with price discounts and pricing schemes," *International Journal of Production Economics*, vol. 117, no. 1, pp. 80–96, 2009.View at: Publisher Site | Google Scholar
7. W.-y. K. Chiang, D. Chhajed, and J. D. Hess, "Direct marketing, indirect profits: a strategic analysis of dual-channel supply-chain design," *Management Science*, vol. 49, no. 1, pp. 1–20, 2003.View at: Publisher Site | Google Scholar
8. S. Panda, N. M. Modak, S. S. Sana, and M. Basu, "Pricing and replenishment policies in dual-channel supply chain under continuous unit cost decrease," *Applied Mathematics and Computation*, vol. 256, pp. 913–929, 2015.View at: Publisher Site | Google Scholar
9. D. Q. Yao and J. J. Liu, "Competitive pricing of mixed retail and e-tail distribution channels," *Omega*, vol. 33, pp. 235–247, 2005.View at: Google Scholar

10. B. Dan, G. Xu, and C. Liu, "Pricing policies in a dual-channel supply chain with retail services," *International Journal of Production Economics*, vol. 139, no. 1, pp. 312–320, 2012.View at: Publisher Site | Google Scholar
11. Q.-H. Li and B. Li, "Dual-channel supply chain equilibrium problems regarding retail services and fairness concerns," *Applied Mathematical Modelling*, vol. 40, no. 15-16, pp. 7349–7367, 2016.View at: Publisher Site | Google Scholar
12. A. Roy, S. S. Sana, and K. Chaudhuri, "Joint decision on EOQ and pricing strategy of a dual channel of mixed retail and e-tail comprising of single manufacturer and retailer under stochastic demand," *Computers & Industrial Engineering*, vol. 102, pp. 423–434, 2016.View at: Publisher Site | Google Scholar
13. A. A. Tsay and N. Agrawal, "Channel conflict and coordination in the e-commerce age," *Production and Operations Management*, vol. 13, pp. 93–110, 2004.View at: Google Scholar
14. Z. Zhang, S. Liu, and B. Niu, "Coordination mechanism of dual-channel closed-loop supply chains considering product quality and return," *Journal of Cleaner Production*, vol. 248, Article ID 119273, 2020.View at: Publisher Site | Google Scholar
15. S. Huang, C. Yang, and H. Liu, "Pricing and production decisions in a dual-channel supply chain when production costs are disrupted," *Economic Modelling*, vol. 30, pp. 521–538, 2013.View at: Publisher Site | Google Scholar
16. Z. Pi, W. Fang, and B. Zhang, "Service and pricing strategies with competition and cooperation in a dual-channel supply chain with demand disruption," *Computers & Industrial Engineering*, vol. 138, 2019.View at: Publisher Site | Google Scholar
17. F. Soleimani, A. Arshadi Khamseh, and B. Naderi, "Optimal decisions in a dual-channel supply chain under simultaneous demand and production cost disruptions," *Annals of Operations Research*, vol. 243, no. 1-2, pp. 301–321, 2016.View at: Publisher Site | Google Scholar
18. T. M. Rofin and B. Mahanty, "Optimal dual-channel supply chain configuration for product categories with different customer preference of online channel," *Electronic Commerce Research*, vol. 18, no. 3, pp. 507–536, 2018.View at: Publisher Site | Google Scholar

19. T. Xiao, T.-M. Choi, and T. C. E. Cheng, “Product variety and channel structure strategy for a retailer-Stackelberg supply chain,” *European Journal of Operational Research*, vol. 233, no. 1, pp. 114–124, 2014. View at: Publisher Site | Google Scholar
20. J. Zhou, R. Zhao, and B. Wang, “Behavior-based price discrimination in a dual-channel supply chain with retailer’s information disclosure,” *Electronic Commerce Research and Applications*, vol. 39, 2020.View at: Publisher Site | Google Scholar
21. I. Duenyas and W. J. Hopp, “Quoting customer lead times,” *Management Science*, vol. 41, no. 1, pp. 43–57, 1995.View at: Publisher Site | Google Scholar
22. K. C. So and J.-S. Song, “Price, delivery time guarantees and capacity selection,” *European Journal of Operational Research*, vol. 111, no. 1, pp. 28–49, 1998.View at: Publisher Site | Google Scholar
23. F. F. Easton and D. R. Moodie, “Pricing and lead time decisions for make-to-order firms with contingent orders,” *European Journal of Operational Research*, vol. 116, no. 2, pp. 305–318, 1999.View at: Publisher Site | Google Scholar
24. P. Keskinocak, R. Ravi, and S. Tayur, “Scheduling and reliable lead-time quotation for orders with availability intervals and lead-time sensitive revenues,” *Management Science*, vol. 47, no. 2, pp. 264–279, 2001.View at: Publisher Site | Google Scholar
25. S. Webster, “Dynamic pricing and lead-time policies for make-to-order systems,” *Decision Sciences*, vol. 33, pp. 579–599, 2002.View at: Google Scholar
26. B. Watanapa and A. Techanitisawad, “Simultaneous price and due date settings for multiple customer classes,” *European Journal of Operational Research*, vol. 166, no. 2, pp. 351–368, 2005.View at: Publisher Site | Google Scholar
27. K. Charnsirisakskul, P. M. Griffin, and P. Keskinocak, “Pricing and scheduling decisions with leadtime flexibility,” *European Journal of Operational Research*, vol. 171, no. 1, pp. 153–169, 2006.View at: Publisher Site | Google Scholar
28. A. Mustafa, B. Ata, and T. Olsen, “Congestion based lead time quotation for heterogenous customers with convex-concave delay costs: optimality of a cost balancing policy based on convex hull functions,” *Operations Research*, vol. 57, no. 3, pp. 753–768, 2009. View at: Google Scholar

29. S. K. Chaharsooghi, M. Honarvar, M. Modarres, and I. N. Kamalabadi, "Developing a two stage stochastic programming model of the price and lead-time decision problem in the multi-class make-to-order firm," *Computers & Industrial Engineering*, vol. 61, no. 4, pp. 1086–1097, 2011.View at: Publisher Site | Google Scholar

30. S. K. Chaharsooghi, M. Honarvar, and M. Modarres, "A multi-stage stochastic programming model for dynamic pricing and lead time decisions in multi-class make-to-order firm," *Scientia Iranica*, vol. 18, no. 3, pp. 711–721, 2011.View at: Publisher Site | Google Scholar

31. K.-Y. Chen, M. Kaya, and Ö. Özer, "Dual sales channel management with service competition," *Manufacturing & Service Operations Management*, vol. 10, no. 4, pp. 654–675, 2008.View at: Publisher Site | Google Scholar

32. H. Xu, Z. Z. Liu, and S. H. Zhang, "A strategic analysis of dual-channel supply chain design with price and delivery lead time considerations," *International Journal of Production Economics*, vol. 139, no. 2, pp. 654–663, 2012.View at: Publisher Site | Google Scholar

33. R. Batarfi, M. Y. Jaber, and S. Zanoni, "Dual-channel supply chain: a strategy to maximize profit," *Applied Mathematical Modelling*, vol. 40, no. 21-22, pp. 9454–9473, 2016.View at: Publisher Site | Google Scholar

34. N. M. Modak and P. Kelle, "Managing a dual-channel supply chain under price and delivery-time dependent stochastic demand," *European Journal of Operational Research*, vol. 272, no. 1, pp. 147–161, 2019. View at: Publisher Site | Google Scholar

35. H. Kurata, D.-Q. Yao, and J. J. Liu, "Pricing policies under direct vs. indirect channel competition and national vs. store brand competition," *European Journal of Operational Research*, vol. 180, no. 1, pp. 262–281, 2007.View at: Publisher Site | Google Scholar

36. G. L. Nemhauser and L. A. Wolsey, *Integer and Combinatorial Optimization*, Wiley-Interscience, New York, NY, USA, 1988.

37. J. Kennedy and R. Eberhart, "Particle swarm optimization," in *Proceedings of 1995 IEEE International Conference on Neural Networks*, pp. 1942–1948, IEEE, Perth, Australia, November 1995. View at: Publisher Site | Google Scholar

38. J. Kennedy, RC. Eberhart, and Y. Shi, *Swarm Intelligence*, Morgan Kaufmann, San Mateo, CA, USA, 2001.

39. S.KozielandZ.Michalewicz,"Evolutionaryalgorithms,homomorphous mappings, and constrained parameter optimization," *Evolutionary Computation*, vol. 7, pp. 19–44, 1999.View at: Google Scholar

40. J. H. Holland, *Adaptation in Natural and Artificial Systems: An Introductory Analysis with Applications to Biology, Control, and Artificial Intelligence*, University of Michigan Press, Ann Arbor, MI, USA, 1975.

CHAPTER 3

Inverting the Multiple-Assisting Tool Network Problem to Solve for Optimality

Robert Rich[1]

[1]Industrial & Systems Engineering, Liberty University, Lynchburg, VA, USA

ABSTRACT

Many network problems deal with the routing of a main tool comprised of several parallel assisting tools. These problems can be found with multi-tool-head routing of CNC machines, waterjets, plasma sprayers, and cutting machines. Other applications involve logistics, distribution, and material handling that require a main tool with assisting tools. Currently no studies exist that optimally route a main tool comprised of and fitted with multiple tools, nor do any studies evaluate the impact of adding additional

Citation: Robert Rich, "Inverting the Multiple-Assisting Tool Network Problem to Solve for Optimality", Advances in Operations Research,vol.2020,Article ID 3515709,13 pages,2020.https://doi.org/10.1155/2020/3515709

capabilities to the tool set. Herein we define the network routing problem for a main tool comprised of multiple secondary tools. We introduce first principles to properly configure the main tool with the appropriate number of supporting tools such that that system is not overstatured. We invert the network geometry to extract the "best case" configuration for toolset configuration to include speed, range, and number of such that the system is lean. Our computational studies reveal that the theorems introduced herein greatly improve the overall system performance without oversaturating it with unused resources. In order to validate experiments, we define a mixed integer program and compare it to our metaheuristics developed for experimentation. Both the MIP and the metaheuristics herein optimally route a main tool with multiple assisting tools as well as the routing of a parcel delivery truck comprised of many drones.

INTRODUCTION

The use of multiple tools consisting of a main tool and several parallel resources in conjunction with the primary tool offers processing efficiencies in a range of applications. Semiconductor manufacturers are looking at ways to allow autonomous vehicles to interact such that overall travel time and traffic-congestion are reduced. Advances in multiheaded CNC and 3D printing-devices are expected to reduce print time significantly. In these designs, the main tool is confined to the primary axis of performance while assisting tool heads operate in parallel to the main tool. From a logistics standpoint, UPS, FedEx, and Amazon are evaluating the use of a single truck that is fitted with multiple parcel delivery drones to achieve greater efficiencies in delivery time. The simplest example of this problem can be visualized as a mother with several children assisting with grocery shopping. The family has one grocery cart, and children are dispatched in different directions around the store to retrieve items back to the cart while the parent routes through the store. It is obvious that there exists a set of optimal routes that the mother and children take that would optimize the total shopping time. However, it is less clear how the underlying relationships between the speed of the mother and children as well as their ranges of travel will also greatly impact any gained efficiencies in shopping. If a child is exceptionally slow, then the parent is better off to carry the child. If the child is exceptionally fast, but does not have the travel range, then the child's speed gives little advantage. Furthermore, that if the mother were to bring all her children into the store (in which case she has more than a dozen), she would be overwhelmed by idle children accompanying the overburdened shopping cart with much complaining due to their unemployed idle time.

So, it is easy to imagine that there exists a set of optimal assisting tool parameters (speed, range, and number of) that properly define the configuration of the overall system. For simplicity, we adopt a parcel delivery truck comprised of and fitted with multiple delivery drones to describe the problem. The main idea here is that there exists a geometric relationship between the size and delivery-density of the total operations area (Figure 1) and the density of truck-drone operations area. Specifically, the size and number of deliveries in the total operational area needs to be proportional to the size and number of deliveries in the truck-drone operations area for the most efficient operations. In words, a single truck-drone operation is described as a truck launching one or more drones, each drone traverses out to make deliveries, and then subsequently rendezvous with the truck at a downstream delivery location. In a single truck-drone operation, we know that there is a max coverage area (denoted by the elliptical circle) that n drones can reach (being constrained by range) and a total number of deliveries that can be performed. To achieve the maximum efficiencies, the truck-drone operation area density needs to be proportional to the total density of the larger delivery space. Since the larger space density is fixed, we adjust the drone parameters (speed, range, and number of) in order to match the truck-drone operations density to the larger delivery density. Thus, we advise to invert the network problem and solve the assisting tool parameters *a priori* such that they fit "*the most*" typical delivery scenarios for the delivery region in order to achieve maximum efficiencies.

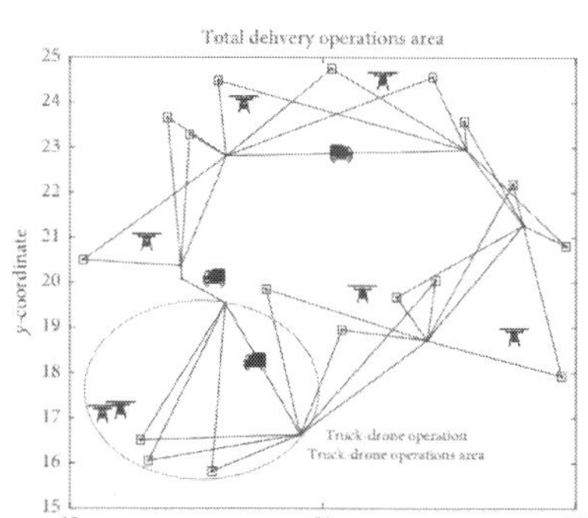

Figure 1. Total delivery operations area proportional to the truck-drone operations area.

In Section (2), we lay out the first principles geometric relationships between drone parameters and the truck-drone operations area. The notion here is that an operations is comprised of several truck-drone routing triangles. Each of these triangles fits within the area of an ellipse defined by truck speed and drone range. Thus, the elliptical area of the truck-drone operation and the number of deliveries within this space must be proportional to the larger delivery space delivery-density.

To support the working theory, we conduct several computer experiments that reveal the production curve and "efficiencies gained" for systems described by the geometric relationships compared to systems defined by standard drone parameters of speed and range found in the literature. Section (3) reveals the results of these experiments.

An evolutionary algorithm (EA) was developed to perform empirical tests of our theory. The EA solves the truck-multidrone network routing problem by minimizing the total cumulative time based on either the max truck time or the max drone time for each set of launch-deliver-rendezvous operations cumulated in the total tour. Figure 1 shows ~optimal routing solution for one of the problem scenarios.

The algorithm is used to test the performance (total delivery time) of a *lean* configured truck-drone/s system to a *standard* configured system. The EA is validated by comparing it to the mixed integer program (MIP) shown in Section (5). The MIP is also validated by comparing it to known solutions and to brute force methods for smaller problem sets. As such, the validated metaheuristic solves problems optimally for nearly all of the smaller problem sets involving ten or less delivery stops and ~optimal for problems as large as one hundred stops.

The sections comprised herein are as follow: Section (2) discusses working theory and theoretical insights. Section (3) reveals the results of the hypothesis testing performed in relation to the fundamental theorems.

Section (4) discusses the literature surrounding the network routing problems. Section (5) defines a tractable version of the mixed integer programming (MIP) used to validate metaheuristics that can be easily adopted to optimization software tools available to researchers. Section (6) formulates EA used to higher dimension truck-multidrone routing problems. Section (7) gives comparative studies: the first comparative study shows the performance of the evolutionary algorithms (EA-1 and EA-2) used to solve the optimal routing for empirical studies. The algorithms are compared to each other and to MIP-solved optimal solutions. The second study compares a *lean* configured system to a standard system configurations found in the literature. Section (8) concludes the research findings.

THEORETICAL INSIGHTS

Truck-Drone Performance Boundaries

The simplest truck-drone example involves a truck and one drone (Figure 2). For this, there are three nodes comprised of a depot v_0 and two delivery locations i and j. If the truck's speed is scaled to one (1) and the drone's speed is a factor of the truck speed, then an optimal configuration exists when the drone's range (κ) equals the drone's speed factor (α). So, as the truck travels out one unit distance from the depot (Figure 2) to a delivery (i), the drone also travels one (κ) units of distance (where α = κ) from the depot to delivery location j for best configuration. The important thing here is that under the "best" configuration both the truck and drone return at the same time such that neither is waiting on the other. So, the best percent time (Π) improvement that the dual system can possibly achieve over a truck-only system (truck having to deliver the two parcels itself and return to the depot tsp) is defined by [1].)us, the best case (Π) of a truck-one-drone over a truck-only system is

$$\Pi = 1 - \left(\frac{2}{2\alpha + 2}\right) = 1 - \left(\frac{1}{a + 1}\right). \quad (1)$$

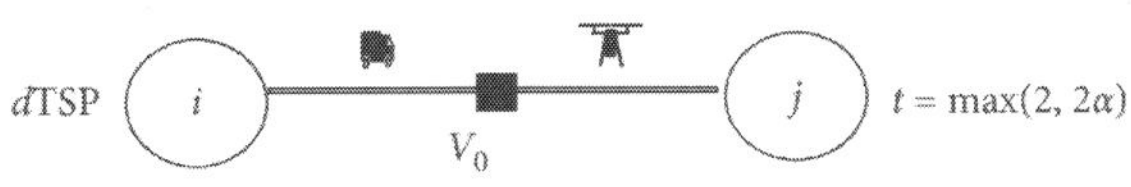

Figure 2. Best case for truck-drone (Agatz et al., 2015).

Truck-Multiple-Drone Performance Boundaries

The more complex case for a truck fitted with multiple drones necessarily involves more than three delivery locations, say (i, j, m, v_0) For this argument, we assume that a well-configured system will involve the drones as much as possible throughout the delivery process. Therefore, if there are N delivery stops to be made during the day, then the "best case" number of deliveries allotted to each resources is (N) divided by the number of assisting drones (v) plus the main tool (truck): (N/(v + 1)). Furthermore, that delivery density plays an important role in determining the proper speed (α) and range (κ) factors in the analysis where total delivery density ρ is defined as the number (N) of required deliveries for the operation divided by the size of the geometric operating or delivery area (A) as in $\rho = N/A$. A densely populated delivery area requires more short-range drones while a sparse delivery density requires longer range and higher speed drones.

In order to construct a case for the *theoretical upper boundary* in potential performance improvement, we examine the geometry of a circle. We surmise that the best case (truck-multidrone) system for delivery time performance Π (over a truck-only) can be determined by launching drones from the depot, sending the drones out to delivery locations along the perimeter of the circle and returning to the depot as shown in (Figure 3). Both drones and truck launch from the depot, traverse out, and return to the depot. Conversely, in an absolute worst case scenario, the truck would have to deliver to all delivery locations. This means that the worst-case truck route having no drones (i.e., hub-spoke route) will have to traverse from depot to edge of circle and then back to the depot.

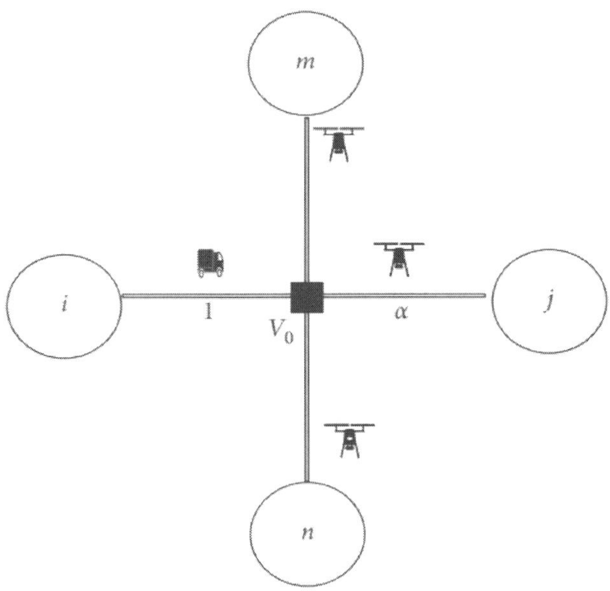

Figure 3. Practical best case truck-multiple drone delivery.

Concretely, there exists a maximum and minimum yield (Π) on potential performance improvement for a truck and multiple drone/s system over the truck-only system. The worst case performance (lower boundary) is the truck having to make all the deliveries itself inside the circle (hub-spoke).

The best case improvement requires the drones to traverse out to the radius of the circle (α); each makes a delivery, and then traverses back to the depot rendezvousing simultaneously with the truck.

Summarizing, if the truck traverses out one unit distance while the drones traverse κ units of distance where ($\kappa = \alpha$), then the max time for any truck or drone is 2 units of time (Figure 3).)us, a worst-case situation would exist if a truck had to visit all v + 1 deliveries on the circle (hub-spoke fashion). In such case, the absolute worst time the truck could encounter would be $2(1 + v\alpha)$ where (v) is the number of deliveries around the circle (thus number of drones) at radius (α) and one delivery made by the truck (1). Conversely, the best case (Π) improvement for truck-multidrone (v) over a truck-only system (Π) is

$$\Pi = 1 - \frac{2}{2 + 2\nu\alpha} = 1 - \frac{1}{1 + \nu\alpha}. \qquad (2)$$

Truck-Multidrone Operations Area

Expanding on the theoretical delivery problem means that trucks and each drone must traverse through the operating space and, therefore, do not return immediately to the depot.

As such, drones travel using a triangular path to launch and deliver and then rendezvous with the truck downstream where the two legs of the drone's triangular path cannot exceed the drone's total range kappa (κ). Since the truck and the drone do not have the same speed,

we have to adopt a configuration using an ellipse (as opposed to a circle) (Figure 4) such that the foci f points on the ellipse denotes the truck's delivery locations, while a triangle formed inside the ellipse (using the foci points and edge of ellipse) denotes any of the potential launch-delivery-rendezvous operation for any of the drones.

Thus, the drones are confined by range to operate within the area of the ellipse where they are launched. They are launched from the first foci point and must then be retrieved at the second foci of the ellipse by the truck in a perfect scenario. Therefore, it is easy to imagine that there exists a proper sizing of the ellipse whereby the truck's speed and distance (from foci to foci) must be aligned with the drone's range, drone's speed, and the number of drones. The number of drones is based on the density of deliver locations within the elliptical area.

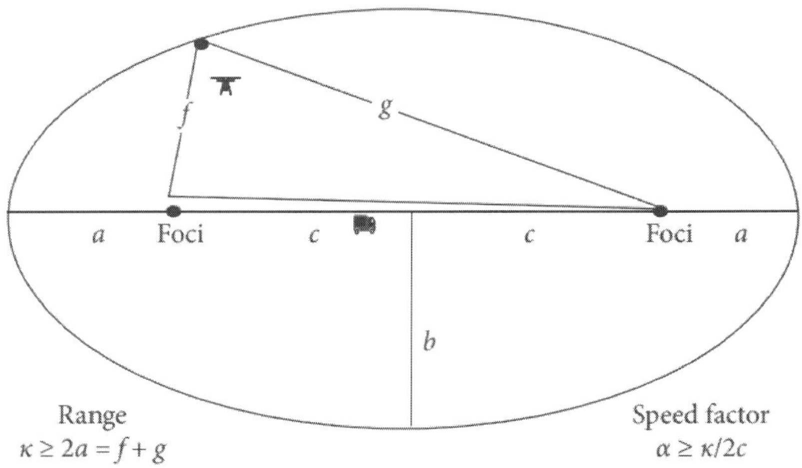

Figure 4. Example of truck-drone delivery inside an ellipse.

The first interrelationship requires that in order for the drone to traverse out to its maximum distance and return back to the truck without any idle time, it is require that the truck-only travel kappa divided by alpha κ/α distance. Therefore, the distance between the two foci is a factor of both drone range and drone speed. For example, if the drone travels twice the speed of the truck ($\alpha = 2$) and the drone can travel a total of 10 km ($\kappa = 10$), then the truck should only travel 5 km before having to retrieve the faster drone having travelled 10 km in the same time period if there is to be no idle time. As such, we limit our elliptical area such that from foci to foci is denoted as (2c) must be equal to (κ/α) for a perfectly lean scenario.

The second interrelationship requires that the total distance that the drone can travel is limited by the drone's range. Therefore, the drone should be limited to twice the axis of the major axis ($\kappa = 2a$). For relaxation of the problem, drone range must be greater than twice the major axis ($\kappa \geq 2a$), and truck travel distance (2c) at speed one is consistent with drone speed factor ($\alpha \geq \kappa/2c$) where the truck speed s_T and drone speed s_D are defined by speed factor alpha $\alpha = s_D/s_T$.

Moreover, delivery density directly reflects the number of drones assigned to deliver within the elliptical area. We can imagine that if given the total delivery density (ρ) for the entire area is comprised of total deliveries (N) and a total area (A), then delivery density would be computed as $\rho = (N/A)\ km^2$. Under a best-case delivery density scenario, the densities of two of the ellipses (ρ') would need to be proportional to the total delivery density for the problem space as in $\rho = \rho'$.)us, we define the truck-drone operations area as that of two overlapping ellipses joined at the foci (minus intersection) as

$$A' = 2\pi ab - \left[\sqrt{\Phi}\,\pi(a-c)^2\right], \qquad (3)$$

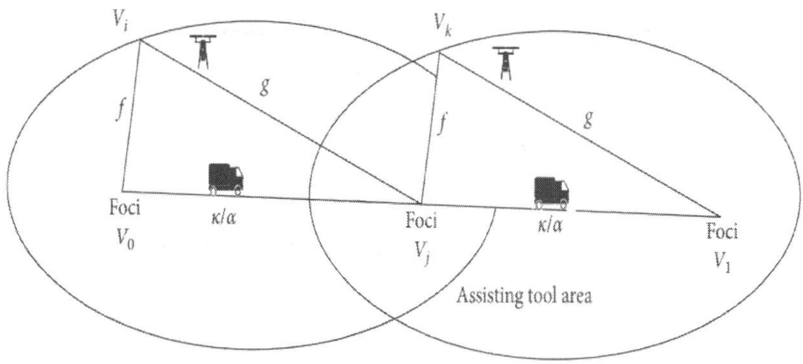

Figure 5. Best-case geometric layout for truck-multiple-drone scenario.

where (a) is the major axis, (b) is the minor axis ($a > b$), (c) is the foci distance from centre to foci, and (Φ) is the golden ratio (~1.618) shown in figure (Figure 5).

Furthermore, based on the interrelationship between truck speed, drone speed, and drone range, the total distance between two foci (2c) is approximately equal to (κ/α). Therefore, we would expect to see N′ deliveries within the dual elliptical area A′.)us, we compute the number of expected deliveries inside any of our elliptical drone operational areas as $N' = A'(\rho)$ for a "practical best-case" scenario.

In summary, we can surmise for practical best-case situation that twice the distance from the centre of the ellipse to the foci is a factor of both drone range (κ) and drone speed (α) as in ($2c = \kappa/\alpha$) so that the truck and drone's operation reduces idle time and that drone range is greater than or equal to the two legs of the triangle ($\kappa \geq 2a$) (Figure 5). In such case, the "practical best case" for the number of drones (v) fitted to the truck is based on N′ deliveries needed to be within the combined area of the two ellipses (A′). Also, since the truck can make three of the total (N′) deliveries, then the number of drones required to be fitted to the truck under "best case" scenario is $v = (N' - 3)$.

Using the area of two ellipses A′ and the appropriate "practical best case" density ρ′ for a given system, the problem can be inverted by knowing a priori the delivery density ρ and then forcing ρ′ to be equal to ρ by selecting proper parameters for range, speed, and number of drones.)us inverting the problem, we solve for the drone range, drone speed, and number of drones that would produce a "practical best case" geometric design for the problem.

Theoretical Summary

In summary, the hypothesis for dvtsp is as follows.

Maximum Theoretical Upper Boundary

There exists a maximum theoretical upper boundary for best possible percent time improvement (Π) over a truck-only system that will never be exceeded without special construction of problem scenario based on the assisting drone's time improvement factor (1 + α]) where α denotes the speed factor and v denotes the number of drones assigned.

$$\Pi < 1 - \frac{1}{1 + \alpha\nu}. \tag{4}$$

Lower Boundary

The *theoretical lower boundary percent* improvement (worst case) for a truck-multiple-drone system over a single truck only solution is Π = 0% based on the fact that a truck can perform all the deliveries by itself and default to a single truck route, thus no time improvement.

Improvement

Improvement Π moves toward the practical upper boundary as the delivery density of the given operating area (ρ) scenario moves toward the calculated theoretical "best case" delivery density (ρ′). As such, the problem can be inverted. As the absolute value between the problem delivery density (ρ) and the "practical best case" delivery density (ρ′) are minimized, the best case (optimal) drone speed factor α, drone range κ, and number of drones v based on the geometric layout of the ellipse emerge. Due to the stochastic nature of most problem scenarios, the optimal drone speed, drone range, and number of drones is a lower boundary optimal.

$$\min|\rho - \rho'|, \tag{5}$$

S.T.

$$\kappa \geq 2a, \tag{6}$$

$$\nu = N' - 3, \tag{7}$$

$$\alpha \geq \frac{\kappa}{2c}, \tag{8}$$

$$N' = A'(\rho), \tag{9}$$

$$\rho = \frac{N}{A}, \tag{10}$$

$$\rho' = \frac{N'}{A'}, \tag{11}$$

$$A' = 2\pi ab - \left[\sqrt{\Phi}\,\pi(a-c)^2\right], \tag{12}$$

$$c^2 + b^2 = a^2, \tag{13}$$

$$\alpha = \frac{s_D}{s_T}, \tag{14}$$

$$s_D > s_T, \tag{15}$$

$$s_{D_{LB}} \leq s_D \leq s_{D_{UB}}, \tag{16}$$

$$\kappa_{LB} \leq \kappa \leq \kappa_{UB}, \tag{17}$$

$$\nu_{LB} \leq \nu \leq \nu_{UB}, \tag{18}$$

$$\rho, \rho' > 0, \alpha, \kappa, \nu \geq 1. \tag{19}$$

For the proper configuration of the minimization problem, let (a) be the major axis, let (b) be the minor axis, and let (c) denote the distance from foci to the centre of the ellipse. Furthermore, let alpha (α) denote the speed factor of drone as related to the truck's speed of one (1). Let kappa (κ) denote the range of the drone whereby the range must be greater than twice the major axis of the ellipse; let nu () be the number of drones assigned to assist the truck. Let the area of one ellipse be $A_{ellipse} = (\pi ab)$ and the combined area of two ellipses ($A_{two} = 2\pi ab$). Since the ellipses are intersecting at a foci point (c), then let the area of the two combined ellipses be reduced by the area of one of the intersected ellipses [$\sqrt{\Phi}\,\pi(a-c)^2$] where (Φ) is the golden ratio (~1.618).

Concretely, equation (5) minimizes the absolute distance between the problem delivery density and the practical best case delivery density of the two ellipses. Equation (6) forces the range of the drone (κ) to be greater than that of the two legs of the triangle inscribed within an ellipse. Equation (7) solves for the optimal number of drones to be fitted to the truck by allowing the truck the opportunity to deliver to three of the delivery vertices found within the area of the two ellipses. Equation (8) guarantees that the interrelationship between truck speed, drone speed, and range are configured such that for each operation, (launch-deliver-rendezvous) the truck and drone arrive back at nearly the same time to reduce potential idle time for either truck or drone. The truck travels (2c), while the drone travels a max of (2a) and the interrelationship gives enough flexibility to solve the problem by establishing that ($2c \geq \kappa/\alpha$). Equation (9) calculates the expected number of deliveries found within the area of two ellipses (A′). Equation (10) defines the delivery density as the number of deliveries per total area of operation. Equation (11) calculates the delivery density of the two ellipses. Equation (12) defines the area of the two ellipses minus their intersecting area. Equation (13) establishes the geometric relationship of the triangle inscribed within an ellipse where (a) is the hypotenuse. Equation (14) solves for drone speed factor alpha as the ratio of the drone's speed to the truck's speed. Equations (15)–(18) are boundary equations for the drone's speed, truck's speed, drone range, and the total number of drones. Equation (19) ensures that delivery density of ellipses is greater than zero while drone range, drone speed factor, and number of drones are greater than one.

HYPOTHESIS TESTING

To test our hypothesis, thousands of random experiments were conducted by holding range (κ), speed factor (α), and number of drones (]) stationary while perturbing number of deliveries (N) over the delivery area (A). In essence, the optimal delivery density (ρ′) is precalculated based on drone parameters (α, κ, v). For each experiment, delivery density (ρ) changes based on changing number of deliveries (N) within the delivery space (A). The experiments are designed to test the performance of a truck multidrone system against a truck-stand-alone (no drones) system. In the experiments, we see that, when the delivery density (along the x-axis) reaches the optimal density line on the graph, then the overall performance of the system improves until it begins to saturate. In words, the denser the deliveries are in a delivery space, the better the system performs until saturation. Using the fundamentals mentioned above, we can now calculate the optimal density (ρ′) at which the system saturates. The figure below (Figure 6) reveals low performance when delivery density is sparse.)e system then begins to improve up to and including the optimal delivery density before the resources begin to saturate, and improvement curve asymptotically reaches its peak performance.

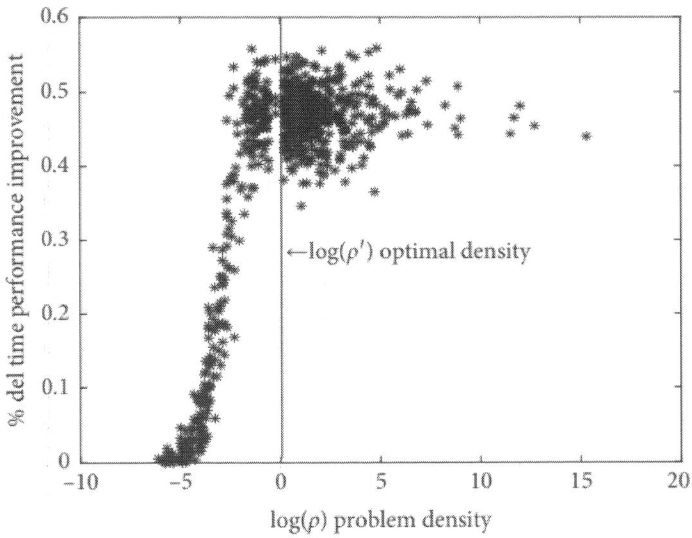

Figure 6. As problem delivery density nears optimal density for the system, performance improves.

The figure shows that the fundamental relationships established in Section 3 accurately predict the system performance. Using the fundamentals as a guide, the experiments revealed that as the delivery density (ρ) approached the optimal density (ρ′) the overall performance of the system improved. Concretely, that as density ρ = N/A moved closer toward ρ′ = N′/A′ the system improved until saturation. As such, we can surmise that the geometric layout of the ellipses comprising the delivery density ρ′ is an accurate depiction of the practical best case geometric configuration for the system. Therefore, the minimization problem established in Section 3 is a "practical best case" method to derive speed factor alpha (α), drone range kappa (κ), and number of drones nu (ν) to be fitted to a truck based on a typical delivery situation.

LITERATURE

A large body of literature exists about the traveling salesman problem (tsp) and the vehicle routing problem (vrp). Many approaches and variations to both can be found in surveys, reports, and papers [2–4]. As a general rule, the vrp problem extents the tsp problem by adding additional constraints. These constraints are comprised of time windows, priorities, range, loiter times, permissible-to-vehicle type route segments, load configurations, traffic patterns, etc. [5, 6]. Originally, Danzig and Ramser [7] investigated the vehicle routing problem. Later, Clarke and Wright [8] proposed an effective greedy heuristic which subsequently followed by several techniques or models involving exact and heuristic approaches to solve the extensions and variations of the vrp. An extensive survey can be found in [9] with exact methods to solve various routing problems.

While many variants to the tsp problem exist in the literature including multivehicle, customer pickup and delivery problem, multiple synchronization constraints [10], multiple depot vehicle scheduling problem [11], and many-to-many milk run routing problem [12], there exists only a handful of studies concerning the *truck-drone* problem. Moreover, the truck multidrone problem is a relatively new paradigm not found in the recent literature. Although the truck-drone problem has been addressed by research thus far, there is no work found on the truck-*n*-drone problem.

Truck-Drone Problem

The operational aspects and the formulation of the truck-drone problem were first shown by Murray and Chu [13]. Shortly thereafter, Agatz et al. [1] considered a close, but slightly altered version of the problem. Murray and Chu identify two ways that drone can be useful in delivery: drones can be launched and recovered from the depot (parallel drone scheduling traveling salesman problem PDSTSP) or drones can assist the truck in a parallel operation (launch-deliver-recover) as in the "flying sidekick tsp" FSTSP. Both Murray and Agatz define the material aspects of the truck-drone problem and discuss the mixed integer formulation for the optimal min-time route. They consider the drone is constrained in range, capacity, and speed. Agatz only considers a slightly altered version of the FSTSP and does not address the PDSTSP. Agatz, like Murray, considers the truck-drone in tandem as a team whereby the truck launches the drone, traverses to a separate delivery location from the drone, and then rendezvous with the drone again. However, the main difference between the approaches is that Agatz et al. [1] proposed that the drone and truck traverse along the road network system, a constraint not enforced by [13]. They do this to facilitate construction of heuristic approaches with approximations that guarantee a bound on the maximum achievable gain of the delivery system over a "truck-only solution." Murray and Chu [13] formally define the *flying sidekick traveling salesman problem (FSTSP)* as an NP-hard problem. Their study suggests a mixed integer programming (MIP) as well as a metaheuristic approach. They also consider a second similar hub-type problem that addresses the case where the customers are close enough to the depot to be serviced directly from the depot by the drone while the truck delivers to farther reach areas. This is denoted as the *parallel drone scheduling TSP* (*PDSTSP*). More recently, Agatz et al. [14] presented an exact solution approach for the truck-drone problem denoted as the TSP-D based on dynamic programming. They conducted experimental studies using the different dynamic programming heuristics which indicated that the dynamic-programming-problem could solve larger problems better than mathematically programming approaches found in the literature.Although existing research shows mathematical formulations and discusses various methods to approach the metaheuristics, the main *gaps* are that no studies reveal details concerning the expected percent improvement for a given configuration of the truck-drone to various delivery densities. Specifically, there are no studies that investigate the tradeoff between number of drones, speed factor, and drone range.

Truck Multiple Drones

Ferrandez et al. [15] introduce a method using *k-means* clustering to cluster deliveries based on distance and then to route the truck around to each of the clusters using a tsp approach. This cluster-first-route-second approach also introduces a genetic algorithm to optimally tsp route the truck through center mass quasinodes created from each of the clusters. In the described scenario, the truck may loiter at any center mass cluster node (hub) while the drone/s deliver to customers within the cluster region using a minimum spanning tree (MST) hub-spoke type of approach. Additionally, analysis was conducted on the tradeoffs between total time and total energy consumed by the truck-drone team as a factor of number of truck stops.

Our study herein is different than Ferrandez et al. [15]. Here, we require that the truck launch multiple drones then proceed to the next delivery location prior to rendezvous. It is an improvement in time over a hub-spoke method. Furthermore, we use an MIP metaheuristic to solve for optimal network routing while minimizing total delivery time. Furthermore, we introduce the concepts of "practical best case" or lean configuration which has not been addressed by any author for any of the main tool/assisting tool problems. Other gaps in the literature are that no research deals squarely with the truck-multidrone problem from a mathematical modeling or a metaheuristic perspective. Moreover, there exists no study that discusses of the tradeoffs or selection of system parameters α, κ, v with changes in the operating area or density of the delivery area. There is obviously a delta time improvement that can be determined using two or more drones or proper selection of range and speed when these are used in conjunction with a truck addressed herein.

MIXED INTEGER PROGRAM (MIP)

The MIP for the truck-multiple (v) drones d traveling salesman problem (dvtsp) solves for the optimal network routing of a truck with multiple] drones while minimizing total time. It was developed and used herein as the basis to analyze and evaluate the effectiveness of the "practical best case" configuration. Material elements of the d]tsp are described as a network graph G = (V, E) where V denotes the customers-delivery-stops and E the edges between stops. Each vertex V = { 1, . . . , |V|} where n = |V|. Edge E is described by two vertices {i, j}, and two edges or an operation is described by two vertices {i, j} for a truck indicating launch and recovery. Furthermore, the binary variables (x_{ij}, $y_{ik}y_{kj}$) denote a truck and the two edges used by a drone in an operation if an operation exists between (i, j). The binary variable $x_{ij} = 1$ if truck traverses edge E(i, j). If a drone is dormant, it is assigned no vertices for a null operation. As in the truck-drone problem, the first and last vertices of an operation must always be the same for the truck and all drones. The distance matrix is denoted as D where $d_{ij} \in D$ is the cost or distance of traversing edge E(i, j). The subtour elimination variable U_j ensures that sequence of j follows i in the event that E(i, j) is part of the solution; as such, it restricts subtour formations. The minsum of max (of max) objective function minimizes the max time of either the truck's time or any of v drone's time for an operation. As such, z_{ij} is used to evaluate each vehicle's maxtime where i, j are the first and last nodes of the two visited nodes for truck (x_{ij}) as well as the first and last of three nodes traveled by any drone as in ($y_{ik}y_{kj}$). Since z_{ij} is greater than or equal to the max time of the truck or any of the v drones, it forces z_{ij} to evaluate the max (of max) of all vehicle's time.

$$\text{minimize } Z = \sum_{i \in V} \max_{j \in V} z_{ij}, \tag{20}$$

S.T.

$$\begin{cases} z_{ij} \geq d_{ij} x_{ij}, \\ z_{ij} \geq \dfrac{y_{iK} y_{Kj} \left(y_{iK} d_{ik} + y_{Kj} d_{Kj} \right)}{\alpha}, \\ \forall i, j, k,\ i, j, k \in V, \end{cases} \tag{21}$$

$$\sum_{i \in k} x_{ik} = \sum_{j \in k} x_{kj}, \quad \forall k \in V, \tag{22}$$

$$\sum_{\substack{i \in V \\ i \neq k}} x_{ik} + \sum_{\substack{i \in V \\ i \neq k}} y_{jk} \geq 1, \quad \forall k \in V, \tag{23}$$

$$\sum_{\substack{i \in V \\ i \neq k}} x_{ki} + \sum_{\substack{j \in V \\ j \neq k}} y_{kj} \geq 1, \quad \forall k \in V, \tag{24}$$

$$\sum_{\substack{i \in V \\ i = k}} x_{ki} = 0, \quad \forall k \in V, \tag{25}$$

$$\sum_{\substack{i \in V \\ i = k}} y_{ki} = 0, \quad \forall k \in V, \tag{26}$$

$$x_{ij} \geq y_{ik} y_{kj}, \quad \forall i, k, j \in V, \tag{27}$$

$$x_{ij} + y_{ij} \leq 1, \quad \forall i, j \in V, \tag{28}$$

$$\underset{\substack{j > 1 \\ j \neq k}}{U_j} \geq U_k + x_{kj} - (N-2)\left(1 - x_{kj}\right) + (N-3) x_{jk}, \quad \forall k, j \in V, \tag{29}$$

$$\underset{k > 1}{U_k} \leq N - 1 - (N-2) x_{1k}, \quad \forall k \in V, \tag{30}$$

$$\underset{k > 1}{U_k} > 1 + (N-2) x_{k1}, \quad \forall k \in V, \tag{31}$$

$$3 \leq \left\lceil \frac{n}{\nu} \right\rceil \leq N \leq n, \tag{32}$$

$$y_{ik}y_{kj}\left(y_{ik}d_{ik} + y_{kj}d_{kj}\right) \leq \kappa, \quad \forall i, k, j \in V, \tag{33}$$

$$d_{ij} \geq 0, \kappa_{\mathrm{LB}} \leq \kappa \leq \kappa_{\mathrm{UB}}, \alpha_{\mathrm{LB}} \leq \alpha \leq \alpha_{\mathrm{UB}}, \quad \forall i, j \in V, \tag{34}$$

$$x_{ij} \in \{0, 1\}, \quad \forall i, j, i, j \in V, \tag{35}$$

$$y_{ij} \in \{0, 1\}, \quad \forall i, j, i, j \in V. \tag{36}$$

In the dvtsp construction, equation (20) minimizes the sum all the max of truck or drone edges (i, j) to all operations described as a truck route $x_{ij} = 1$ and potentially a triangle drone operation described by ($y_{ik}y_{kj} = 1$). Equation (21) forces the max (worst case) of either a truck edge or the drones' two-edges to be in the resultant minimization calculation for objective Z.)e distance for a truck edge (i, j) is denoted as $d_{ij}x_{ij}$ whereas the drone's triangular operation distance is denoted by $y_{ik}y_{kj}(d_{ik}y_{ik} + d_{kj}y_{kj})/\alpha$ where α is the drone speed as a factor of truck speed s where s = 1.

In essence, this equation serves as a mini-sum of max (of max) construction. Equation (22) constrains the truck portion of the route x_{ij} = 1 to make a circuit or at a minimum manifests a sub-tour. Equations (23) and (24) ensure each city has at least one truck or one drone visiting the city. Equation (23) establishes that each city must be visited (entered), while (24) constrains that each city visited must then be exited. Equations (25) and (26) prohibit any truck or drone from loitering as in xii or y_{ii}. Equation (27) requires that any drone operation triangle $y_{ik}y_{kj}$ has an associated truck launch and recovery x_{ij}. Equation (28) forces that any drone operation edge is not already a truck edge.

Equation (29) forces a utility variable U_j and U_k to properly sequence truck route segments x_{jk} such that sequence number j follows k so that no subtour is formed in the truck sequence of the truck portion of the route. Equation (30) and (31) work in conjunction with equation (29) to set vertex (20) as first and vertex N | (3 ≤ n/] ≤ N ≤ n) as the last of the truckroute segments. Equation (32) sets the minimum number of truck cities visited to be greater than or equal to three (3 ≤ n/] ≤ N ≤ n) but less than total nodes available, n. Equation (33) limits any drone operation to be less than the constrained range of the drone (κ) if those two edges (i, k) and (k, j) are used as a drone operation $y_{ik}y_{kj} = 1$. Equations (34)–(36) set minimum requirements for distance matrix, range, and speed factor as well as sets truck variable x_{ij} and drone variable y_{ij} as binary integers.

EVOLUTIONARY ALGORITHM

The tournament-based evolutionary algorithm (EA) here adopts a *cluster-during-routing* approach to solve the truck-drone problem. More accurately, it assigns both *truck* and *drone* labels during the routing process. Currently, there are no other algorithms found in the literature. The algorithm denoted as EA1 creates a population matrix of randomly permuted routes whereby each node in a tour is evaluated as a potential drone-delivery node unless that node is out of range. Since a population of many randomly generated tours is evaluated simultaneously, any node not within drone range is autoassigned and labelled *truck*; otherwise, the algorithm labels it drone.

Evolutionary Algorithm (1) Steps

The EA randomly permutes a population P of m tours where each tour denoted as a genome sequence (1, 2, . . . , n) for n delivery nodes in the tour. It determines the fitness for each population member (tours) based on total tour delivery time. All fitness times are saved for seed tournament.)e total population is then divided into groups of five tours each to conduct a set of seed tournaments. For each of the groups, the best member within the seed group (of the five) is chosen as the single gene to mutate for the remaining four members of the seed group. Gene mutation (tour mutation) first copies the fittest member of the group of five within the seed tournament to replace the four less fit members. Each of the four less fit members (now identical to the fittest) is then slightly mutated to improve fitness. For each of the four, mutations are comprised of only one of (a) randomly selecting and swapping two nodes within the tour, (b) reverse ordering of the tour between two nodes, (c) sliding a tour segment down between nodes to left or right, and (d) replacing the last node in the tour with any other node.)e algorithm iterates repeatedly until convergence, or a terminating condition is met based on a predetermined budget, tolerance, or a saturation found in improvements (Algorithm 1).

Algorithm 1. Evolutionary Algorithm (EA1).

```
DATA : Random Population matrix tours where Route_i ∈ Population
RESULT : Route_opt ~ optimal tour ∈ Population of n delivery stops
FOR iterations in budjet LOOP
  FOR each (i) population member LOOP
      Route_i ⟵ a tour within the populations of tours
      launch ⟵ Route_i (1) first node;
      rendezvous ⟵ Route_i (2) second node;
      candidate ⟵ Route_i (3) third node;
      drone count ⟵ 0;
      route time ⟵ 0;
    WHILE candidate ≤ total delivery nodes DO
      truck op time ⟵ get truck time (launch, rendezvous);
      drone op time ⟵ get drone time (launch, candidate, rendezvous);
```

```
drone op dist ⟵ get drone distance (launch, candidate, rendezvous);
IF candidate ≥ total delivery nodes THEN
  IF drone count ≥ number drones avail.OR drone op dist. > drone range THEN
    Max op time = max (truck op time, drone op time, multiple drone op time);
    route time ⟵ route time + max op time + truck op time;
    route time ⟵ route time + get truck time (candidate depot);
  ELSE (drone makes last delivery)
    Max op time = max (truck op time, drone op time, multiple drone op time);
    Route time ⟵ route time + max op time + get truck time (rendezvous, depot);
  END IF
  BREAK while loop;
END IF
IF drone count ≥ number drone avail OR drone dist. > drone range THEN

  Max op time ⟵ max (truck op time, multiple drone op time);
  Route time ⟵ route time + max op time;
  Launch ⟵ rendezvous;
  rendezvous ⟵ candidate;
  candidate ⟵ min (Route_i(candidate + 1), total number stops);
  drone count ⟵ 0;
  multi drone op time ⟵ 0;
ELSE (assign delivery to drone)
  max op time ⟵ max (truck op time, drone op time, multiple drone op time);
  route time ⟵ route time + max op time
  candidate ⟵ Route_i(candidate + 1);
  drone count ⟵ drone count + 1;
  multi drone op time(drone count) ⟵ drone op time;
END IF
  List of all Route Times (p) ⟵ route time;

    END WHILE
  END FOR (each population member)
    Population ⟵ randomly shuffle among the tours in population, keep routes intact
  FOR each of five tours in population LOOP through, keep tournament winners, mutate losers
    Best time, Id, Best Route ⟵ Get Fittest Member tour of tournament of 5;
    Overwrite each of the four less fit tours with the fittest member;
    Mutate first of the four less fit tours by random swap;
    Mutate second of the four less fit tours by random segment slide;
    Mutate third of the four less fit tours by random segment flip;
    Mutate fourth of the four less fit tours by swapping last node in tour with any other
    Do nothing for fifth tour; keep the fittest tour intact;
  END FOR mutations and tournament winners
    Population ⟵ update old Population with new Population mutations and winners
END FOR total budget exhausted
RETURN overall best tour in population (time, route) and graph route;
```

Evolutionary Algorithm (2)

The second metaheuristic denoted as EA2 is nearly identical to EA1, but differs by performing a cost check before assigning a drone to the job. Concretely, EA2 performs a calculation to determine if it is more efficient to assign the truck or the next drone. If the truck is more efficient, the greedy algorithm assigns the truck and foregoes assigning a drone.

Both algorithms use an approach similar to simulated annealing. A seed tournament genetic algorithm's strength lies in the ability to overcome local optima by retaining multiple paths (or seeds) that simultaneously advance toward optimization. This ideology is especially critical for network routing problems. Furthermore, because there are multiple members of the population within a seed tournament, the algorithm allows for various mutation methodologies to be performed on the members of the seed tournament. In this case, the *random swap* (*or pairwise swap*), *flip*, and *slide* mutations have proven to be robust, fast, and extremely accurate for problems involving permutations whereby order matters.

The performance of the evolutionary algorithm is based on the underlying theoretical principles: (a) by initializing a relatively large population (i.e., *5n*) of randomly permuted tours, multiple tracks are maintained toward optimal convergence (seed tournaments). These multiple, but different, paths slowly converge and, thus, increase the probability of an optimal convergence. (b) By saving the fittest gene in a seed and then slightly perturbing (mutating) the best gene (tour) found in the seed group ensure the solution never gets worse while promoting improvements at each iteration. (c) By autoassigning the drone to any "within range" node, the use of the drone is maximized throughout the routing process while simultaneously reducing the truck's overall tour length. The risk of assigning the wrong node to a drone is mitigated by multiple seeds. (d) Multiple path (seed) random search is much faster than having to calculate the greediness or the exactness of each neighborhood within reach as in other algorithms. Therefore, the algorithm relies on computational speed and iterations without the burden of unnecessary calculations.

Both EA1 and EA2 use a *route-during-clustering* approach. This approach has several advantages over other algorithms. All algorithms found in the literature with the exception of [16] use route-first, cluster-second or a cluster-first, route-second approach. This results in a two-phase algorithmic process whereby the first phase solves for routing (as in travelling salesperson routing (tsp) or minimum spanning tree, (mst) routing); the second phase relying extensively on solutions found in first phase performs swaps between truck or drone delivery. The best known algorithms for *tsp* (or mst) routing are on the order of $O(n^2 2^n)$ where n is the number of nodes in the route. The second phase in a *route-first-cluster-second* algorithm relies heavily on the initial routing solutions being in the vicinity of a truck-drone solution. As such, the second phase is equally complex as the first phase with the disadvantage of being locked into a potentially suboptimal routing scheme especially when drone configurations change.

Conversely, the route-during-clustering algorithm performs the routing and the selection of the truck or drone for the next operational move, segment of the route. As such, the complexity of the algorithm is based on the number of iterations (n^2), the population size (4n), and the number of nodes (n) found in the tour. In such case, the algorithm is consistent with the order of $O(n^2 \times 4n \times n)$ or $O(4n^4)$. A computer with an Intel core (i5) and 1.70 GHz CPU, comprised of 50 nodes and 3 drones, requires approximately ~25 million iterations or ~35 seconds for convergence.

The overall complexity of the truck-multidrone problem denoted as (d] tsp) can be analysed from a brute force methodology. To evaluate every potential solution for a ten node problem would require slightly less than $((10!/2) \times 2^{10})$ permutations or approximately ~1.86e + 09 routes and binary drone assignments. Every solution requires two genes, a gene describing the order of the tour (10!/2) and a gene to describe every possible truck or drone assignments 210. For larger problem sizes, say n = 50 would require $((50!/2) \times 2^{50}) = 6.8486e+79$ iterations translating to approximately 2.8e61 seconds computer time or ~7.1e53 years on a 1.70 GHz CPU. For such method, the number of iterations to accomplish would exceed the time in the universe.

COMPARISON STUDIES

Two different computational studies are conducted here. The first study analyses the performance of the metaheuristics against the optimal solution based on MIP and brute force solution. The second study analyses the performance of the system model parameters (α, κ, v) against arbitrarily (standard) assigned system variables.

In the first study, the performance of the two metaheuristics described in Section 6 is analysed. The two metaheuristics are compared against themselves as well as compared with the guaranteed optimal results obtained from the mixed integer program (MIP). The comparison study analyses four different performance factors: (1) the average delta delivery-time of the metaheuristic versus the optimal delivery time found using the MIP, (2) the max delta (delivery-time) between the metaheuristic versus the optimal time of MIP, and (3) the total number of metaheuristic experiments that resulted in optimal results out of ten studies, and the average computer solution time for the metaheuristic to compute the optimal solution.

The second study analyses the performance of properly (optimally) configured system parameters (α, κ, v) as described in Section 2 (theoretical insights: system model) against standard parameters found in the literature

($\alpha = 2$, $\kappa = 10$, $v = 3$). In many cases, the literature arbitrarily chooses system parameters as drone speed factor set at two times the speed of the truck ($\alpha = 2$) and an arbitrary drone range set at ten kilometres ($\kappa = 10$) based on current technical capabilities. The system comparison study here analyses three performance metrics: (1) the total deliverytime obtained by a properly configured (lean) system versus the total delivery-time obtained by arbitrary system parameters (α, κ, v), (2) the overall utilization of the drones throughout the delivery: a properly configured (lean) system versus the utilization of the drones from a system with arbitrary parameters, and (3) the under or oversaturation of the system determined by the average wait or idle time (by the truck or by drone) for all operations.

Computational Study 1: Metaheuristic Performance to Optimal Solution

For the sake of simplicity, a plane of (x, y) coordinates is uniformly randomly generated from the Cartesian coordinate system with Euclidean distance between the nodes. Concretely, all nodes for problem comparisons herein were sampled from the uniform distribution from {0, 1, 2, …, 30}. For each metaheuristic experiment, 10 runs were conducted in order to average the results. Results obtained were based on delta from optimal, whereby delta is defined as the difference between delivery-time (and final route) obtained from the metaheuristic versus the delivery-time (and final route) obtained by an optimal guarantee process (i.e., MIP).

$$\Delta = \frac{\text{objective value heuristic} - \text{optimal objective value}}{\text{optimal objective value}}. \tag{37}$$

The metaheuristic experiments were run on a Windows 10 operating system, Intel® Core™ i5-835OU CPU @ 1.70 GHz and 4 GB of memory. The main purpose of the experiments was to determine the accuracy of the metaheuristics denoted by delta (Δ or deviation from optimal) and total computer-solution time (*elapse time*). Since optimization can only be guaranteed (compared to MIP) by smaller job sizes, the job size for experiments ranged from 10 to 15 jobs (nodes). Table (Table 1) below shows results for 10 randomly generated instances with 10 and 15 nodes of each instance type. For each experiment (10 runs), the delta from optimal is averaged (avg.), the max deviation from optimal is determined (max), the total number of experiments found to be optimal (#opt out of 10), and the total computer-solution time is given (*elapse time*) for comparison.

Table 1. Comparison of EA1 and EA2 to optimal solutions.

	From optimal						Comp. performance time	
	EA-I			EA-2			EA-1	EA-2
Sire	kg	Max	#opt	Avg.	Max	#opt ,	Elapse time (s)	Elapse time (s)
10 nodes	0.00	0.00	10/10	0.00	0.00	111,10	23	25
15 node.	0.03	0.03	9/10	0.04	0.04	9/10	32	41

Note: algorithm performances arc compared to known optimal solulions (brute force, 4EP). Rends: averaged over ten MICHMelat. Standard drone parameters: Delivery nodes sampled from random uniform distribution, drone speed factor α = 2, drone range κ = 10, number drones assigned truck v = 4.Results indicate that both EA-1 and EA-2 are capable of solving the truck-multidrone problem near *optimal*. The EA-1 slightly outperformed EA-2 due to the higher utilization of drones. It is estimated that depending on the configuration of the system parameters to the operational spaced, EA-2 is efficient when there are more drones than necessary fitted to the truck or the system is highly saturated with resources.

Computational Study 2: Lean System Parameters Compared to Standard Parameters

Lean System Parameters Compared to Standard Parameters.)e second study compares the performance of a lean or properly configured system parameters (α, κ, v) against an arbitrarily set of system parameters (α = 2, κ = 10, v = 3) often found in the literature. For the sake of simplicity, an experiment was drawn from a uniform random distribution of the (x, y) plane comprised of Euclidean distance between each of the coordinate delivery nodes. Since the experiments are designed to compare the performance of a lean system to an arbitrary system, the performance parameters selected for comparison are delivery-time ti, utilization util., and truck or drone wait time. Each experiment was comprised of ten (10) runs in order to obtain an averaged set of results. The delivery area was perturbed between 1 km^2 to 1000 km^2 to evaluate various delivery densities for 30 delivery sites. Table 2 shows the results of six experiments each with ten runs comparing the arbitrary system parameters to the lean system parameters derived from the nonlinear optimization in Section 3. The evolutionary algorithm EA-1 (Section 7) was used to optimally route and cluster the truck and n-drones throughout the delivery area to arrive at the optimal delivery times for each run for each experiment. EA-1 was also used to obtain average utilization (of drones) and the average wait time for either the truck or the drone for each operation within each run. The average of the ten runs is reported in Table 2.

Table 2. Lean system performance results. System performance Comparing geometric-based optimal drone parameters to standard drone parameters

Exp.	Size operating area (km²)	Standard parameters (α=2, κ=10, v=3)			Optimal system parameters (α, κ, and v)		
		Avg. utiL	Avg. wait	Avg. time	Avg. utiL	Avg. wait	Avg. time
1	1	1.00	0.063	2.48	0.98	0.05	1.08
2	25	1.00	0.286	11.82	0.99	0.24	7.91
3	100	0.95	0.93	25.03	0.86	1.76	21.13
4	300	0.75	2.03	51.52	0.64	2.10	47.10
5	700	0.22	2.73	102.94	0.29	3.30	93.24
6	1000	0.24	2.22	139.47	0.29	3.16	103.31

For each of the six experiments, ten runs were conducted for each delivery area (km^2) and results were averaged. Each experiment was comprised of 30 delivery nodes. Arbitrary system parameters ($\alpha=2$, $\kappa=10$, $v=3$) were compared to lean system parameters determined by nonlinear-optimization equations found in Section 2. Experiment results show that by adopting the lean system parameters, the overall delivery time is consistently reduced. Results show delivery time reductions for each experiment (56%, 33%, 15%, 8%, 9% and 26%). It is noted that for at least two experiments, the arbitrary configuration was close to the calculated lean configuration; thus, no major improvements were expected for experiments 4 and 5. An expected tradeoff is utilization. In poorly configured systems, the drones are either highly utilized or rarely utilized; thus, they range from bottleneck to saturated resources. The lean configuration slightly relieves the bottleneck system and unsaturates low utilization configurations. Average wait time for each operation slightly increased due to the stochastic nature of within the experiment design. In this case, the drones often found themselves advance of the truck in operations and therefore at an increased opportunity to accumulate wait time.

CONCLUSIONS

The current literature lacks any usable information in terms of the highest yield regions of the design space (speed, range, and number drones) for a main tool with multiple assisting tools; specifically, herein, we analysed a truck-multiple-drone configuration. Few, if any, research articles address a tractable, usable MIP for the truck-multidrone problem for quick transfer and testing within standard optimization tools (IBM® ILOG CPLEX, Lindo® Lingo). Therefore, any business wishing to evaluate the use of a single truck fitted with multiple drones has a way neither to guide the selection of drone parameters nor to evaluate the efficiencies of those selections without developing scientific experiments. The "practical best case" geometric approach shown herein inverts the problem space and solves for practical best case drone speed, drone range, and number of drones given a general delivery scenario. Concretely, it serves as a basis to address practical design decisions regarding the proper configurations to achieve the maximum desired yields for a system. It serves as a way to screen out low performing designs while giving the decision maker a Pareto front of potential configuration solutions that achieve a lean alignment. Moreover, both the MIP and metaheuristics developed herein for empirical studies accurately solve for the optimal truck-multidrone routing, thus giving the business ample tools for evaluating generalized scenarios.

The simplified version of the MIP is not found in any other study; it is useful and an easily implemented metaheuristic necessary to solve for the optimal route and optimal time for the truck-multiple-drone for smaller problem sets. For larger problem sets, the single chromosome evolutionary algorithm (EA-1) is best in class metaheuristic to test various test-case scenarios. Both EA-1 and EA-2 were modelled as functions within the MATLAB® development environment language, and the files were made available at Mathworks® file exchange (dvtsp_ga_basic) for evaluation and general purpose use/testing. As far as we could surmise, our EA is the only algorithm available for such problems found in the literature or in an open resource environment.

In conclusion, this research answers the questions of expected efficiencies in time that would be expected given a truck-multidrone configuration as well as finding "what is the proper configuration for a truck-multidrone situation." It gives business a foundation to evaluate a variety of configurations against a typical daily last-mile parcel-delivery scenario. The work also opens several additional questions for future research. These questions tend to be toward the total energy savings and the energy saturation for different truck-multiple-drone generalized situations the business may encounter. Future research studies a fleet of trucks each fitted with the optimal number and type of drones.

DATA AVAILABILITY

Genetic algorithms and metaheuristics are available online in MATLAB after publication.

CONFLICTS OF INTEREST

The authors declare that they have no conflicts of interest.

AUTHORS' CONTRIBUTIONS

(1) R. R. conceived the presented idea, developed the system theory, performed the computations, developed the theoretical formalism, performed the analytic calculations, performed the numerical simulations, verified the analytical methods of the system theory, mixed integer program and metaheuristics, wrote all sections of the paper including all the tables and graphs, conducted the fundamental experiments, conducted the performance comparison for evolutionary algorithms 1 and 2 against optimal solution, and conducted the lean comparison tests to the arbitrarily configured system.

REFERENCES

1. N. Agatz, P. Bouman, and M. Schmidt, "Optimization approaches for the traveling salesman problem with drone," *Transportation Science*, vol. 52, no. 4, pp. 965–981, 2018.
2. B. Golden, S. Raghavan, and E. Wasil, "The vehicle routing problem: latest advances and new challenges," in *Operations Research/Computer Sciences Interfaces*, vol. 43, Springer, Berlin, Germany, 2008.
3. P. Toth and D. Vigo, *The Vehicle Routing Problem SIAM Monographs on Discrete Mathematics and Applications*, SIAM Monographs on Discrete Mathematics and Applications, Philadelphia, PA, USA, 2002.
4. S. N. Kumar and R. Panneerselvam, "A survey on the vehicle routing problem and its variants," *Intelligent Information Management*, vol. 4, no. 3, pp. 66–74, 2012.
5. P. Toth and D. Vigo, *Vehicle Routing: Problems, Methods, and Applications*, Society for Industrial and Applied Mathematics, Philadelphia, PA, USA, 2nd edition, 2014.
6. C. E. Miller, A. W. Tucker, and R. A. Zemlin, "Integer programming formulation of traveling salesman problems," *Journal of the ACM (JACM)*, vol. 7, no. 4, pp. 326–329, 1960.
7. G. Dantzi and J. Ramser, "The truck dispatching problem," *Management Science*, vol. 6, no. 1, pp. 80–91, 1959.
8. G. Clarke and J. W. Wright, "Scheduling of vehicles from a central depot to a number of delivery points," *Operations Research*, vol. 12, no. 4, pp. 568–581, 1964.
9. M. Desrochers, J. K. Lenstra, and M. W. P. Savelsbergh, "A classification scheme for vehicle routing and scheduling problems," *European Journal of Operational Research*, vol. 46, no. 3, pp. 322–332, 1990.
10. M. Drexl, "Synchronization in vehicle routing-A survey of VRPs with multiple synchronization constraints," *Transportation Science*, vol. 46, no. 3, pp. 297–316, 2012.
11. A.-S. Pepin, G. Desaulniers, A. Hertz, and D. Huisman, "A comparison of five heuristics for the multiple depot vehicle scheduling problem," *Journal of Scheduling*, vol. 12, no. 1, pp. 17–30, 2009.
12. Y. Lin, Z. Bian, S. Sun, and T. Xu, "A two stage simulated annealing algorithm for the many-to-many milk run routing problem with pipleline inventory cost," *Mathematical Problems in Engineering*, vol. 2015, Article ID 428925, 22 pages, 2015.

13. C. C. Murray and A. G. Chu, "The flying sidekick traveling salesman problem: optimization of drone-assisted parcel delivery," *Transportation Research Part C: Emerging Technologies*, vol. 54, pp. 86–109, 2015.
14. P. Bouman, N. Agatz, and M. Schmidt, "Dynamic programming approaches for the traveling salesman problem with drone," September 2017, https://ssrn.com/abstract=3035323.
15. S. Ferrandez, T. Harbison, T. Weber, R. Sturges, and R. Rich, "Optimization of a truck-drone in tandem delivery network using *K*-means and genetic algorithm," *Journal of Industrial Engineering and Management*, vol. 9, no. 2, pp. 379–383, 2016.
16. R. Rich, "Inverting the truck-drone network problem to find best case configuration," *Advances in Operations Research*, vol. 2020, Article ID 4053983, 10 pages, 2020.

CHAPTER 4

A Dynamic Active-Set Method for Linear Programming

Alireza Noroziroshan, H. W. Corley, Jay M. Rosenberger
IMSE Department, The University of Texas at Arlington, Arlington, USA

ABSTRACT

An efficient active-set approach is presented for both nonnegative and general linear programming by adding varying numbers of constraints at each iteration. Computational experiments demonstrate that the proposed approach is significantly faster than previous active-set and standard linear programming algorithms.

Keywords:- Constraint Optimal Selection Techniques, Dynamic Active-Set Methods, Large-Scale Linear Programming, Linear Programming

Citation: Alireza Noroziroshan, H. W. Corley, Jay M. Rosenberger (2015) A Dynamic Active-Set Method for Linear Programming. American Journal of Operations Research,05,526-535. doi: 10.4236/ajor.2015.56041.

INTRODUCTION

Consider the linear programming problem

(P) $\text{Max } z = \boldsymbol{c}^T \boldsymbol{x}$ (1)

s.t.

$\boldsymbol{Ax} \le \boldsymbol{b}$ (2)

$x \geq 0,$ (3)

where x and c are the n-dimensional column vectors of variables and objective coefficients, respectively, and z represents the objective function. The matrix A is an m × n matrix $[a_{ij}]$ with row vectors $\boldsymbol{a}_1, \cdots, \boldsymbol{a}_m$; $\boldsymbol{b}$ is an m-dimensional column vector; and 0 is an n-dimensional column vector of zeros. The non-polynomial simplex methods and polynomial interior-point barrier-function algorithms illustrate the two different approaches to solve problem P. There is no single best algorithm [1] . For any existing approach, there is a problem instance for which the developed method performs poorly [2] [3] . However, interior point methods do not provide efficient post-optimality analysis, so the simplex algorithm is the most frequently used approach [2] , even for sparse large scale linear programming problems where barrier methods perform extremely well. In fact, the simplex method has been called "the algorithm that runs the world" [4] ; yet it often cannot efficiently solve the large scale LPs required in many applications.

In this paper we consider both the general linear program (LP) and the special case with $\boldsymbol{a}_i \ge \boldsymbol{0}$ and $\boldsymbol{a}_i \ne \boldsymbol{0}$, $\forall i = 1, \cdots, m$; $\boldsymbol{b} > \boldsymbol{0}$; and $\boldsymbol{c} > \boldsymbol{0}$, , which is called a nonnegative linear program (NNLP). NNLPs have some useful properties that simplify their solution, and they model various practical applications such as determining an optimal driving route using global positioning data [5] and updating airline schedules [6] , for example. We propose active-set methods for LPs and NNLPs. Our approach divides the constraints of problem P into operative and inoperative constraints at each iteration. Operative constraints are those active in a current relaxed subproblem P_r, $r = 1, 2, \cdots$, of P, while the inoperative ones are constraints of the problem P not active in P_r. In our active-set method we iteratively solve P_r, $r = 1, 2, \cdots$, of P after adding one or more violated inoperative constraints from (2) to P_{r-1} until the solution $\boldsymbol{x}_r^*$ to P_r is a solution to P.

Active-set methods have been studied by Stone [7] , Thompson et al. [8] , Myers and Shih [9] , and Curet [10] , for example. More explicitly,

Adler et al. [11] suggested a random constraint selection rule, in which a violated constraint was randomly added to the operative set. Bixby et al. [1] developed a sifting method for problems having wide and narrow structure. In effect, the sifting method is an active-set method for the dual. Zeleny [12] used a constraint selection rule called VIOL here, which added the constraint most violated at each iteration. Mitchell [13] used a multi-cut version of the VIOL for an interior point cutting plane algorithm. Corley et al. [14]

developed a cosine simplex algorithm where a single violated constraint maximizing $\cos(\boldsymbol{a}_i,\boldsymbol{c}) = \frac{\boldsymbol{a}_i\boldsymbol{c}}{\|\boldsymbol{a}_i\|\|\boldsymbol{c}\|}$ was

added to the operative set. Junior and Lins [15] used a similar cosine criterion to determine an improved initial basis for the simplex algorithm.

References [6] [16] and [17] are the most directly related to the current work. The constraint optimal selection technique (COST) RAD was introduced in [6] and [16] as a constraint selection metric for NNLPs, and then generalized in [17] to GRAD for LPs. In RAD and GRAD, an initial problem P_0 is formulated from P such that all variables are bounded by at least one constraint, which may be an artificial bounding constraint $\boldsymbol{c}^T\boldsymbol{x} \leq M$ for sufficiently large M. Multiple violated constraints are then added to problems P_r, $r = 0,1,\cdots$ according to the constraint selection metric RAD or GRAD. The contribution here is to improve significantly the speed of RAD and GRAD by varying the number of added constraints at each P_r according to an empirically derived function estimating the effectiveness of that iteration.

The paper is organized as follows. The constraint selection metric and dynamic active-set conjunction with RAD and GRAD is given in Section 2. In Section 3, we describe the problem instances and CPLEX preprocessing settings. Section 4 contains computational experiments for both NNLPs and LPs. Then the performance of the new methods is compared to the previous ones as well as the CPLEX simplex, dual simplex, and barrier method. In Section 5, we present conclusions.

A DYNAMIC ACTIVE-SET APPROACH

The purpose of our dynamic active-set method is to add violated constraints to problem P_r more effectively than in [6] and [17] . In COST RAD of [6] for NNLPs we use the constraint selection metric

$$\mathrm{RAD}(\boldsymbol{a}_i, b_i, \boldsymbol{c}) = \frac{\boldsymbol{a}_i^T \boldsymbol{c}}{b_i} \tag{4}$$

to order constraints from highest to lowest value of RAD as a geometric heuristic for determining the constraints most likely to be binding at optimality. Moreover, at each iteration in [6] we add violated constraints in order of decreasing RAD until the added constraints contain non-zero coefficients for all variables. In similar fashion for COST GRAD of [17] , we use the constraint selection metric

$$\mathrm{GRAD}(\boldsymbol{a}_i, b_i, \boldsymbol{c}) = \sum_{j=1, c_j>0}^{n} \frac{a_{ij} c_j}{b_i^+} - \sum_{j=1, c_j<0}^{n} \frac{-a_{ij}}{b_i^+}, \tag{5}$$

where

$$b_i^+ = \begin{cases} b_i - \min\limits_{k=1,\cdots,m} [b_k] + \varepsilon & \text{if } \min\limits_{k=1,\cdots,m} [b_k] < 0 \\ b_i & \text{Otherwise} \end{cases} \tag{6}$$

for a small positive constant ε and $i^* \in \operatorname{argmax}\left(\mathrm{GRAD}(\boldsymbol{a}_i, b_i, \boldsymbol{c}) : \boldsymbol{a}_i^T \boldsymbol{x}_r^* > b_i, i \notin \mathrm{OPERATIVE}\right)$ However, in COST GRAD, violated constraints are added until every column of A has at least one positive and one negative value.

In this paper we propose a dynamic method that adds a varying number of constraints to P_r that depends on the progress made at P_{r-1}. No equality constraints are considered here, but any equality constraints can be included in P_0. An active-set function is defined to compensate for the lack of progress in P_{r-1} by adding more violated constraints at P_r. The algorithm stops when the solution $\boldsymbol{x}_r^*$ to P_r is the optimal solution to P. Of course, one could simply add all violated constraints at any one iteration. However, the proposed dynamic method attempts to balance the number of iterations required to eliminate all constraints violation, while still maintaining an efficient active-set method.

Dynamic Active-Set Approach for NNLP

The dynamic active-set approach developed for solving NNLPs is as follows. Constraints are initially ordered by the RAD constraint selection metric (4). To construct P_0, we choose constraints from (2) in descending order of RAD until each variables x_j has an $a_{ij} > 0$ in the coefficient matrix of P_0. We say the variables are covered by the constraints of the initial problem P_0. P_0 is then solved by the primal simplex to achieve an initial solution $\boldsymbol{x}_0^*$

. Now let γ_0 be the number of constraints in the original problem P and, in general, let γ_r be the number of constraints of problem P violated by x_r^*. At each iteration r, the total number of violated constraints γ_r is computed, and the improvement percentage is calculated by

$$\omega_r = \max\left\{0, \left(\frac{\gamma_{r-1} - \gamma_r}{\gamma_{r-1}}\right)\right\} \times 100, \forall r = 1, 2, \cdots, \tag{7}$$

where $\omega_r > 0$ represents the percent of improvement made in reducing the total number of violated constraints at iteration r. Next, with [.] denoting the greatest integer function, let

$$\varphi_{r+1} = \left[\varphi_r * \left(1 + \log\left(101 - \omega_r\right)\right)\right], 0 \le \omega_r \le 100, \forall r = 1, 2, \cdots, \quad \backslash(8)$$

Where $\varphi_1 = 100$. The value ϕ_r is an upper bound on the possible number of non-operative violated constraints that can be added at active-set iteration $r = 1, 2, \cdots$ The actual number added is $\min\{\varphi_{r+1}, \gamma_r\}$. The active-set iterations terminate when $\gamma_r = 0$ and therefore $\omega_r = 100$. Equation (8) was developed from the results of com- putational experiments.

Pseudocode for the dynamic active-set NNLPs is as follows.

Step 1—Identify constraints to initially bound the problem.

1: $\boldsymbol{a}^* \leftarrow 0$

2: while $\boldsymbol{a}^* \ngtr 0$ do

3: Let $i^* \in \operatorname{argmax}_{i \notin \text{BOUNDING}} \text{RAD}(\boldsymbol{a}_i, b_i, \boldsymbol{c})$

4: if $\exists j | a_j^* = 0$ and $a_{i*j} > 0$ then

5: $\text{BOUNDING} \leftarrow \text{BOUNDING} \cup \{i^*\}$

6: end if

7: $\boldsymbol{a}^* \leftarrow \boldsymbol{a}^* + \boldsymbol{a}_{i*}$

8: Optimized ← false

9: end while

Step 2—Using the primal simplex method, obtain an optimal solution x_0^* for the initial bounded problem P_0

$$\text{maximize } z = \boldsymbol{c}^T \boldsymbol{x}$$

$$\text{subject to } \boldsymbol{a}_i^T \boldsymbol{x} \le b_i \ \forall i \in \text{BOUNDING}$$

$$\boldsymbol{x} \ge \boldsymbol{0}$$

Step 3—Perform the following iterations until an optimal solution to problem P is found.

$\varphi_1 \leftarrow \#100$
$r \leftarrow 0$
$\gamma_0 \leftarrow \#rows$
$while(Optimized = false)do$
  $r \rightarrow r+1$
  $if\left\{\boldsymbol{a}_i^T \boldsymbol{x}_r^* > b_i, \forall i = 0,\cdots,rows\right\} then$ calculate γ_r
    Calculate $\omega_r = \max\left\{0, \left(\frac{\gamma_{r-1}-\gamma_r}{\gamma_{r-1}}\right)\right\} * 100$
  $if\, 0 \le \omega_r < 100\ then\ \varphi_{r+1} = \left[\varphi_r * \left(1+\log\left(101-\omega_r\right)\right)\right]$
    $Let\ i^* \in \text{argmax}_{i \notin \text{OPERATIVE}}\left\{\text{RAD}\left(\boldsymbol{a}_i, b_i, \boldsymbol{c}\right) : \boldsymbol{a}_i^T \boldsymbol{x}_r^* > b_i\right\}$
  for ($i = 0$ to $\min\{\varphi_{r+1}, \gamma_r\}$) OPERATIVE $\leftarrow$ OPERATIVE $\cup \{i^*\}$ *end*
    Solve the following P_r by the dual simplex method to obtain $\boldsymbol{x}_r^*$
  else if ($\omega_r = 100$) *then Optimized* $\leftarrow$ *true*// $\boldsymbol{x}_r^*$ is an optimal solution to *P*
    end if
  else Optimized $\leftarrow$ *true*// $\boldsymbol{x}_r^*$ is an optimal solution to *P*.
  end if
end while

Dynamic Active-Set Approach for LP

The dynamic active-set approach for solving LP similar to the one for NNLPs. We construct P_0 by choosing a number of constraints ρ_1 from (2) in descending order of GRAD from (2) until no variable x_j is left without at least either a positive or a negative coefficient. In addition, we include an artificial bounding constraint $\mathbf{1}^T \boldsymbol{x} \le M$. If $\rho_1 < 100$, then set $\rho_1 = 100$. Then P_0 is solved to obtain an initial solution $\boldsymbol{x}_0^*$. As in Section 2.1, it is initially assumed that all constraints are violated $(\gamma_0 = m)$. Then the relative improvement percent ω_r is calculated by (7) for P_r and P_{r+1}. Now let

$$\rho_{r+1} = \left[\rho_r * \log\left(101-\omega_r\right)\right], 0 \le \omega_r \le 100, \forall r = 1,2,\cdots, \tag{9}$$

where the value ρ_r is an upper bound on the possible number of non-operative violated constraints that can be added at active-set iteration $r = 1,2,\cdots$. The actual number added is $\min\{\rho_{r+1}, \gamma_r\}$. As ω decreases, ρ_{r+1} increases in (9) to add more violated constraints to P_{r+1}. The algorithm stops at 100% reduction in the number of violated constraints.

Pseudocode for the dynamic active-set for LPs is as follows.

Step 1—Identify constraints to initially bound the problem.

$a^* \leftarrow 0$
while $a^* \not> 0$ *do*
  Let $i^* \in \text{argmax}_{i \notin \text{BOUNDING}} \text{GRAD}(a_i, b_i, c)$
  if $\exists j | a_j^* = 0$ *and* $a_{i*j} > 0$ *or* $a_{i*j} < 0$ *then*
    $\text{BOUNDING} \leftarrow \text{BOUNDING} \cup \{i^*\}$
  end if
  $a^* \leftarrow a^* + a_{i*}$
Optimized $\leftarrow$ *false*
end while

Step 2—Using the primal simplex method, obtain an optimal solution x_0^* for the initial bounded problem P_0 given by

$$\text{maximize } z = c^T x$$

$$\text{subject to } a_i^T x \le b_i \ \forall i \in BOUNDING$$

$$\mathbf{1}^T x \le M$$

$$x \ge \mathbf{0}$$

Step 3—Perform the following iterations until an optimal solution to problem P is found.

$\rho_1 \leftarrow \text{Max}\{\#\text{BOUNDING}, 100\}$
$r \leftarrow 1$
$\gamma_0 \leftarrow \#rows$
while $(Optimized = false)$ *do*
  $r \leftarrow r + 1$
if $\{a_i^T x_r^* > b_i, \forall i = 0, \cdots, rows\}$ *then* calculate γ_r
  Calculate $\omega_r = \max\left\{0, \left(\dfrac{\gamma_{r-1} - \gamma_r}{\gamma_{r-1}}\right)\right\} * 100$
  if $0 \le \omega_r < 100$ *then* $\rho_{r+1} = \lfloor \rho_r * \log(101 - \omega_r) \rfloor$
  Let $i^* \in \text{argmax}_{i \notin \text{OPERATIVE}} \{\text{GRAD}(a_i, b_i, c) : a_i^T x_r^* > b_i\}$
   for $(i = 0$ to $\min\{\rho_{r+1}, \gamma_r\})$ $\text{OPERATIVE} \leftarrow \text{OPERATIVE} \cup \{i^*\}$ *end*
   Solve the following P_r by the dual simplex method to obtain x_r^*
 end if
 else Optimized $\leftarrow$ *true*// x_r^* is an optimal solution to P.
 end if
end while

PROBLEM INSTANCES AND CPLEX PREPROCESSING

Four sets of NNLPs used in [6] are considered to evaluate the performance of the developed algorithm. Each problem set contains five problem instances for 21 different density levels and for varying ratios of (m constraints)/(n variables) from 200 to 1. Each set contains 105 randomly generated NNLPs with various densities p ranging from 0.005 to 1. Randomly generated real numbers between 1 and 5, 1 and 10, 1 and 10 were assigned to the elements of A, b and c respectively. To avoid having a constraint in the form of an upper bound on a variable, each constraint is required to have at least two non-zero a_{ij}. For general LP, a problem set containing 105 randomly generated by Saito et al. [17] is compared with the dynamic approach of this paper. These LP problems contain 1000 variables (n) and 200,000 constraints (m), with various densities ranging from 0.005 to 1 and the randomly generated a_{ij} ranging between −1 and −5 or between 1 and 5.

Two parameters that CPLEX uses for solving linear programming are PREIND (preprocessing pre-solve indicator) and PREDUAL (preprocessing dual). As described in [17] and [6] , when parameter setting PREIND = 1 (ON), the preprocessing pre-solver is enabled and both the number of variables and the number of constraints is reduced before any type of algorithm is used. By setting PREIND = 0 (OFF) the pre-solver routine in CPLEX is disabled. PREDUAL is the second preprocessing parameter in CPLEX. By setting parameter PREDUAL = 0 (ON) or −1 (OFF), CPLEX automatically selects whether to solve the dual of the original LP or not. Both are used with the default settings for the CPLEX primal simplex method, the CPLEX dual simplex method, and the CPLEX barrier method. Neither CPLEX pre-solver nor PREDUAL parameters were used in any part of the developed dynamic active-set methods for NNLPs and LPs.

COMPUTATIONAL EXPERIMENTS

The computations were performed on an Intel Core (TM) 2 Duo X9650 3.00 GHz with a Linux 64-bit operating system and 8 GB of RAM. The developed methods use IBM CPLEX 12.5 callable library to solve linear programming problems. The dynamic RAD and dynamic GRAD are compared with the previously developed COST RAD and COST GRAD, respectively, as well as VIOL, the CPLEX primal simplex method, the CPLEX dual simplex method, and the CPLEX barrier method.

Computational Results for NNLP

Table 1 illustrates the performance comparison between dynamic RAD method and the previously defined constraint selection technique COST RAD on Set 1 to Set 4 for various dimensions of the matrix A used in [6] . Both methods are compared with the CPLEX barrier method (interior point), the CPLEX primal simplex method, and the CPLEX dual simplex method. The worst performance occurs at m/n ratio of 200, where on average, dynamic

Table 1. Results from dynamic RAD and COST RAD for set 1-set 4, (random, NNLP $a_{ij} = 1 - 5$, $b_i = 1 - 10$, $c_j = 1 - 10$).

		Dynamic RAD$^+$				COST RAD$^+$			
	n	1000	3163	10,000	14,143	1000	3163	10, 000	14, 143
	m	200, 000	63,246	20,000	14,143	200, 000	63, 246	20, 000	14, 143
	m/n	200	20	2	1	200	20	2	1
Den-sity	No	CPU time, sec^{++}							
0.005	1	2.02	30.88	108.52	126.93	2.10	30.82	108.70	127. 55
0.006	2	2.47	32.19	106.48	113.68	2.42	31.48	104.87	114. 03
0.007	3	2.63	30.39	96.61	104.34	2.65	29.41	92.45	104. 18
0.008	4	2.46	31.03	90.14	89.24	2.54	30.63	88.20	90.73
0.009	5	2.67	28.89	82.66	86.57	2.78	30.10	83.53	85.21
0.01	6	2.73	28.22	75.46	83.66	2.79	27.81	77.90	80.43
0.02	7	2.88	23.17	45.55	49.82	3.09	24.69	47.63	49.95
0.03	8	2.83	17.97	33.85	37.35	3.22	20.49	36.68	38.33
0.04	9	2.92	15.24	29.23	28.98	3.33	19.06	32.74	32.53
0.05	10	2.97	14.10	24.83	26.37	3.34	16.97	28.23	28.59
0.06	11	2.86	11.93	23.38	24.45	3.20	14.94	27.58	27.27
0.07	12	2.94	11.21	20.38	21.08	3.41	14.88	23.59	23.79
0.08	13	2.87	10.25	19.47	21.43	3.32	13.57	23.44	24.19

0.09	14	3.05	9.33	19.43	20.49	3.38	12.67	23.09	23.80
0.1	15	3.20	9.33	18.03	18.78	3.39	12.92	22.93	20.85
0.2	16	4.39	8.07	14.86	16.50	4.30	11.09	18.87	20.31
0.3	17	5.26	8.19	13.77	15.27	4.97	10.58	18.11	19.46
0.4	18	6.40	9.19	14.32	15.60	5.76	12.31	18.55	18.88
0.5	19	7.80	9.84	14.33	15.97	6.98	11.92	18.00	19.89
0.75	20	10.86	11.91	14.55	16.26	8.26	12.01	17.19	18.06
1	21	12.93	12.01	12.61	14.58	8.39	12.20	17.71	18.50
Average		4.24	17.30	41.83	45.11	3.98	19.07	44.28	46.98

$^{+}$Used CPLEX preprocessing parameters of presolve = off and predual = off. $^{++}$Average of 5 instances of LPs at each density.

RAD is 8% faster than COST RAD for densities less than 0.2 and 18% slower for densities above 0.2. When the density increases, dynamic RAD shows an increase in computation time more than that of COST RAD. On the other hand, for an m/n ratio of 20 the CPU times decrease with an increase in density. For higher densities above 0.01, dynamic RAD is more efficient and takes less computation times than COST RAD.

On average, dynamic RAD is 10% more efficient than COST RAD. For an m/n ratio of 2 at densities higher than 0.009, the data show that COST RAD starts taking significantly more time than dynamic RAD. Dynamic RAD was 5.5% faster than COST RAD over all densities and 21% faster on average for densities above 0.5.

For an m/n ratio of 1 with densities greater than 0.01, dynamic RAD is about 8% more efficient than COST RAD. On average, dynamic RAD is superior performance to COST RAD for problem sets 2, 3, and 4.

Table 2 from [6] is presented to provide an immediate comparison of the developed dynamic RAD method with the standard CPLEX solvers.

A reporting limit of 3000 seconds was used. On average, the CPU times for dynamic RAD were faster than any of the CPLEX solvers across all densities and ratios. However, CPLEX barrier methods show smaller CPU times when ratio m/n = 20 and the density is less than or equal to 0.01.

Table 2. Results from the CPLEX primal, the dual simplex, and the barrier method for set 1-set 4, (random, NNLP $a_{ij} = 1 - 5$, $b_i = 1 - 10$, $c_j = 1 - 10$) [6] .

		Primal⁻				Dual⁻				Barrier⁻			
	n	1000	3163	10,000	14,143	1000	3163	10,000	14,143	1000	3163	10,000	14,143
	m	200,000	63,246	20,000	14,143	200,000	63,246	20,000	14,143	200,000	63,246	20,000	14,143
	m/n	200	20	2	1	200	20	2	1	200	20	2	1
Density	No					CPU time, sec⁺⁺							
0.005	1	7.01	71.02	228.51	309.83	54.84	762.62	1597.24	1169.04	2.36	14.52	240.17	650.83
0.006	2	10.36	77.28	245.60	291.07	60.29	803.97	1607.16	2413.42	2.39	16.30	224.08	666.54
0.007	3	12.98	75.84	219.72	265.09	91.39	876.85	1483.20	1702.47	3.04	18.34	233.55	671.56
0.008	4	15.72	82.01	206.45	239.30	100.06	912.75	1445.54	1236.76	3.90	20.70	232.38	668.82
0.009	5	19.25	80.35	196.72	216.23	114.95	898.99	1375.73	427.95	4.76	22.66	232.23	649.26
0.01	6	21.92	78.50	182.47	216.60	123.49	912.63	1252.05	436.31	5.53	24.29	228.76	650.30
0.02	7	39.90	78.80	118.28	127.59	203.08	963.66	807.29	362.34	17.13	32.08	242.54	711.26
0.03	8	45.42	79.75	98.02	108.60	217.18	1207.76	545.91	723.98	28.79	45.03	266.90	727.61
0.04	9	50.30	78.78	89.75	88.32	248.75	1489.40	450.08	539.92	41.50	62.28	292.15	806.80
0.05	10	55.16	78.92	81.09	82.14	256.49	1746.46	418.69	519.50	53.72	81.32	327.01	837.67
0.06	11	60.34	77.49	77.28	78.27	251.39	2124.31	378.71	409.47	67.58	100.48	359.53	897.58
0.07	12	62.07	78.93	70.44	70.37	251.74	2446.69	310.89	544.15	84.70	125.49	401.72	948.01

0.08	13	62.92	76.96	70.21	69.81	264.48	2799.62	307.25	388.94	99.51	149.37	454.01	1038.86
0.09	14	66.57	79.07	71.46	72.37	258.14	2523.03	718.04	427.95	119.26	186.06	495.28	1153.31
0.1	15	71.00	74.57	67.43	62.64	287.36	2251.10	267.14	436.31	138.67	207.54	539.64	1194.56
0.2	16	87.49	83.12	64.38	62.99	294.39	1450.82	201.73	362.34	379.68	691.77	1298.76	2529.97
0.3	17	94.57	77.91	67.14	66.61	341.44	1280.71	175.16	267.16	657.45	1333.29	2418.75	b
0.4	18	99.33	78.46	73.58	71.48	384.10	1236.30	146.09	233.39	985.86	2076.09	b	b
0.5	19	111.30	84.30	86.50	75.62	427.16	1173.49	133.49	208.65	1350.82	b	b	b
0.75	20	128.26	99.35	115.00	102.51	410.98	1056.18	132.25	181.95	b	b	b	b
1	21	207.55	94.09	393.54	145.96	375.89	411.19	148.90	165.45	b	b	b	b
Average		63.30	80.26	134.46	134.45	238.93	1396.60	662.03	626.55	n/a	n/a	n/a	n/a

‾Used CPLEX preprocessing parameters of presolve = ON and predual = Auto; ++Average of 5 instances of LPs at each density; [b]Runs with CPU times > 3000 s are not reported.

Computational Results for LP

Table 3 shows computational results for the CPLEX primal simplex method, the dual simplex method, and the interior point barrier method for the general LP problem set used in [17] . CPU times for COST GRAD and VIOL using both the multi-cut technique and dynamic approaches are presented for comparison. Dynamic GRAD is stable over the range of densities. In addition, its performance is superior to multi-cut GRAD for every problem instance. Average CPU times for GRAD using multi-cut method and dynamic approach are 43.87 and 24.57 seconds, respectively, a 42% improvement in computation time. Average computation times for GRAD and VIOL using dynamic approach are 24.57 seconds vs. 33.82 seconds, respectively.

It should be noted that GRAD captures more information than VIOL in higher densities to discriminate between constraints. Interestingly, when the dynamic active-set is used for both GRAD and VIOL, their CPU

Table 3. Comparison of computation times of CPLEX solvers, GRAD, and VIOL using both dynamic active-set and multi- cut method on general LP problem set (random LP with 1000 variables and 200,000 constraints [17]).

		Constraint selection metric+				CPLEX--		
No	Den-sity	GRAD	VIOL	GRAD	VIOL	Primal	Dual	Barrier
		Multi-cut method		Dynamic active-set				
		CPU time, sec++						
1	0.005	9.85	12.31	7.96	9.26	40.99	23.05	2.39
2	0.006	11.48	14.50	9.44	11.11	84.56	35.52	2.62
3	0.007	13.36	14.21	10.60	12.58	128.65	48.62	3.79
4	0.008	14.24	14.67	12.09	13.00	183.70	61.56	4.93
5	0.009	15.41	15.32	12.25	14.57	212.79	75.34	6.06
6	0.01	16.55	17.09	14.10	15.33	256.70	92.11	7.33
7	0.02	24.74	22.24	20.93	21.79	396.55	205.25	15.86
8	0.03	27.84	24.30	22.91	26.21	460.01	295.18	26.63
9	0.04	30.55	24.47	23.87	29.52	602.73	350.86	35.26
10	0.05	37.59	28.72	28.52	33.57	617.29	396.10	46.76
11	0.06	34.29	26.58	26.86	33.80	656.22	438.92	59.55
12	0.07	37.46	28.05	26.91	34.34	729.43	465.61	71.65
13	0.08	36.28	26.29	25.54	33.46	739.21	510.10	82.98
14	0.09	37.97	27.74	24.60	33.21	823.11	521.89	94.01
15	0.10	39.50	28.30	25.99	35.61	956.17	554.29	108.03
16	0.20	56.26	36.64	27.97	41.28	1456.41	759.66	280.09
17	0.30	60.93	42.40	28.41	40.68	1664.83	900.12	527.05
18	0.40	74.58	56.97	33.39	52.19	2033.10	1057.27	760.07

19	0.50	85.02	71.35	36.85	54.68	1925.32	1334.80	1076.40
20	0.75	113.02	116.78	39.44	59.53	2232.88	1571.28	2132.53
21	1.00	144.35	173.02	57.22	104.58	2301.76	1717.25	3267.10
Av-er-age		43.87	39.14	24.57	33.82	881.07	543.56	410.05

+Used CPLEX preprocessing parameters of presolve = off and predual = off. $1^T x \leq M = 10^{10}$ was used as the bounding constraint; ++Average of 5 instances of LPs at each density; --Used CPLEX preprocessing parameters of presolve = ON and predual = Auto.

times are significantly faster than the same metrics with the multi-cut method. GRAD using the multi-cut technique takes the longest computation time in comparison to others at higher densities.

Unlike the proposed dynamic approach, the LP algorithm COST GRAD requires checking the signs of the nonzero a_{ij} and therefore more computation time for higher densities.

The efficiency of VIOL decreases significantly with increasing density. On average, dynamic GRAD is approximately 35 times faster than the CPLEX primal simplex, 21 times faster than the CPLEX dual simplex, and 17 times faster the CPLEX barrier linear programming solvers without preprocessing.

The superior overall performance of GRAD using dynamic approach is apparent across all densities in general LP set.

For comparison purposes, Table 4 shows GRAD and VIOL computation times when a fixed number of violated constraints is added at each iteration. Adding a fixed number of constraints is examined for both GRAD and VIOL. At densities below 0.03, dynamic GRAD takes less CPU time than the fixed-cut approach. GRAD

Table 4. Comparison of computation times of GRAD using dynamic active-set and fixed cut method on general LP problem set (random LP with 1000 variables and 200,000 constraints [17]).

Constraint selection metric[+]											
No	Den-sity	GRAD	VIOL		GRAD				VIOL		
		Dynamic active-set			Fixed number of constraints						
				100	500	1000		100	500	1000	
				CPU times, sec[++]							
1	0.005	7.96	9.26		14.58	10.04	8.05		10.49	8.82	11.56
2	0.006	9.44	11.11		18.75	13.14	9.48		12.61	10.57	14.45
3	0.007	10.60	12.58		20.32	13.24	10.67		13.87	12.22	15.01
4	0.008	12.09	13.00		23.57	12.63	12.05		14.97	12.82	16.22
5	0.009	12.25	14.57		23.16	13.60	12.32		15.88	14.00	18.35
6	0.01	14.10	15.33		26.04	14.83	13.59		17.72	15.35	19.26
7	0.02	20.93	21.79		36.49	21.27	20.38		23.55	22.70	28.35
8	0.03	22.91	26.21		38.40	22.33	22.30		25.40	26.07	34.22
9	0.04	23.87	29.52		38.48	22.68	23.19		25.63	27.51	36.21
10	0.05	28.52	33.57		46.34	27.77	28.69		29.67	32.25	41.66
11	0.06	26.86	33.80		40.35	24.47	26.26		27.12	30.36	40.53
12	0.07	26.91	34.34		41.91	26.05	27.92		28.88	32.69	42.09
13	0.08	25.54	33.46		37.80	24.61	26.36		26.64	31.62	42.58
14	0.09	24.60	33.21		37.71	25.01	28.18		27.38	32.19	43.69
15	0.1	25.99	35.61		39.30	25.54	28.00		29.12	33.81	46.08
16	0.2	27.97	41.28		41.66	29.36	33.48		33.54	40.20	57.02
17	0.3	28.41	40.68		38.05	28.25	34.05		32.88	41.53	59.39
18	0.4	33.39	52.19		41.45	33.58	41.14		39.98	50.68	74.25
19	0.5	36.85	54.68		42.40	36.86	46.14		44.68	56.76	81.71
20	0.75	39.44	59.53		45.88	40.36	50.28		52.67	69.07	101.71

21	1	57.22	104.58	48.44	46.14	57.55	61.59	78.23	114.15
Average		24.57	33.82	35.29	24.37	26.67	28.30	32.36	44.69

[+]Used CPLEX preprocessing parameters of presolve = off and predual = off. $1^T x \leq M = 10^{10}$ was used as the bounding constraint; [++]Average of 5 instances of LPs at each density.

[*]Corresponding author.

CONCLUSION

In this paper, dynamic active-set methods have been proposed for both NNLPs and LPs. In particular, these new approaches were compared to existing methods for problems with various sizes and densities. On average, dynamic RAD shows superior performance over COST RAD for the NNLP problem sets 2, 3, and 4. In the LP problem set, dynamic GRAD significantly outperformed the COST GRAD as well as the CPLEX primal simplex and the dual simplex. In this LP problem set, however, the barrier solver did outperform all methods for densities up to 0.03. In addition, dynamic GRAD outperformed a dynamic version of VIOL, which was a standard method in column generation and decomposition methods.

REFERENCES

1. Bixby, R.E., Gregory, J.W., Lustig, I.J., Marsten, R.E. and Shanno, D.F. (1992) Very Large-Scale Linear Programming: A Case Study in Combining Interior Point and Simplex Methods. Operations Research, 40, 885-897. http://dx.doi.org/10.1287/opre.40.5.885
2. Rosenberger, J.M., Johnson, E.L. and Nemhauser, G.L. (2003) Rerouting Aircraft for Airline Recovery. Transportation Science, 37, 408-421. http://dx.doi.org/10.1287/trsc.37.4.408.23271
3. Todd, M.J. (2002) The Many Facets of Linear Programming. Mathematical Programming, 91, 417-436. http://dx.doi.org/10.1007/s101070100261
4. Elwes, R. (2012) The Algorithm That Runs the World. New Scientist, 215, 32-37. http://dx.doi.org/10.1016/S0262-4079(12)62078-8
5. Dare, P. and Saleh, H. (2000) GPS Network Design: Logistics Solution Using Optimal and Near-Optimal Methods. Journal of Geodesy, 74, 467-478. http://dx.doi.org/10.1007/s001900000104
6. Saito, G., Corley, H.W., Rosenberger, J.M., Sung, T.-K. and Noroziroshan, A. (2015) Constraint Optimal Selection Techniques (COSTs) for Nonnegative Linear Programming Problems. Applied Mathematics and Computation, 251, 586-598. http://dx.doi.org/10.1016/j.amc.2014.11.080
7. Stone, J.J. (1958) The Cross-Section Method, an Algorithm for Linear Programming. DTIC Document, P-1490, 24.
8. Thompson, G.L., Tonge, F.M. and Zionts, S. (1996) Techniques for Removing Nonbinding Constraints and Extraneous Variables from Linear Programming Problems. Management Science, 12, 588-608. http://dx.doi.org/10.1287/mnsc.12.7.588
9. Myers, D.C. and Shih, W. (1988) A Constraint Selection Technique for a Class of Linear Programs. Operations Research Letters, 7, 191-195. http://dx.doi.org/10.1016/0167-6377(88)90027-2
10. Curet, N.D. (1993) A Primal-Dual Simplex Method for Linear Programs. Operations Research Letters, 13, 233-237. http://dx.doi.org/10.1016/0167-6377(93)90045-I
11. Adler, I., Karp, R. and Shamir, R. (1986) A Family of Simplex Variants Solving an m× d Linear Program in Expected Number of Pivot Steps Depending on d Only. Mathematics of Operations Research, 11, 570-590. http://dx.doi.org/10.1287/moor.11.4.570

12. Zeleny, M. (1986) An External Reconstruction Approach (ERA) to Linear Programming. Computers & Operations Research, 13, 95-100. http://dx.doi.org/10.1016/0305-0548(86)90067-5
13. Mitchell, J.E. (2000) Computational Experience with an Interior Point Cutting Plane Algorithm. SIAM Journal on Optimization, 10, 1212-1227. http://dx.doi.org/10.1137/S1052623497324242
14. Corley, H.W., Rosenberger, J., Yeh, W.-C. and Sung, T.K. (2006) The Cosine Simplex Algorithm. The International Journal of Advanced Manufacturing Technology, 27, 1047-1050. http://dx.doi.org/10.1007/s00170-004-2278-1
15. Junior, H.V. and Lins, M.P.E. (2005) An Improved Initial Basis for the Simplex Algorithm. Computers & Operations Research, 32, 1983-1993. http://dx.doi.org/10.1016/j.cor.2004.01.002
16. Corley, H.W. and Rosenberger, J.M. (2011) System, Method and Apparatus for Allocating Resources by Constraint Selection. US Patent No. 8082549.
17. Saito, G., Corley, H.W. and Rosenberger, J. (2012) Constraint Optimal Selection Techniques (COSTs) for Linear Programming. American Journal of Operations Research, 3, 53-64. http://dx.doi.org/10.4236/ajor.2013.31004

CHAPTER 5

The Sliding Gradient Algorithm for Linear Programming

Hochung Liu, Peizhuang Wang

College of Intelligence Engineering and Mathematics, Liaoning Technical University, Fuxin, China.

ABSTRACT

The existence of strongly polynomial algorithm for linear programming (LP) has been widely sought after for decades. Recently, a new approach called Gravity Sliding algorithm [1] has emerged. It is a gradient descending method whereby the descending trajectory slides along the inner surfaces of a polyhedron until it reaches the optimal point. In R^3, a water droplet pulled by gravitational force traces the shortest path to descend to the lowest point. As the Gravity Sliding algorithm emulates the water droplet trajectory, it exhibits strongly polynomial behavior in R^3. We believe that it could be a strongly polynomial algorithm for linear programming in R^n too.

Citation: Liu, H. and Wang, P. (2018) The Sliding Gradient Algorithm for Linear Programming. American Journal of Operations Research, 8, 112-131. doi: 10.4236/ajor.2018.82009.

In fact, our algorithm can solve the Klee-Minty deformed cube problem in only two iterations, irrespective of the dimension of the cube. The core of gravity sliding algorithm is how to calculate the projection of the gravity vector g onto the intersection of a group of facets, which is disclosed in the same paper [1]. In this paper, we introduce a more efficient method to compute the gradient projections on complementary facets, and rename it the Sliding Gradient algorithm under the new projection calculation.

Keywords:- Linear Programming, Mathematical Programming, Complexity Theory, Optimization

INTRODUCTION

The simplex method developed by Dantiz [2] has been widely used to solve many large-scale optimizing problems with linear constraints. Its practical performance has been good and researchers have found that the expected number of iterations exhibits polynomial complexity under certain conditions [3] [4] [5] [6] . However, Klee and Minty in 1972 gave a counter example showing that its worst case performance is $O(2^n)$ [7] . Their example is a deliberately constructed deformed cube that exploits a weakness of the original simplex pivot rule, which is sensitive to scaling [8] . It is found that, by using a different pivot rule, the Klee-Minty deformed cube can be solved in one iteration. But for all known pivot rules, one can construct a different deformed cube that requires exponential number of iterations to solve [9] [10] [11] . Recently, the interior point method [12] has been gaining popularity as an efficient and practical LP solver. However, it was also found that such method may also exhibit similar worse case performance by adding a large set of redundant inequalities to the Klee-Minty cube [13] .

Is it possible to develop a strongly polynomial algorithm to solve the linear programming problem, where the number of iterations is a polynomial function of only the number of constraints and the number of variables? The work by Barasz and Vempala shed some light in this aspect. Their AFFINE algorithm [14] takes only $O(n^2)$ iterations to solve a broad class of deformed products defined by Amenta and Ziegler [15] which includes the Klee-Minty cube and many of its variants.

In certain aspect, the Gravity Sliding algorithm [1] is similar to the AFFINE algorithm as it also passes through the interior of the feasible region. The main difference is in the calculation of the next descending vector. In the gravity falling approach, a gravity vector is first defined (see Section 3.1 for details). This is the principle gradient descending direction where other descending directions are derived from it. In each iteration, the algorithm first computes the descending direction, then it descends from this direction until it hits one or more facets that forms the boundary of the feasible region. In order not to penetrate the feasible region, the descending direction needs to be changed. The trajectory is likened a water droplet falling from the sky but is blocked by linear planar structures (e.g. the roof top structure of a building) and needs to slide along the structure. The core of gravity sliding algorithm is how to calculate the projection of the gravity vector g onto the intersection of a group of facets. This projection vector lies on the intersection of the facets and hence lies on the null space defined by these facets. Conventional approach is to compute the null space first and then find the projection of g onto this null space. An alternative approach is disclosed in [1] which operates directly from the subspace formed by the intersecting facets. This direct approach is more suitable to the Gravity Sliding algorithm. In this paper, we further present an efficient method to compute the gradient projections on complementary facets and also introduce the notion of selecting the steepest descend projection among a set of candidates. With these refinements, we rename the Gravity Sliding algorithm as the Sliding Gradient algorithm. We have implemented our algorithm and tested it on the Klee-Minty cube. We observe that it can solve the Klee-Minty deformed cube problem in only two iterations, irrespective of the dimension of the cube.

This paper is organized as follows: Section 2 gives an overview of the Cone-Cutting Theory [16] , which is the intuitive background of the Gravity Sliding algorithm. Section 3 discusses the Sliding Gradient algorithm in details. The pseudo-code of this algorithm is summarized in Section 4 and Section 5 gives a walk-through of this algorithm using the Klee-Minty as an example. This section also discusses the practical implementation issues. Finally, Section 6 discuss about future work.

CONE-CUTTING PRINCIPLE

The cone-cutting theory [16] offers a geometric interpretation of a set of inequality equations. Instead of considering the full set constraint equations in a LP problem, the cone-cutting theory enables us to consider a subset of equations, and how an additional constraint will shape the feasible region. The geometric insight forms the basis of our algorithm development.

Cone-Cutting Principle

In an m-dimension space R^m , a hyperplane $y^T\tau$=cy cuts R^m into two half spaces. Here τ is the normal vector of the hyperplane and c is a constant. We denote the positive half space $\{y||y^T\tau \geq c\}$ the accepted zone of the hyperplane and the negative half space where $\{y||y^T\tau<c\}$ is rejected zone. Note that the normal vector τ points to accepted zone area and we call the hyperplane with such orientation a facet α:(τ,c). When there are m facets in R^m and $\{\tau_1,\tau_2,\cdots,\tau_m\}$ are linear independent, this set of linear equations has a unique solution which is a point V in R^m. Geometrically, $\{\alpha_1,\alpha_2,\cdots,\alpha_m\}$ form a cone and V is the vertex of the cone. We now give a formal definition of a cone, which is taken from [1] .

Definition 1. Given m hyperplanes in R^m, with rank $r(\alpha_1,\cdots,\alpha_m)$=m and intersection V, $C=C(V;\alpha_1,\cdots,\alpha_m)=\alpha_1\cap\cdots\cap\alpha_m$ is called a cone in R^m. The area $\{y\,|\,y^T\tau_i\geq c_i\,(i=1,2,\cdots,m)\}$ is called the accepted zone of C. The point V is the vertex and α_j is the facet plane, or simply the facet of C.

A cone C also has m edge lines. They are formed by the intersection of (m − 1) facets. Hence, a cone can also be defined as follows.

Definition 2. Given m rays $R_j=\{V+tr_j|0\leq t<+\infty\}(j=1,\cdots,m)$ shooting from a point V with rank $r(r_1,\cdots,r_m)=m$, $C=C(V;r_1,\cdots,r_m)=c[R_1,\cdots,R_m]$, the convex closure of m rays is called a cone in R^m. R_j is the edge, r_j the edge direction, and $R_j^+=\{V+tr_j\,|\,\infty<t<+\infty\}$ the edge line of the cone C.

The two definitions are equivalent. Furthermore, P.Z. Wang [11] has observed that R_i^+ and α_i are opposite to each other for i=1,⋯,m. Edge-line R_i^+ is the intersection of all C-facets except α_i, while facet α_i is bounded by all C-edges except R_i^+. This is the duality between facets and edges. For i=1,⋯,m, $\{\tau_i, R_i\}$ is called a pair of cone C.

It is obvious that $r_j^T\tau_i=0\ \ (\text{for}\ \ i\neq j)$ since r_j lies on α_i . Moreover, we have

$$r_i^T\tau_i\geq 0\,(\text{for}\ i=1,\cdots,m) \tag{1}$$

Cone Cutting Algorithm

Consider a linear programming (LP) problem and its dual:

$$\text{(Primary):}\ \ \max\left\{\tilde{c}^T x\,|\,Ax\leq b;\,x\geq 0\right\} \tag{2}$$

$$\text{(Dual):}\ \ \min\left\{y^T b\,|\,y^T A\geq \tilde{c};\,y\geq 0\right\} \tag{3}$$

In the following, we focus on solving the dual LP problem. The standard simplex tableau can be obtained by appending an m×m identity matrix $I_{m\times m}$ which represents the slack variables as shown below:

$$\begin{bmatrix} a_{11} & \cdots & a_{1n} & 1 & \cdots & 0 & b_1 \\ \vdots & \ddots & \vdots & \vdots & \ddots & \vdots & \vdots \\ a_{m1} & \cdots & a_{mn} & 0 & \cdots & 1 & b_m \\ \tilde{c}_1 & \cdots & \tilde{c}_n & 0 & \cdots & 0 & 0 \end{bmatrix}$$

We can construct a facet tableau whereby each column is a facet denoted as α_j:(τ_j,c_j) , where $\alpha_j:(\tau_j,c_j)$, where $\tau_i=(a_{1i},a_{2i},\cdots,a_{mi})^{\mathrm{T}}$ and

$$c_i=\begin{cases}\tilde{c}_i & \text{for } 1\le i\le n \\ 0 & \text{for } n<i\le m+n\end{cases} \tag{4}$$

The facet tableau is depicted as follow. The last column $(b_1,b_2,\cdots,b_m,0)^{\mathrm{T}}$ is not represented in this tableau.

$$\begin{matrix} \alpha_1 & \alpha_2 & \cdots & \alpha_{m+n} \end{matrix}$$
$$\begin{bmatrix} \tau_1 & \tau_2 & \cdots & \tau_{m+n} \\ c_1 & c_2 & \cdots & c_{m+n} \end{bmatrix}$$

When a cone $C=C(V;\alpha_1,\cdots,\alpha_m)=C(V;r_1,\cdots,r_m)$ is intersected by another facet α_j, the ith edge of the cone is intersected by α_j at certain point q_{ij} . We call α_j cuts the cone C and the cut points q_{ij} can be obtained by the following equations:

$$\boldsymbol{q}_{ij}=\boldsymbol{V}+t_i\boldsymbol{r}_i \quad \text{where } t_i=\left(c_i-\boldsymbol{V}^{\mathrm{T}}\boldsymbol{\tau}_j\right)\big/\boldsymbol{r}_i^{\mathrm{T}}\boldsymbol{\tau}_j \tag{5}$$

The intersection is called real if $t_i\ge 0$ and fictitious if $t_i<0$. Cone cutting greatly alters the accepted zone, as can be seen from the simple 2-dimension example as shown in Figures 1(a)-(e). In 2-dimension, a facet α:(τ,c) is a line. The normal vector τ is perpendicular to this line and points to the accepted zone of this facet. Furthermore, a cone is formed by two non-parallel facets in 2-dimension. Figure 1(a) shows such a cone C(V;α_1,α_2) . The accepted zone of the cone is the intersection of the two accepted zones of facets α_1 and α_2. This is represented by the shaded area A in Figure 1(a). In Figure 1(b), a new facet α_3 intersects the cone at two cut points q_{13} and q_{23}. They are both real cut points. Since the arrow of normal vector τ_3 points to the same general direction of the cone, V lies in the rejected zone of α_3 and we say α_3 rejects V. Moreover, the accepted zone of α_3 intersects with the accepted zone of the cone so that the overall accepted zone is reduced to the shaded area marked as B. In Figure 1(c), τ_3 points to the opposite direction. α_3 accepts V and the overall accepted zone is confined to the area marked as C.

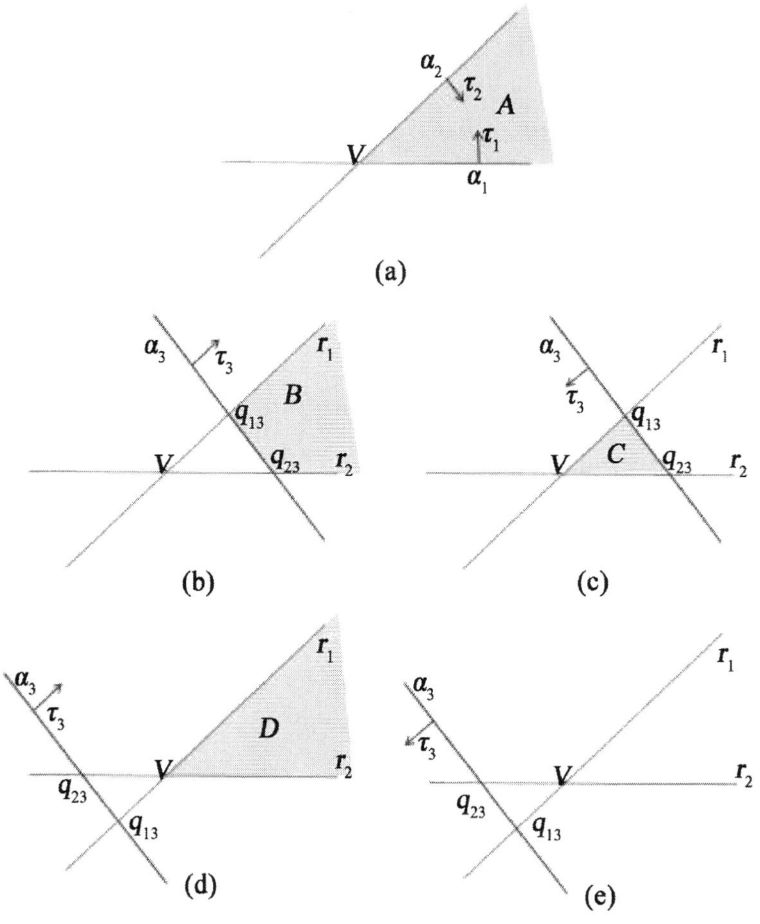

Figure 1. Accepted zone area of a cone and after it is cut by a facet.

As the dual feasible region $\mathcal{D}$ of a LP problem must satisfy all the constraints, it must lie within area C. In Figure 1(d), α_3 cuts the cone at two fictitious points. Since τ_3 points to the same direction of the cone, V is accepted by α_3. However, the accepted zone of α_3 covers that of the cone. As a result, α_3 does not contribute to any reduction of the overall accepted zone area, and so it can be deleted for further consideration without affecting the LP solution. In Figure 1(e), τ_3 points to the opposite direction of the cone. The intersection between the accepted zone of α_3 and that of the cone is an empty set. This means that the dual feasible region $\mathcal{D}$ is empty and the LP is infeasible. This is actually one of the criteria that can be used for detecting infeasibility.

Based on this cone-cutting idea, P.Z. Wang [16] [17] have developed a cone-cutting algorithm to solve the dual LP problem. Each cone is a combination of m facets selected from (m + n) choices. Let D denotes the index set of facets of C, (i.e. if $\Delta(i)=j$, then $\tau_{\Delta(i)}=\tau_j$). The algorithm starts with an initial coordinate cone C_o, then finds a facet to replace one of the existing facet α_{out} thus forming a new cone. This process is repeated until an optimal point is found. The cone-cutting algorithm is summarized in Table 1 below.

This algorithm finds a facet that rejects V the least as the cutting facet in steps 2 and 3. This facet cuts the edges of the cone at m points. In step 4 and 5, the real cut point q_{I*} that is closest to the vertex V is identified. This becomes the vertex of a new cone.

Table 1. Cone-cutting algorithm.

steps	**Input:** A, b **and** $\tilde{c}$ ***Output:*** ***either V as the optimal point & $V^T b$ as optimal value or declare the LP problem as infeasible.***
0	$V = V_0 = (0,0,\cdots,0)$ % the coordinate cone; $\Delta(j) = n + j \ for \ j = 1,\cdots,m$ $r_1 = (1,0,\cdots,0), r_2 = (0,1,\cdots,0),\cdots, r_m = (0,0,\cdots,1)$;
1	**while** (true)
2	% For all facets not in C & reject V, find one that rejects V the least. $\bar{\Delta} = \{1:(m+n)\} \setminus \Delta$; % $\bar{\Delta}$ are facets not in C $j^* = \arg\min_j \{e_j \mid e_j = (V^T \tau_{\bar{\Delta}(j)} - c_{\bar{\Delta}(j)}), j \in \bar{\Delta}\}$
3	**if** $e_{j^*} \geq 0$ return ($\boldsymbol{V}$ & $\boldsymbol{V}^T\boldsymbol{b}$) as optimal vertex & optimal value **else** $J^* = \bar{\Delta}(j^*)$; % j^* is the facet index to enter
4	$t_i = \dfrac{c_{J^*} \boldsymbol{V}^T \boldsymbol{\tau}_{J^*}}{\boldsymbol{r}_i^T \boldsymbol{\tau}_{J^*}}; i = 1,\cdots,m; \ i = 1,\cdots,m \}; \ \boldsymbol{q}_i = \boldsymbol{V} + t_i \boldsymbol{r}_i$;
5	**if** $(t_i < 0 \ \forall i = 1,\cdots,m)$ **return**("LP is infeasible") **else** % For all the real cuts $\boldsymbol{q}_{i,}$, find the $\boldsymbol{q}_{I^*}$ that is closest to $\boldsymbol{V}$ $I^* = \arg\min_i \{\boldsymbol{q}_i^T \boldsymbol{b} \mid t_i > 0\}$ % I^* is the facet index to leave

% Form new cone C_{k+1} by updating $\boldsymbol{V}_{k+1}$; edge vectors and facets

6 $V_{k+1} = \boldsymbol{q}_{I^*}$; $r_i = \begin{cases} r_i & i = I^* \\ sign(t_i)[\boldsymbol{q}_i - V_{k+1}] & i \neq I^* \end{cases}$

$\Delta(I^*) = J^*$;

7 **end**

This new cone retains all the facets of the original cone except that the cutting facet replaces the facet corresponding to the edge I^*. Yet the edge I^* is retained but the rest of the edges must be recomputed as shown in step 6. Amazingly, P.Z. Wang shows that when b>0, this algorithm produces exactly the same result as the original simplex algorithm proposed by Dantz [2] . Hence, the cone-cutting theory offers a geometric interpretation of the simplex method. More significantly, it inspires the authors to explore new approach to tackle the LP problem.

SLIDING GRADIENT ALGORITHM

Expanding on the cone-cutting theory, the Gravity Sliding Algorithm [1] was developed to find the optimal solution of the LP problem from a point within the feasible region $\mathcal{D}$. Since then, several refinements have been made and they are presented in the following sections.

Determining the General Descending Direction

The feasible region $\mathcal{D}$ is a convex polyhedron formed by constraints $(y^{\mathrm{T}}\tau_j \geq c_j; 1 \leq j \leq n+m)$, and the optimal feasible point is at one of its vertices. Let $\Omega = \{V_i \mid V_i^{\mathrm{T}}\tau_j \geq c_j; j = 1, \cdots, n+m\}$ be the set of feasible vertices. The dual LP problem (3) can then be stated as: $\min\{V_i^{\mathrm{T}}\boldsymbol{b} \mid V_i \in \Omega\}$. As $V_i^{\mathrm{T}}\boldsymbol{b}$ is the inner-product of vertex V_i and b, the optimal vertex V^* is the vertex that yields the lowest inner-product value. Thus we can set the principle descending direction g_0 to be the opposite of the b vector (i.e. g_0=−b) and this is referred to as the gravity vector. The descending path then descends along this principle direction inside $\mathcal{D}$ until it reaches the lowest point in $\mathcal{D}$ viewed along the direction of b. This point is then the optimal vertex V^* .

Circumventing Blocking Facets

The basic principle of the new algorithm can be illustrated in Figure 2. Notice that in 2-dim, a facet is a line. In this figure, these facets (lines) form a closed polyhedron which is the dual feasible region $\mathcal{D}$. Here the initial point P_0 is inside $\mathcal{D}$. From P_0, it attempts to descend along the g_0=−b direction. It can go as far as P_1 which is the point of intersection between the ray $R=P_0+tg_0$ and the facet α_1. In essence, α_1 is blocking this ray and hence it is called the blocking facet relative to this ray. In order not to penetrate $\mathcal{D}$, the descending direction needs to change from g_0 to g_1 at P_1, and slides along g_1 until it hits the other blocking facet α_2 at P_2. Then it needs to change course again and slides along the direction g_2 until it hits P_3. In this figure, P_3 is the lowest point in this dual feasible region $\mathcal{D}$ and hence it is the optimal point V^*.

It can be observed from Figure 2 that g_1 is the projection of g_0 onto α_1 and g_2 is the projection of g_0 onto α_2. Thus from P_1, the descending path slides along α_1 to reach P_2 and then slides along α_2 to reach P_3. Hence we call this algorithm Sliding Gradient Algorithm. The basic idea is to compute the new descending direction to circumvent the blocking facets, and advance to find the next one until it reaches the bottom vertex viewed along the direction of b.

Let σtσt denotes the set of blocking facets at the tth iteration. From an initial point P_0 and a gradient descend vector g_0, the algorithm iteratively performs the following steps:

1) compute a gradient direction g_t based on σ_t. In this example, the initial set of blocking facets σ_0 is empty and g_0=−b.

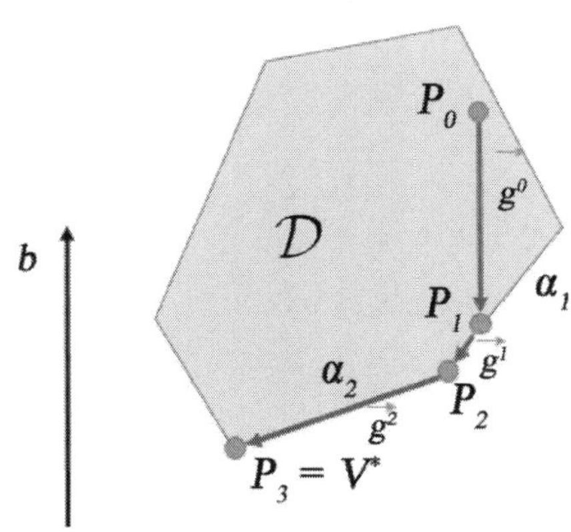

Figure 2. Sliding gradient illustration.

2) move P_t to P_{t+1} along g_t where P_{t+1} is a point at the first blocking facet.

3) Incorporate the newly encountered blocking facet to σ_t to form σ_{t+1}.

4) go back to step 1.

The algorithm stops when it cannot find any direction to descend in step (1). This is discussed in details in Section 3.6 where a formal stopping criterion is given.

Minimum Requirements for the Gradient Direction g_t

For the first step, the gradient descend vector g_t needs to satisfy the following requirements.

Proposition 1. gtgt must satisfy $(g_t)^T g_0 \geq 0$ so that the dual objective function $y^T b$ will be non-increasing when y move from P_t to P_{t+1} along the direction of g_t.

Proof. Since $(g_t)^T g_0 \geq 0$, g_t aligns to the principle direction of g_0. As $P_{t+1} = P_t + tg_t$, P_{t+1} moves along the principle direction of g_0 when t>0.

Since $P_{t+1}^T b = P_t^T b + t(g_t)^T b = P_t^T b - t(g_t)^T g_0$, $P_{t+1}^T b \leq P_t^T b$ when $(g_t)^T g_0 \geq 0$.

END

This means that if $(g_t)^T g_0 \geq 0$, then P_{t+1} is "lower than" P_t when viewed along the b direction.

Proposition 2. If $P_0 \in \mathcal{D}$, g_t must satisfy $\left(\tau_{\sigma_t(j)}\right)^T g_t \geq 0$ for all $j \in \sigma_t$ to ensure that P_{t+1} remains dual feasible (i.e. $P_{t+1} \in D$).

Proof. If for some j, $\left(\tau_{\sigma_t(j)}\right)^T g_t < 0$, this means that g_t is in the opposite direction of the normal vector of facet $\alpha_{\sigma_t(j)}$ so a ray $Q = P_t + tg_t$ will eventually penetrate this facet for certain positive value of t. This means that Q will be rejected by $\alpha_{\sigma_t(j)}$ and hence Q is no longer a dual feasible point. END

Maximum Descend in Each Iteration

To ensure that $P_{t+1} \in D$, we need to make sure that it won't advance too far. The following proposition stipulates the requirement.

Proposition 3. Assuming that $\mathcal{D}$ is non-empty and $P_0 \in D$. If g_t satisfies Propositions 1 and 2; and not all $g_t^T \tau_j = 0$ for $j = 1, \cdots, m+n$, then $P_{t+1} \in \mathcal{D}$ provided that the next point P_{t+1} is determined according to (6) below:

$$P_{t+1} = P_t + t_{j^*} g_t \qquad (6)$$

where $j^* = \arg\min_j \left\{ t_j \mid t_j = \dfrac{c_j - P_t^T \tau_j}{g_t^T \tau_j}; t_j > 0; j = 1, \cdots, m+n \right\}$.

Proof. The equation for a line passing through P along the direction g is P+tg. If this line is not parallel to the plane (i.e. $g^T\tau \neq 0$), it will intersect a facet $\alpha:(\tau,c)$ at a point Q according to the following equation:

$$Q = P + tg \text{ where } t = \left(c - P^T\tau\right)/g^T\tau \tag{7}$$

We call t the displacement from P to Q. So $t_j = \dfrac{c_j - P_t^T\tau_j}{g_t^T\tau_j}$ is the displacement from P_t to α_j. The condition $t_j > 0$ ensures that P_{t+1} moves along the direction g_t but not the opposite direction. j^* is the smallest of all the displacements thus α_{j^*} is the first blocking facet that is closest to P_t.

To show that $P_{t+1} \in D$, we need to show $P_{t+1}^T\tau_j - c_j \geq 0$ for $j = 1,\cdots,m+n$. Note that

$$P_{t+1}^T\tau_j - c_j = \left(P_t + t_{j^*}g_t\right)^T\tau_j - c_j = \left(P_t^T\tau_j - c_j\right) + t_{j^*}g_t^T\tau_j.$$

Since $P_t \in D$, $\left(P_t^T\tau_j - c_j\right) \geq 0$ for $j = 1,\cdots,m+n$, so we need to show that $t_{j^*}g_t^T\tau_j \geq 0$ for $j = 1,\cdots,m+n$.

The displacements t_j can be split into two groups. For those displacements where $t_j < 0$, $\dfrac{c_j - P_t^T\tau_j}{g_t^T\tau_j} = t_j < 0$ so $g_t^T\tau_j = \left(c_j - P_t^T\tau_j\right)/t_j = -\left(c_j - P_t^T\tau_j\right)/k_1$, where $k_1 = -t_j$ is a positive constant. Since tj*>0tj*>0.

$$t_{j^*}g_t^T\tau_j = -\frac{t_{j^*}}{k_1}\left(c_j - P_t^T\tau_j\right) = k_2\left(P_t^T\tau_j - c_j\right) \geq 0$$ since $P_t \in D$ & $k_2 > 0$.

For those displacements where $t_j \geq 0$, we have that t_{j^*} is the minimum of all t_j in this group. Let k_3 be the ratio between t_{j^*} and t_j. Obviously, $k_3 = \dfrac{t_{j^*}}{t_j} \leq 1$

$$\begin{aligned}
&\left(P_t^T\tau_j - c_j\right) + t_{j^*}g_t^T\tau_j \\
&= \left(P_t^T\tau_j - c_j\right) + k_3 t_j g_t^T\tau_j = \left(P_t^T\tau_j - c_j\right) + k_3\frac{c_j - P_t^T\tau_j}{g_t^T\tau_j}g_t^T\tau_j \\
&= \left(P_t^T\tau_j - c_j\right) - k_3\left(P_t^T\tau_j - c_j\right) \geq 0
\end{aligned}$$

So $P_{t+1} \in D$. END

If $g_t^T\tau_j = 0$, g_t is parallel to α_j. Unless all facets are parallel to g_t, Proposition 3 can still find the next descend point P_{t+1}. If all facets are parallel to g_t, this means that facets are linearly dependent with each other. The LP problem is not well formulated.

Gradient Projection

We now show that the projection of g_0 onto the set of blocking facets σ_t satisfies the requirements of Proposition 1 and 2. Before we do so, we discuss the projection operations in subspace first.

Projection in Subspaces

Projection is a basic concept defined in vector space. Since we are only interested in the gradient descend direction of g_t but not the actual location of the projection, we can ignore the constant c in the hyperplane $\{y||y^T\tau=c\}$. In other words, we focus on the subspace $V(\tau)$ spanned by τ and its null space $N(\tau)$ rather than the affine space spanned by the hyperplanes.

Let Y be the vector space in $\mathbb{R}^m$, $V(\tau)=\{y\,|\,y=t\tau; t\in\mathbb{R}\}$ and its corresponding null space is $N(\tau)=\{x\,|\,x^Ty=0 \text{ for } x\in Y \text{ and } y\in V(\tau)\}$. Extending to k hyperplanes, we have $V(\tau_1,\tau_2,\cdots,\tau_k)=\{y\,|\,y=\sum_j^k t_j\tau_j; t_j\in R\}$ and the null space is $N(\tau_1,\tau_2,\cdots,\tau_k)=\{x\,|\,x^Ty=0 \text{ for } x\in Y$ and $y\in V(\tau_1,\tau_2,\cdots,\tau_k)\}$. It can be shown that $N(\tau_1,\tau_2,\cdots,\tau_k)=N(\tau_1)\cup N(\tau_2)\cup\cdots\cup N(\tau_k)$. Since $V(\tau_1,\tau_2,\cdots,\tau_k)$ and $N(\tau_1,\tau_2,\cdots,\tau_k)$ are the orthogonal decomposition of the whole space Y, a vector g in R^m can be decomposed into two components: the projection of g onto $V(\tau_1,\tau_2,\cdots,\tau_k)$ and the projection of g onto $N(\tau_1,\tau_2,\cdots,\tau_k)$. We use the notation $g\downarrow_{[\tau_1,\tau_2,\cdots,\tau_k]}$ and $g\downarrow_{\alpha_1\cap\cdots\cap\alpha_k}$ to denote them and they are called direct projection and null projection respectively.

The following definition and theorem were first presented in [1] and is repeated here for completeness.

Let the set of all subspaces of $Y=R^m$ be $\mathcal{N}$, and let $\mathcal{O}$ stand for 0-dim subspace, we now give an axiomatic definition of projection.

Definition 3. The projection defined on a vector space Y is a mapping

$$*\downarrow_{\#}: Y\times\mathcal{N}\to Y$$

where * is a vector in Y, # is a subspace X in $\mathcal{N}$ satisfying that

(N.1) (Reflectivity).

For any $g\in Y, g\downarrow_Y=g$;

(N.2) (Orthogonal additivity).

For any $g\in Y$ and subspaces $X,Z\in N$, if X and Z are orthogonal to each other, then $g\downarrow_X+g\downarrow_Z=g\downarrow_{X+Z}$, where X+Z is the direct sum of X and Z.

(N.3) (Transitivity).

For any g∈Y and subspaces $X, Z \in \mathcal{N}$, $(g\downarrow_X)\downarrow_{X\cap Z} = g\downarrow_{X\cap Z}$,
(N.4) (Attribution).
For any g∈Yg∈Y and subspace $X \in \mathcal{N}$, $g\downarrow_X \in X$, and especially,
(N.5) For any g∈Y and subspace $X \in \mathcal{N}$, $g^T g\downarrow_X \geq 0$.

A convention approach to find $g\downarrow_{\alpha_1\cap\cdots\cap\alpha_k}$ is to compute it directly from the null space $N(\tau_1,\tau_2,\cdots,\tau_k)$. We now show another approach that is more suitable to our overall algorithm.
Theorem 1. For any g∈Y, we have

$$g\downarrow_{\alpha_1\cap\cdots\cap\alpha_k} = g - \sum_i^k g^T o_i \tag{8}$$

where $\{o_1,\cdots,o_k\}$ are an orthonormal basis of subspace $V(\tau_1,\tau_2,\cdots,\tau_k)$.

Proof. Since $\alpha_1\cap\cdots\cap\alpha_k$ and $[\tau_1,\tau_2,\cdots,\tau_k]$ are the orthonormal decomposition of Y, according to (N.2) and (N.1) we have

$$g = g\downarrow_{[\tau_1,\cdots,\tau_k]} + g\downarrow_{\alpha_1\cap\cdots\cap\alpha_k}.$$

According to (N.2) and (N.4), the first term becomes

$$g\downarrow_{[\tau_1,\cdots,\tau_k]} = g^T o_1 + \cdots + g^T o_k.$$

Hence (8) is true. END

The following theorem shows that the projection of g_0 onto the set of all blocking facets σ_t always satisfies Propositions 1 and 2. First, let us simplify the notation and use σ to represent σ_t in the following section and $g_0\downarrow_\sigma$ to stand for $g_0\downarrow_{\alpha_{\sigma(1)}\cap\cdots\cap\alpha_{\sigma(k)}}$ where $k=|\sigma|$ where k=|σ| is the number of elements in σ.

Theorem 2. $(\tau_{\sigma(j)})^T(g_0\downarrow_\sigma)=0$ and $g_0^T(g_0\downarrow_\sigma)\geq 0$ for all $j=1,\cdots,k$.

Proof. Since $g_0\downarrow_\sigma$ lies on the intersection of $(\alpha_{\sigma(1)}\cap\cdots\cap\alpha_{\sigma(k)})$, it lies on each facet $\alpha_{\sigma(j)}$ for j=1,···,k. Thus $g_0\downarrow_\sigma$ is perpendicular to the normal vector of $\alpha_{\sigma(j)}$ (*i.e.* $(\tau_{\sigma(j)})^T(g_0\downarrow_\sigma)=0$). So it satisfies Proposition 2.

According to (N.5), $g_0^T(g_0\downarrow_\sigma)\geq 0$. So it satisfies proposition 1 too. END

As such, $g_0\downarrow_\sigma$, the projection of g0g0 onto all the blocking facets, can be adopted as the next gradient descend vector g_t. Hence, $g_0\downarrow_\sigma$, the projection of g0g0 onto all the blocking facets, can be adopted as the next gradient descend vector g_t.

Selecting the Sliding Gradient

In this section, we explore other projection vectors which also satisfy Propositions 1 and 2. Let the jth complement blocking set σ_j^c be the blocking set σ excluding the jth element; i.e. $\sigma_j^c = \alpha_1 \cap \cdots \cap \alpha_{j-1} \cap \alpha_{j+1} \cap \cdots \cap \alpha_k$. We examine the

projection $g_0 \downarrow_{\sigma_j^c}$ for j=1,⋯,k. Obviously, $g_0^{\mathrm{T}}\left(g_0 \downarrow_{\sigma_j^c}\right) \geq 0$ according to (N.5) as $g_0 \downarrow_{\sigma_j^c}$ is a projection of g_0. So if $\left(\tau_{\sigma(j)}\right)^{\mathrm{T}}\left(g_0 \downarrow_{\sigma_j^c}\right) \geq 0$ for all j∈σ,

it satisfies Proposition 2 and hence is a candidate for consideration. For all the candidates, including $g_0\downarrow_\sigma$, which satisfy this proposition, we can compute the inner product of each candidate with the initial gradient descend vector g_0, (i.e. $g_0^{\mathrm{T}}\left(g_0 \downarrow_{\sigma_j^c}\right)$) and select the maximum. This inner product is a measure of how close or similar a candidate is to g_0 so taking the maximum means getting the steepest descend gradient. Notice that if a particular $g_0 \downarrow_{\sigma_j^c}$ is selected as the next gradient descending vector, the corresponding α_j is no longer a blocking facet in computing $g_0 \downarrow_{\sigma_j^c}$. Thus α_j needs to be removed from σ_t to form the set of effective blocking facets σ_t^*. The set of blocking facets for the next iteration σ_{t+1} is σ_t^* plus the newly encountered blocking facet. In summary, the next gradient descend vector g_t is:

$$g_t = \max\left(g_0^{\mathrm{T}}\left(g_0 \downarrow_\sigma\right), g_0^{\mathrm{T}}\left(g_{[1]}\right), g_0^{\mathrm{T}}\left(g_{[2]}\right), \cdots, g_0^{\mathrm{T}}\left(g_{[k]}\right)\right) \tag{9}$$

where $g_{[j]} = g_0 \downarrow_{\sigma_j^c}; j \in \sigma$ with $\left(\tau_{\sigma(j)}\right)^{\mathrm{T}}\left(g_0 \downarrow_{\sigma_j^c}\right) \geq 0$ and $k = |\sigma|$.

The effective blocking set σ_t^* is

$$\sigma_t^* = \begin{cases} \sigma_t & \text{if } g_{t+1} = g_0 \downarrow_\sigma \\ \sigma_t \setminus \{\alpha_j\} & \text{if } g_{t+1} = g_0 \downarrow_{\sigma_j^c} \end{cases}. \tag{10}$$

At first, this seems to increase the computation load substantially. However, we now show that once $g_0\downarrow_\sigma$ is computed, $g_0 \downarrow_{\sigma_j^c}$ can be obtained efficiently.

Computing the Gradient Projection Vectors

This session discusses a method of computing $g\downarrow_\sigma$ and $g\downarrow_{\sigma_j^c}$ for any vector g. According to (8), $g\downarrow_\sigma = g\downarrow_{\alpha_1\cap\cdots\cap\alpha_k} = g - \sum_i^k g^{\mathrm{T}} o_i$. The orthonormal basis $\{o_1, o_2, \cdots, o_k\}$ can be obtained from the Gram Schmidt procedure as follows:

$$o_1 = \tau_1;\ o_1 = o_1 / |o_1| \tag{11}$$

$$o_2 = \tau_2 - \tau_2^{\mathrm{T}} o_1;\ o_2 = o_2 / |o_2| \tag{12}$$

Let us introduce the notation $a\downarrow b$ to denote the projection of vector a onto vector b. We have $a\downarrow b = \left(\frac{a^{\mathrm{T}}b}{b^{\mathrm{T}}b}\right)b$, then $o_2 = \tau_2 - \tau_2 \downarrow o_1$ as $\left(o_1^{\mathrm{T}} o_1\right) = 1$. Likewise,

$$o_j = \tau_j - \sum_{i=1}^{j-1} \tau_j \downarrow o_i;\ o_j = o_j / |o_j|. \tag{13}$$

Thus from (8),

$$g\downarrow_\sigma = g - \sum_i^k g^{\mathrm{T}} o_i = g - \sum_{i=1}^k g\downarrow o_i. \tag{14}$$

After evaluating $g\downarrow_\sigma$, we can find $g\downarrow_{\sigma_j^c}$ backward from j=k to 1. Firstly,

$$g\downarrow_{\sigma_k^c} = g - \sum_{i=1}^{k-1} g\downarrow o_i = g\downarrow_\sigma + g\downarrow o_k. \tag{15}$$

Likewise, it can be shown that

$$g\downarrow_{\sigma_j^c} = g - \sum_{i=1}^{j-1} g\downarrow o_i - \sum_{i=j+1}^{k} g\downarrow o_i^{(j)} \text{ for } j = 2 \text{ to } k. \tag{16}$$

The first summation is projections of g onto existing orthonormal basis o_i. Each term in this summation has already been computed before and hence is readily available. However, the second summation is projections on new basis $o_i^{(j)}$. Each of these basis must be re-computed as the facet α_j is skipped in σ_j^c. Let

$$T_k = g + g\downarrow o_k;\ S_k = 0 \tag{17}$$

$$T_j = T_{j+1} + g \downarrow o_j;\ S_j = \sum_{i=j+1}^{k} g \downarrow o_i^{(j)}. \tag{18}$$

Then we can obtain $g \downarrow_{\sigma_j^c}$ recursively from j=k,k−1,···,1 by:

$$g \downarrow_{\sigma_j^c} = T_j - S_j. \tag{19}$$

To compute $o_i^{(j)}$, some of the intermediate results in obtaining the orthonormal basis can also be reused.

Let $\mu_{j,1}=0$ for all j=2,···,k and $\mu_{j,i} = \mu_{j,i-1} + \tau_j \downarrow o_i$ for i=1,···, j−1, then we have

$$o_j = \tau_j - \mu_{j,j-1};\ o_j = o_j / \left| o_j \right| \text{ for } j = 2,\cdots,k. \tag{20}$$

The intermediate terms $\mu_{j,i}$ can be reused in computing $o_m^{(j)}$ as follows:

$$o_m^{(j)} = \tau_m - \mu_{m,j-1} - \sum_{i=j+1}^{m-1} \tau_m \downarrow o_i^{(j)};\ o_m^{(j)} = \frac{o_m^{(j)}}{\left| o_m^{(j)} \right|} \text{ for } m = +1,\cdots,k. \tag{21}$$

By using these intermediate results, the computation load can be reduced substantially.

Termination Criterion

When a new blocking facet is encountered, it will be added to the existing set of blocking facets. Hence both σ_t and σ_t^* will typically grow in each iteration unless one of the $g_0 \downarrow_{\sigma_j^c}$ is selected as g_t. In this case, $\alpha_{\sigma(j)}$ is deleted from σ_t according to (10). The following theorem, which was first presented in [1] shows that when $|\sigma_t^*| = m$, the algorithm can stop.

Theorem 3 (Stopping criterion) Assuming that the dual feasible region $\mathcal{D}$ is non-empty, let $P_t \in D$ and is descending along the initial direction $g_0=-b$; let $|\sigma_t^*|$ be the number of effective blocking facets in σ_t^* at the tth iteration. If $|\sigma_t^*| = m$ and the rank $r(\sigma_t^*) = m$, then P_t is a lowest point in the dual feasible region $\mathcal{D}$.

Proof. If $|\sigma_t^*| = m$ and the rank $r(\sigma_t^*) = m$, then the m facets in σ_t^* form a cone C with vertex $V=P_t$. Since the rank is m, its corresponding null space contains only the zero vector. So $g_0 \downarrow_\sigma = g_0 \downarrow_{\alpha_{\sigma(1)} \cap \cdots \cap \alpha_{\sigma(k)}} = 0$.

As mentioned about the facet/edge duality in Section 2, for j=1,⋯,m, edge-line R_i^+ is the intersection of all C-facets except α_i. That means $R_i^+ = \sigma_i^c$. Since an edge-line is a 1-dimensional line, the projection of a vector

g_0 onto R_i^+ equals to $\pm r_i$ and hence $g_t \downarrow_{\sigma_i^c} = g_t \downarrow_{R_i^+} = \pm r_i$. Since $g_t \downarrow_{\sigma_i^c}$ are projections of g_0, according to (N.5), $g_0^T \left(g_t \downarrow_{\sigma_i^c} \right) \geq 0$.

Since $|\sigma_t^*| = m$, it means that $g_t \downarrow_{\sigma_i^c}$ does not satisfy Proposition 2 for all i=1,⋯,m. Otherwise, one of the $g_t \downarrow_{\sigma_i^c}$ would have been selected as the next gradient descend vector and, according to (10), it would be deleted from σ_t^* and hence $|\sigma_t^*|$ would be less than m. This means that at least one of j∈σ has a value $\tau_j^T \left(g_t \downarrow_{\sigma_i^c} \right) < 0$. However, for all $k \neq i$, $g_t \downarrow_{\sigma_i^c}$ is in the null space of α_k so $\tau_k^T \left(g_t \downarrow_{\sigma_i^c} \right) = 0$. This leaves $\tau_i^T \left(g_t \downarrow_{\sigma_i^c} \right) < 0$. If $g_t \downarrow_{\sigma_i^c} = r_i$. If $g_t \downarrow_{\sigma_i^c} = r_i$, then $\tau_i^T \left(g_t \downarrow_{\sigma_i^c} \right) = \tau_i^T r_i < 0$. This contradicts to the fact that $\tau_i^T r_i \geq 0$ in (1). Therefore, $g_t \downarrow_{\sigma_j^c} = -r_i$. Since $g_0^T \left(g_t \downarrow_{\sigma_i^c} \right) \geq 0$, $g_0^T (-r_i) = (-g_0)^T r_i \geq 0$. Note that

$(-g_0)^T r_i$ means that edge r_i is in opposite direction of g_0. As this is true for all edges, there is no path for g_t to descend further from this vertex. It is obvious that the vertex V is the lowest point of C when viewed in the b direction.

Since P_t is dual feasible, and V is a vertex of $\mathcal{D}$. Cone C coincides with the dual feasible region $\mathcal{D}$ in a neighborhood N of V, it is obvious that P_t is the lowest point of $\mathcal{D}$ when viewed in the b direction. END

In essence, when the optimal vertex V^* is reached, all the edges of the cone will be in opposite direction of the gradient vector $g_0 = -b$. There is no path to descend further so the algorithm terminates.

THE PSEUDO CODE OF THE SLIDING GRADIENT ALGORITHM

The entire algorithm is summarized as follows in Table 2.

Step 0 is the initialization step that sets up the tableau and the starting point P. Step 2 is to find a set of initial blocking facets σ in preparation of step 4. In the inner loop, Step 4 calls the Gradient Select routine. It computes $g_0 \downarrow_\sigma$ and $g_0 \downarrow_{\sigma_j^c}$ in view of σ using Equations (11) to (21) and select the best gradient vector g according to (9). This routine not only returns g but also the effective blocking facets σ^* and $g_0 \downarrow_{\sigma_j^c}$ for subsequent use. Theorem 3 states that when the size of σ^* reaches m, the optimal point is reached. So when it does, step 5 returns the optimal point and the optimal value to the calling routine. Step 6 is to find the closest blocking facet according to (6). Because P lies on every facets of σ, $t_j = 0$ for $j \in \sigma$. Hence, we only need to compute those t_j where j□σ. The newly found blocking facet is then included in σ in step 7 and the inner loop is repeated until the optimal vertex is found.

Table 2. The sliding gradient algorithm.

steps	**Input:** A, b and $\tilde{c}$; **and** $\mathcal{D}$ **is non-empty and** P_0 **is inside** $\mathcal{D}$ **Output:** *OptPt* & *OptVal*
0	Construct the Facet Tableau $[\tau]$ from A, b and $\tilde{c}$ $P = P_0$; $g_0 = -b$
1	$d_j = P^T \tau_j - c_j$ for $j = 1, \cdots, m+n$ }
2	$\sigma = \{ j \| \|d_j\| < \delta \}$; % δ is a small constant $\sigma^* = \sigma$; % σ^* is set of blocking facet
3	**while** (true)
4	**if** $\sigma \neq \phi$ % compute $g, g\downarrow_{\sigma_j^c}$ and update σ^* $[g, g\downarrow_{\sigma_j^c}, \sigma^*]$ = Gradient Select($g_0, \sigma, [\tau], P, c$) **else** $g = g_0$
5	**if** $\|\sigma^*\| == m$ $OptPt = P; OptVal = P^T b$; return (*OptPt, OptVal*) **else** $\sigma = \sigma^*$
6	$t_j = \dfrac{c_j P^T \tau_j}{g^T \tau_j}$; for $j \in \{1, 2, \cdots, m+n\} \setminus \sigma$ $j^* = \arg\min_j \{t_j \| t_j > 0\}$; $Q = P + t_{j^*} g$; $P = Q$
7	$\sigma = \sigma^* \cup \{ i \| \|t_i - t_{j^*}\| < \epsilon \}$ % update blocking facets
8	**end**

IMPLEMENTATION AND EXPERIMENTAL RESULTS

Experiment on the Klee-Minty Problem

[1]Other derivations of the Klee-Minty formulas have also been tested and the same results are obtained.

We use the Klee-Minty example presented in [18] [1] to walk through the algorithm in this section. An example of the Klee-Minty Polytope example is shown below:

$$\max 2^{m-1}x_1 + 2^{m-2}x_2 + \cdots + 2x_{m-1} + x_m .$$

Subject to

x_1									$\leq$	5^1
$4x_1$	+	x_2	+						$\leq$	5^2
$8x_1$	+	$4x_2$	+	x_3					$\leq$	5^3
$\vdots$									$\vdots$	$\vdots$
$2^m x_1$	+	$2^{m-1}x_2$	+	$2^{m-2}x_3$	+	$\cdots$		x_m	$\leq$	5^m

For the standard simplex method, it needs to visit all 2^{m-1} vertices to find the optimal solution. Here we show that, with a specific choice of initial point P^0, the Sliding Gradient algorithm can find the optimal solution in two iterations—no matter what the dimension m is.

To apply the Sliding Gradient algorithm, we first construct the tableau. For an example with m=5 , the simplex tableau is:

1					5
4	1				25
8	4	1			125
16	8	4			625
32	16	8	4	1	3125
16	8	4	2	1	

The b vector is $b=[5,25,125,625,3125]^T$. After adding the slack variables, the facet tableau becomes:

a_1	a_2	a_3	a_4	a_5	a_6	a_7	a_8	a_9	a_{10}
1	0	0	0	0	1	0	0	0	0
4	1	0	0	0	0	1	0	0	0
8	4	1	0	0	0	0	1	0	0
16	8	4	1	0	0	0	0	1	0
32	16	8	4	1	0	0	0	0	1
16	8	4	2	1	0	0	0	0	0

Firstly, notice that α_5 and α_{10} have the same normal vector (i.e. $\tau_5=\tau_{10}$) so we can ignore α_{10} for further consideration. This is true for all value of m.

If we choose $P_0=Mb$, where M is a positive number (e.g. M=100), It can be shown that P_0 is inside the dual feasible region. The initial gradient descend vector is: $g_0=-b$.

With P_0 and g_0 as initial conditions, the algorithm proceeds to find the first blocking facet using (6). The displacements t_j for each facet can be found by:

$$t_j=\frac{c_j-P_0^{\mathrm{T}}\tau_j}{g_0^{\mathrm{T}}\tau_j}=\frac{c_j}{-b^{\mathrm{T}}\tau_j}-\frac{Mb^{\mathrm{T}}\tau_j}{-b^{\mathrm{T}}\tau_j}=\frac{c_j}{-b^{\mathrm{T}}\tau_j}+M=M-\frac{c_j}{b^{\mathrm{T}}\tau_j}.$$

With P_0 and g_0 as initial conditions, the algorithm proceeds to find the first blocking facet using (6). The displacements t_j for each facet can be found by:

$$t_j=\frac{c_j-P_0^{\mathrm{T}}\tau_j}{g_0^{\mathrm{T}}\tau_j}=\frac{c_j}{-b^{\mathrm{T}}\tau_j}-\frac{Mb^{\mathrm{T}}\tau_j}{-b^{\mathrm{T}}\tau_j}=\frac{c_j}{-b^{\mathrm{T}}\tau_j}+M=M-\frac{c_j}{b^{\mathrm{T}}\tau_j}. \qquad (22)$$

We now show that the minimum of all displacements is t_m.

First of all, at = m, $\tau_m=[0,\cdots,0,1]^{\mathrm{T}}$, $c_m=1$ and $b_m=5^m$, so $t_m=M^{-5}-m$.

For $m<j\leq 2m-1$, $c_j=0$, so $t_j=M>t_m$.

For $1\leq j<m$, $c_j=2^{m-j}$, and the elements of τ_j are:

$$\tau_{ij}=\begin{cases}0 & \text{if } i<j\\ 1 & \text{if } i=j\\ 2^{i-j+1} & \text{if } j<i\leq m\end{cases}.$$

The 2nd term of Equation (22) can be re-written as:

$$\frac{c_j}{\boldsymbol{b}^{\mathrm{T}}\boldsymbol{\tau}_j}=\frac{1}{\boldsymbol{b}^{\mathrm{T}}\left(\frac{\boldsymbol{\tau}_j}{c_j}\right)}=\frac{1}{\boldsymbol{b}^{\mathrm{T}}\left(\frac{\boldsymbol{\tau}_j}{2^{m-j}}\right)}$$

The inner product of the denominator is:

$$\boldsymbol{b}^{\mathrm{T}}\left(\frac{\tau_j}{2^{m-j}}\right)=\sum_{i=1}^{m}b_i\left(\frac{\tau_{ij}}{2^{m-j}}\right)=\sum_{i=1}^{m-1}b_i\left(\frac{\tau_{ij}}{2^{m-j}}\right)+b_m\left(\frac{2^{m-j+1}}{2^{m-j}}\right)=\sum_{i=1}^{m-1}b_i\left(\frac{\tau_{ij}}{2^{m-j}}\right)+2b_m$$

Since all the elements in the b vector and the ττ are positive, the summation is a positive number. Thus

$$\boldsymbol{b}^{\mathrm{T}}\left(\frac{\tau_j}{2^{m-j}}\right)=\sum_{i=1}^{m-1}b_i\left(\frac{\tau_{ij}}{2^{m-j}}\right)+2b_m>b_m$$

Since the value of the denominator is bigger than $b_m=5^m$, we have

$$\frac{1}{\boldsymbol{b}^{\mathrm{T}}\left(\frac{\tau_j}{2^{m-j}}\right)}<5^{-m}$$

So

$$t_j=M-\frac{c_j}{\boldsymbol{b}^{\mathrm{T}}\boldsymbol{\tau}_j}=M-\frac{1}{\boldsymbol{b}^{\mathrm{T}}\left(\frac{\tau_j}{2^{m-j}}\right)}>M-5^{-m}=t_m\,.$$

Hence t_m is the smallest displacement. For the case of m=5, their values are shown in the first row (first iteration) of the following Table 3.

Thus α_m is the closest blocking facet. Hence, $\sigma_1^*=\sigma_1=\{\alpha_m\}$. For the next iteration,

$$\boldsymbol{P}_1=\boldsymbol{P}_0+t_m\boldsymbol{g}_0=M\boldsymbol{b}+\left(M-5^{-m}\right)\left(-\boldsymbol{b}\right)=M\boldsymbol{b}-M\boldsymbol{b}+5^{-m}\boldsymbol{b}=5^{-m}\boldsymbol{b}\,.$$

The gradient vector g_1 is g_0 projects onto α_m. Because $\tau_m=[0,\cdots,1]$ is already an orthonormal vector, we have according to(8)

$$\boldsymbol{g}_1=\boldsymbol{g}_0-\left(\boldsymbol{g}_0^{\mathrm{T}}\boldsymbol{\tau}_m\right)\boldsymbol{\tau}_m=\boldsymbol{g}_0-\left[0,\cdots,-5^m\right]^{\mathrm{T}}=\left[-5,-5^2,\cdots,-5^{m-1},0\right]^{\mathrm{T}}.$$

In other words, g_1 is the same as −b except that the last element is zeroed out. Using P_1 and g_1, the algorithm proceeds to the next iteration and evaluates the displacements t_j again. For j=m+1 to $2m^{-1}$, since $c_j=0$ and τ_j is a unit vector with only one non-zero entry at the jth element,

$$t_j=-\frac{\boldsymbol{P}_1^{\mathrm{T}}\boldsymbol{\tau}_j}{\boldsymbol{g}_1^{\mathrm{T}}\boldsymbol{\tau}_j}=-\frac{5^{-m}\boldsymbol{b}^{\mathrm{T}}\boldsymbol{\tau}_j}{\boldsymbol{g}_1^{\mathrm{T}}\boldsymbol{\tau}_j}=-5^{-m}\frac{b_j}{-b_j}=5^{-m}\quad\text{for } m+1\le j\le 2m-1\,.$$

Thus the displacements t_m to t_{2m-1} have the same value of 5^{-m}.
For 1≤j<m, we have:

$$t_j = \frac{c_j - P_1^T \tau_j}{g_1^T \tau_j} = \frac{c_j - 5^{-m} b^T \tau_j}{g_1^T \tau_j} = 5^{-m} \frac{5^m c_j - b^T \tau_j}{g_1^T \tau_j}.$$

As mentioned before, g_1 is the same as −b except that the last element is zero, we can express $b^T \tau_j$ in terms of $g_1^T \tau_j$ as follows:

$$b^T \tau_j = -g_1^T \tau_j + b_m \tau_{mj}.$$

The numerator then becomes:

$$5^m c_j - b^T \tau_j = 5^m c_j + g_1^T \tau_j - b_m \tau_{mj}.$$

Since $c_j = 2^{m-j}$, $c_j = 2^{m-j}$ and $\tau_{mj} = 2^{m-j+1}$, , substituting these values to the above equation, the numerator becomes

$$5^m c_j - b^T \tau_j = 5^m 2^{m-j} - 5^m 2^{m-j+1} + g_1^T \tau_j = g_1^T \tau_j - 5^m 2^{m-j}.$$

Thus

$$t_j = 5^{-m} \frac{5^m c_j - b^T \tau_j}{g_1^T \tau_j} = 5^{-m} \left(1 - \frac{5^m 2^{m-j}}{g_1^T \tau_j}\right).$$

Table 3. Displacement values titi in each iterations for m=5.

	t_1	t_2	t_3	t_4	t_5	t_6	t_7	t_8	t_9
1	99.9999	99.9999	99.9999	99.9998	99.9997	100	100	100	100
2	0.0018	0.0018	0.0018	0.0035	0	0.00032	0.00032	0.00032	0.00032

Notice that all elements in g_1 are negative but all of τ_j are positive. So the inner product $g_1^T \tau_j$ is a negative number. As a result, the last term inside the bracket is a positive number which makes the whole value inside the bracket bigger than one and hence $t_j > 5^{-m}$ for $1 \le j < m-1$. Moreover, t_m is zero as g_1 lies on α_m. The actual displacement values for the case of m=5 are shown in the second row of Table 3.

Since t_m to t_{2m-1} have the same lowest displacement value, all of them are blocking facets so $\sigma_2^* = \sigma_2 = \{\alpha_m\} \cup \{\alpha_{m+1}, \cdots, \alpha_{2m-1}\} = \{\alpha_m, \alpha_{m+1}, \cdots, \alpha_{2m-1}\}$. Also,

$$P_2 = P_1 + t_{m+1} g_1 = 5^{-m} b + 5^{-m} \left[-5, -5^2, \cdots, -5^{m-1}, 0\right]^T = [0, 0, \cdots, 0, 1]^T.$$

Now $|\sigma_t^*| = m$, so P_2 has reached a vertex of a cone. According to Theorem 3, the algorithm stops. The optimal value is $P_2^T b = 5^m$, which is the last element of the b vector.

Thus with a specific choice of the initial point P_0=Mb, the Sliding Gradient algorithm can solve the Klee-Minsty LP problem in two iterations, and it is independent of m.

Issues in Algorithm Implementation

The Sliding Gradient Algorithm has been implemented in MATLAB and tested on the Klee-Minty problems and also self-generated LP problems with random coefficients. As a real number can only be represented in finite precision in digital computer, care must be taken to deal with the round-off issue. For example, when a point P lies on a plane $y^T\tau=c$, the value $d=P^T\tau-c$ should be exactly zero. But in actual implementation, it may be a very small positive or negative number. Hence in step 2 of the aforementioned algorithm, we need to set a threshold δ so that if $|d|<\delta$, we regard that point P is laid on the plane. Likewise for the Klee-Minty problem, this algorithm relies on the fact that in the second iteration, the displacement values t_i for i=m+1 to 2m−1 should be the same and they should all be smaller than the values of t_j for j=1 to m−1. Due to round-off errors, we need to set a tolerant level to treat the first group to be equal and yet if this tolerant level is set too high, then it cannot exclude members of the second group. The issue is more acute as m increases. It will require higher and higher precision in setting the tolerant level to distinguish these two groups.

CONCLUSIONS AND FUTURE WORK

We have presented a new approach to tackle the linear programming problem in this paper. It is based on the gradient descend principle. For any initial point inside the feasible region, it will pass through the interior of the feasible region to reach the optimal vertex. This is made possible by projecting the gravity vector to a set of blocking facets and using that as descending vector in each iteration. In fact, the descending trajectory is a sequence of line segments that hug either a single blocking facet or the intersections of them, and each line segment is advancing towards the optimal point. It should be noted that there is no parameters (such as step-size, ..., etc.) to tune in this algorithm although one needs to take care of numerical round-off issue in actual implementation.

This work opens up many areas of future research. On the one hand, we are extending this algorithm so that it can relax the constraint of starting from a point inside the feasible region. Promising development has been achieved in this area though more thorough testing on obscure cases need to be carried out.

On the theoretical front, we are encouraged that, from the algorithm walk-through on the Klee-Minty example, this algorithm exhibits strongly polynomial complexity characteristics. Its complexity does not appear to depend on the bit sizes of the LP coefficients. However, more rigorous proof is needed and we are working towards this goal.

ACKNOWLEDGEMENTS

The authors wish to thank all his friends for their valuable critics and comments on the research. Special thanks are given to Prof. Yong Shi, Prof. Sizong Guo for their supports. This study is partially supported by the grants (Grant Nos. 61350003, 70621001, 70531040, 90818025) from the Natural Science Foundation of China, and grant (Grant No. L2014133) from Department of Education of Liaoning Province.

CONFLICTS OF INTEREST

The authors declare no conflicts of interest.

REFERENCES

1. Wang, P.Z., Lui, H.C., Liu, H.T. and Guo, S.C. (2017) Gravity Sliding Algorithm for Linear Programming. Annals of Data Science, 4, 193-210. https://doi.org/10.1007/s40745-017-0108-1
2. Dantzig, G.B. (1963) Linear Programming and Extensions. Princeton University Press, Princeton. https://doi.org/10.1515/9781400884179
3. Megiddo, N. (1987) On the Complexity of Linear Programming. In: Bewley, T., Ed., Advances Economic Theory, 5th World Congress, Cambridge University Press, Cambridge, 225-268. https://doi.org/10.1017/CCOL0521340446.006
4. Megiddo, N. (1984) Linear Programming in Linear Time When the Dimension Is Fixed. Journal of ACM, 31, 114-127. https://doi.org/10.1145/2422.322418
5. Smale, S. (1983) On the Average Number of Steps of the Simplex Method of Linear Programming. Mathematical Programming, 27, 251-262. https://doi.org/10.1007/BF02591902
6. Todd, M. (1986) Todd, Polynomial Expected Behavior of a Pivoting Algorithm for Linear Complementarity and Linear Programming Problems. Mathematical Programming, 35, 173-192. https://doi.org/10.1007/BF01580646
7. Klee, V. and Minty, G.J. (1972) How Good Is the Simplex Method. In: Shisha, O., Ed., Inequalities III, Academic Press, New York, 159-175.
8. Chvatal, V. (1983) Linear Programming. W.H. Freeman and Company, New York.
9. Goldfarb, D. and Sit, W. (1979) Worst Case Behavior of the Steepest Edge Simplex Method. Discrete Applied Mathematics, 1, 277-285. https://doi.org/10.1016/0166-218X(79)90004-0
10. Jeroslow, R. (1973) The Simplex Algorithm with the Pivot Rule of Maximizing Im- provement Criterion. Discrete Mathematics, 4, 367-377. https://doi.org/10.1016/0012-365X(73)90171-4
11. Zadeh, N. (1980) What Is the Worst Case Behavior of the Simplex Algorithm? Technical Report 27, Dept. Operations Research, Stanford University, Stanford.
12. Karmarkar, N.K. (1984) A New Polynomial-Time Algorithm for Linear Programming. Combinatorica, 4, 373-395. https://doi.org/10.1007/BF02579150

13. Deza, A., Nematollahi, E. and Terlaky, T. (2008) How Good Are Interior Point Methods? Klee-Minty Cubes Tighten Iteration-Complexity Bounds. Mathematical Programming, 113, 1-14.
14. Barasz, M. and Vempala, S. (2010) A New Approach to Strongly Polynomial Linear Programming.
15. Amenta, N. and Ziegler, G. (1999) Deformed Products and Maximal Shadows of Polytopes. Contemporary Mathematics, 223, 57-90. https://doi.org/10.1090/conm/223/03132
16. Wang, P.Z. (2011) Cone-Cutting: A Variant Representation of Pivot in Simplex. Information Technology & Decision Making, 10, 65-82.
17. Wang, P.Z. (2014) Discussions on Hirsch Conjecture and the Existence of Strongly Polynomial-Time Simplex Variants. Annals of Data Science, 1, 41-71. https://doi.org/10.1007/s40745-014-0005-9
18. Greenberg, H.J. (1997) Klee-Minty Polytope Shows Exponential Time Complexity of Simplex Method. University of Colorado at Denver, Denver.

Section 2: Graph-based Methods and Techniques

CHAPTER 6

The Route Planning on Campus Bus in H University

Zhengang Shu, Xiaoli Wu

School of Business and Administration, South China University of Technology, Guangzhou, China.

ABSTRACT

In recent years, the university development is at the expansion period. As the important part of the overall planning process, campus traffic planning gets highly consideration as to the university builders. The optimization of the transportation system will bring convenience and safe experience to the school staffs. This paper analyzes the current situation of campus traffic organization in H university campus, and establishes a trip map to the traffic network. Finally, based on the basic principles of traffic planning, we designed campus bus routes and sites. By using of cost analysis, we also design the optimal starting frequency, which optimizes the campus traffic system.

Citation: Shu, Z. and Wu, X. (2018) The Route Planning on Campus Bus in H University. American Journal of Industrial and Business Management, 8, 473-486. doi: 10.4236/ajibm.2018.83031.

Keywords:- Campus Traffic Planning, Route Network, Cost Optimization

INTRODUCTION

Since the new century, the rise of college enrollment rate has made the number of teachers and students in colleges and universities continually increase.

When the campus constantly enriches and perfects its education environment, it also results in some problems such as overlong traffic routes among functional areas in campus, travel inconvenience and so on. Aiming at the mixed-traffic of many kinds of traffic inside the campus, Lin Zhen (2007) [1. researches and puts forward some measures, including campus road network transformation, mixed traffic separation, parking layout optimization, pedestrian space setting and guidance of external traffic. According to the type of spatial layout of college campus, Xu Junbin (2012) [2. divides it into four categories: annular space layout, strip space layout, grid space layout and free-form space layout.

Aiming at the actual situation of road traffic in our country, Yang Dongyuan and Liu Zhiqian (2008) [3. respectively analyze the factors influencing the stop time of bus stop and establish the corresponding empirical formula.

When studying the stop time of stations, Mao BaoLi (2009) [4. has considered the effect of vehicles from outside of the campus on buses and analyzed the effect of traffic flow and running speed on the stop time of bus stop, and he also has researches the queuing time of outbound buses in-depth.

In order to solve this problem, H University (Wushan Campus) opened Campus Bus Line 2 in April 2015 and since being opened for more than two years, it has improved travel efficiency of teachers and students in campus to a certain extent.

H University now have two bus campus routes : Line 1 and Line 2, Line 1 service between North area and South Gate, it mainly to meets the travel needs of North area students and faculty; Line 2 service between South area and the East area, it mainly meets the travel needs of staff in the South area as well as East area. However, there still exist many shortcomings. This paper adopts questionnaire survey to summarize the shortcomings of the current campus bus system, allocates the traffic volume of the campus road network based on Wardrop user equilibrium theory model, and finally designs a new bus route map, and uses cost analysis method to establish a new bus operation schedule.

SURVEY ABOUT CAMPUS TRAVEL DEMAND AND PROBLEM ANALYSIS

Analysis on Campus Status

H University covers an area of more than 2940 thousand square meters. The campus is divided into two campuses, and the Wushan Campus is located in the Shipai College Area of Tianhe District, Guangzhou City, and is the main campus of this study. Wushan Campus covers an area of 182.6 hectares (including military land in south front door) and is founded in the 1930s, and it has experienced development and construction in different stages of more than seventy years before and after liberation, and formed architectural styles in different stages and distinct layout of function division in the process of constant evolution.

There are three main parts of people in the campus: full-time students, faculty members and administrative staff. According to the statistics report of 2014, North Campus has 20,723 undergraduates, 13,276 doctorial candidates and postgraduate students and 1531 overseas students, totaling 43,862. It is expected to reach about 5000 people by the year of 2020: 5590 faculty members, and according to the coefficient of dependent 3.0, the faculty members (including family members) are about 16,000. In addition, there are about 1000 administrative staffs and the total of faculty members is about 19,000.

The planning of campus traffic is the subsystem of the whole college campus planning and construction. The division of functional area in each campus is conducive to the implementation of the campus traffic organization planning, and the planning and design of traffic organization should also make coordination with various functional areas. The relationship between functional areas is carefully analyzed, and the division of functional area in North Campus is confirmed with the combination of use properties of campus land as well as landform. At present, it is divided into five functional areas: central area, north area, west area, south area and east area, as shown in Figure 1. The distance between each functional area is shown in Table 1.

Questionnaire

This questionnaire survey is carries out from April 5th to 10th, 2017, and has completed 425 valid questionnaires, including 143 in the west area, 127 in the north area and 155 in other areas. It mainly investigates the characteristics, demand and intention of the present situation of campus travel.

From Table 2, it can be seen that the users of roads are mainly divided into pedestrians, motor vehicles and non-motor vehicles.

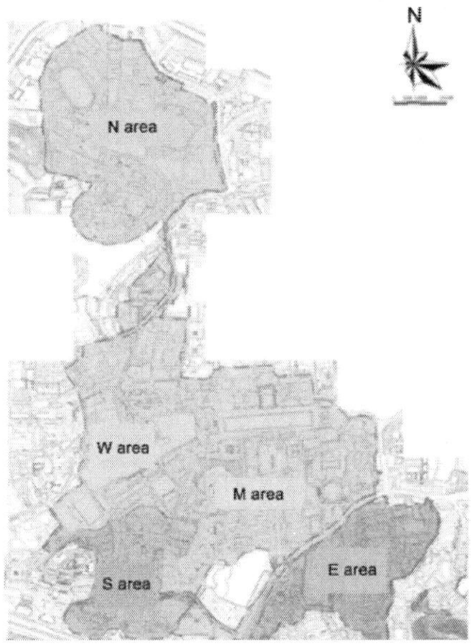

Figure 1. Partition map of H University (H University capital construction department).

Table 1. The distance between each functional area.

Distance (km)	N area	W area	M area	S area	E area
N area	-	1.5	1.2	2	2
W area	-	-	0.6	0.5	0.9
M area	-	-	-	0.9	0.8
S area	-	-	-	-	0.7
E area	-	-	-	-	-

As the main pedestrians of campus traffic behavior, it includes students, faculty members and social visitors, whose travel modes are walk, bike, electric car, campus bus and car. According to the investigation, the travel modes of the students are mainly walk and bike; the travel modes of faculty members and social visitors are mainly motor vehicle.

According to suitable distance for walking, people will feel tired after walking more than 500 m. Therefore, the travel of more than 500 m in the school is defined as long-distance travel, and that of less than 500 m is short-distance travel. The different travel distances frequency is shown in Table 3.

From the above table, it can be seen that there is slight difference between administrative staffs and teachers. Long-distance travel and short-distance travel of students on weekday respectively are 3.3 times a day and 4.2 times a day, while they significantly reduce on weekend. Therefore, this paper focuses on analyzing the travel characteristics of students and teachers during the workdays.

According to the survey, in Figure 2, the travel peak of students on campus is mainly concentrated in the time of 8 a.m. before the class and 12:00 noon after class.

Table 2. Main travel modes on campus of campus traffic (%).

	The car	School bus	Electric car	Bike	Walk
Student	0	6	0	32	62
Faculty member	60	1	8	12	19
Visitor	73	5	0	2	20

Data collection is based on questionnaire survey.

Table 3. Frequency distribution of different travel distances (sub/d).

	Administrative staff	Teachers	Students (working day)	Students (weekend)
Long-distance travel frequency	2.9	3.3	3.3	2
short-distance travel frequency	3.1	2.8	4.2	2.7

Data collection is based on questionnaire survey.

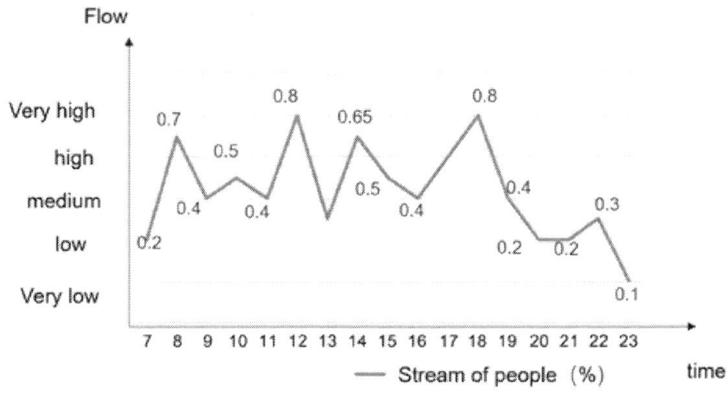

Figure 2. Campus people flow peak chart (Data collection is based on questionnaire survey).

From 11:30 to 12:00, it is the time period when students intensively go for the lunch; from 14:00 to 14:30, it is the peak of attending classes in the afternoon, which is basically the same as the peak in the morning; from 17:00 to 18:00, it is the time period when students intensively go for dinner, and after that, the traffic behavior of students is return to the dormitory for rest; From 18:30 to 19:00 is the main travel time of self-study in the evening; After 21:30, it is a time period when students return to the dormitory to sleep in succession.

Analysis on Travel Problem

The travel problem of this campus are as followings :

- Can't meet the demand of travel in peak hour and Campus Bus Line 1 is the only bus routes for the travel in the north area. in the peak hours of attending and finishing classes, it can't meet the travel demands of students and staffs in north area and also can't meet the general travel demand of some people, such as from residential area in south area to other areas, from students' dormitories in west area to outside school as well as infirmary, residential area of staff in east area to public teaching buildings.
- The distance of Campus Bus Line 1 between Zhongshan Station Baibu Station is less than 200 m, and the most area covered by Baibu Station also can be covered by the area of Zhongshan Station.

- There is a cross between South Changjiang Road to North Changjiang Road and Dongguan Village Road, and it is a downhill from South Changjiang Road to the crossroad, which easily cause dangers. Roads on campus are mainly coexistence of people and vehicles, and the roads are not wide enough with a result of traffic jam.
- Problems of campus bus itself. Campus Bus Line 1 has a low punctuality rate, unfixed cycling time and higher interval service time. Newly-built Campus Bus Line 2 mainly serves teaching staff with few seats, and its departure frequency is less than half of the prescribed one, and many students are not able to enjoy the benefits brought by the new opening line.

CAMPUS TRANSPORTATION MODEL CONSTRUCTION

Travel and Attraction Distribution of H University

Travel and attraction survey is to investigate the starting point and destination of people's travel on campus, and the purpose is to analyze the current situation of the traffic flow relations between different traffic areas [5. and provide basic data for future traffic volume forecast and check. According to the functional layout of different buildings within the campus, it is divided into 12 traffic areas, as shown in Table 4 and Figure 3.

Demand Analysis

Traffic area and road network are the simulation of the actual road network in the form of data, which is the important basis of traffic model.

Table 4. Overview of campus functional areas.

Serial number	Main functions of community	Number of people/ floor area (person/m^2)
1	student dormitory in north area	8218/303,570
2	teaching area	135/81,939
3	postgraduate dormitory	3667/150,893
4	sports field in west area	234/58,402
5	teachers' residential area in south area	1893/412,845
6	student dormitory in west area	7756/223,876
7	college administration laboratory building.	1834/467,323
8	administrative office	561/156,783
9	library	175/20,654
10	cultural and sports center, kindergarten	785/74,593
11	dormitory group in east area	3852/258,313
12	school office	283/40,738

Data collection is based on questionnaire survey.

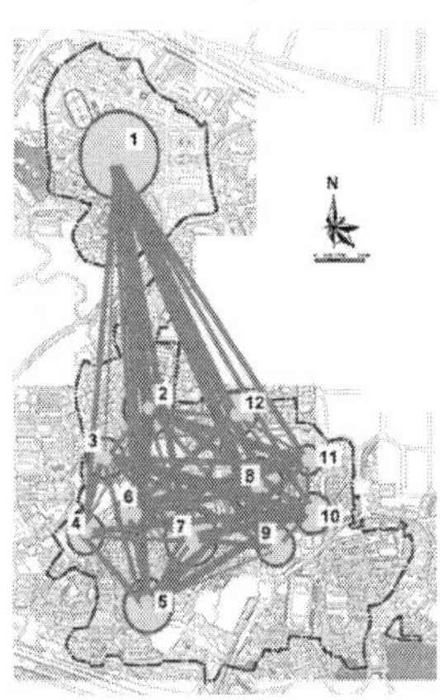

Figure 3. Campus travel flow map.

The size and boundary of traffic area, the scope of road network and the section parameters directly affect the accuracy and authenticity of the model. The road network model is shown in Figure 4. Section 1-22 is the shared road of pedestrians and motor vehicles, of which section 15 is an important pedestrian passageway. According to the characters of campus travel, it can assume the future travel purpose and travel frequency of teachers and students remain the same, and original unit method is used to respectively calculate the original unit of generation and attraction, and then according to the value of travel generation and attraction gotten through the forecast of attributes such as generation original unit and attraction original unit as well as population and area, allocate travels of each community to road network and travel allocation model to calculate the traffic volume in each section and finally get the data of each section. The Wardrop Equilibrium Theory model is expressed as follows:

When $L_k^{pq} > 0,\ c_k^{pq} = c^{pq},\ \forall k \in K^{pq},\ \forall pq \in \Omega$ (1)

When $L_k^{pq} = 0,\ c_k^{pq} \geq c^{pq},\ \forall k \in K^{pq},\ \forall pq \in \Omega$ (2)

$\sum_{k \in K} L_k^{pq} - t^{pq} = 0,\ \forall pq \in \Omega$ (3)

When $L_k^{pq} \geq 0,\ \forall k \in K^{pq},\ \forall pq \in \Omega$ (4)

In the formula, L_k^{pq} is the traffic volume of OD towards pq in the kth path;

c_k^{pq} is the travel time of D towards pq in the kth path;

c^{pq} is the travel time of D towards pq in the shortest path;

t^{pq} is the traffic volume of OD towards pq in shortest path.

From Table 5, it shows the 1 2, 3, 4, 5, 8, 9, 11, 12, 13, 14, 15, 20, 21 road sections have large flow, average daily traffic volume reach from 8000 to 30,000.

Figure 4. Campus road network model.

Table 5. Student traffic volume distribution results in each section (person-time/d).

Section of serial number	Traffic volume (bidirectional)	Section of serial number	Traffic volume (bidirectional)
1	16,063	12	32,040
2	14,242	13	15,166
3	15,052	14	19,348
4	14,522	15	9348
5	10,837	16	5766
6	5287	17	3919
7	0	18	5254
8	8837	19	6715
9	29,810	20	9110
10	2456	21	2181
11	12,519	22	5102

CAMPUS BUS DESIGN SCHEME

According to the construction of road network in the previous chapter, we consider to improve the deficiency of existing traffic on campus. The improvement of scheme follows the principles to design, and on the basis of new scheme, analyze the characteristics of the route optimization and finally conduct the optimization of operation project towards newly designed route plan, including departure time and the optimization of departure frequency, to maximize efficiency of public transportation.

Route Design Optimization Principles

1) Follow the principle that route covers main passenger flow corridor

   The design of campus traffic route must be consistent with the main flow of people. The people flow on campus OD is obtained by investigating the people's intention to travel on campus. From the analysis on campus traffic network, it can be known that Campus Bus Line 1 covers from the north area to the public teaching buildings, administrative buildings and the flow direction of south front door, however, there is no bus route to cover from the north area to infirmary, from dormitories in west area to Wushan subway and from travels in south area and residential area to college buildings and public teaching area. Thus, when designing route, it should consider these main routes.

2) Principle of moderate length of traffic route

The length of the route should be designed within a reasonable range so that the bus system can be better organized and operated. On the one hand, if the distance between the routes is longer, the average cycling time of the bus will become longer, the bias ratio of time when vehicles arrive the station will increase, and there will also e many problems when arranging departure frequency; on the other hand, if the route is too short, the users it covers are fewer and the vehicle's cycling time is short, so that the passenger's intention to ride will decrease and the economy is not good. The relevant data suggest the route length is appropriate when the bus runs 5 - 15 km among 20 - 30 min. This paper recommends to adopt the standard of 6 - 10 km.

3) Principle that traffic route covers densely populated areas

Use personnel density heat map to show the data of builders in each area obtained through survey. Heat map adopts EXCEL plug-in and POWERMAP painting. Label west campus, central area, south area and roads in east area or density of dormitory personnel.

4) Principle of lower linear coefficient.

The design of the bus route should also pay attention to the influence of the linear coefficient. In terms of a bus route, the straight line surely is best, just like Campus Bus Line 1 which exists in H University. When designing the round route, the linear coefficient should also be paid attention to. If the linear coefficient is higher, users' satisfaction with the route will be reduced, and the travel time of users will be greatly increased. It is generally believed that the linear coefficient should be between 1 and 1.3.

5) Principle of satisfying the actual environment

When design traffic route, the important aspect which should be paid attention to is the geographical environment of the campus.

The terrain of H University is undulating, and its local slope is more than 8%, so when considering route design, it should be considered that the driven routes of the vehicle lie in altitude traverse with slight ups and downs as far as possible.

Design of Bus Stops

The bus stop setting on campus is a systematic and complicated work. At the beginning of the study about bus stop setting, stations need to be classified clearly firstly. Bus stopover station is the node of transit network. Travelers use the starting point and end of bus, which provides service for the stop of public transport and passengers' getting on or off the bus. The performance of bus stop directly affects the service efficiency of public transport, even the traffic capacity of the whole road. Bus stop can be divided into two basic forms of station: original and terminal station and midway station according to function.

1) Design of original and terminal station

   The original and terminal station is the beginning and end of the bus line, which provides service for the arrival and departure of operating vehicles and passengers' getting on or off the bus. At the same time, it also is the place where traffic dispatchers organize the operation of vehicles, drivers have a rest and vehicles are maintained. Thus, the original and terminal station should be close to parking lots of public transport or maintenance factories and pay attention to be near the existing stations, in order to be easy to transfer.

2) Design of midway station

   The midway station should be set on the nodal points of main passenger flow that the public transport route passes, and the average distance between stations is 200 m - 300 m. The station distance of campus population density selects the lower limiting value, and the station distance of the marginal area and the area of people flow choose the upper limiting value. The selection of site should also satisfy the three main functions of vehicle: safe stop, convenient passage and convenient transportation.

Campus Traffic Route Design

According to the campus traffic route network map and the campus elevation map, we can get the elevation map of the main people flow on campus.

1) The north section of the Hubin, the section of Songshan and the north section of Zhujiang are all at a lower altitude, and also are in the main traffic section.
2) The main routes of travel in south area and the travel routes of residential area in east area have a single direction with little fluctuation in altitude.
3) The selection of original and terminal station is the station of south front gate, and there is a bus maintenance station in the south front gate, and can be transferred to the original Line 1.

After fully considering travel demands of people on campus, geographical factors and design principle of bus route and station, set up a route (as shown in Figure 5) for south front door—Zhongshanxiang—College of Business—No. 15 building (Zhujiang East Road)—complex buildings in west area (crossing of the Yellow River)—No.17 press building—Xi'er Village (crossing of Duxiufeng)—Lijiang Road—stadium in area—Research Two—Research Four—Research Five—No.31 building—Yifu Science Museum—Architectural Design Institute—the dormitory of the national defense students—East eight—Cultural and Sports Center—kindergarten—918 intersection (tunnel portal)—the south gate.

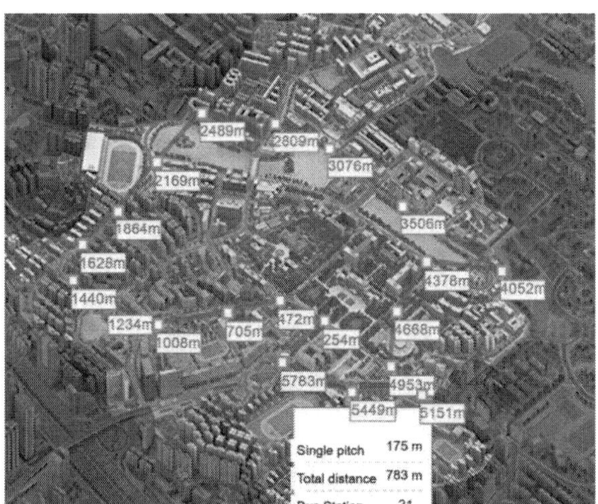

Figure 5. The average distance travelled of the new route.

The route has 21 bus stops in total (Figure 5), and the original and terminal station is located at the south gate terminal, a place where is convenient for drivers to have a rest and the dispatching and maintenance of buses. The total length of this route is about 6000 m, and the average length of station is

200 m. According to the bus speed standard: 30 km/h, the time required for a bus to travel around is 25 min.

THE DESIGN OF THE OPERATING SCHEDULE OF THE CAMPUS BUS

The departure frequency of departure is closely related to the traffic cost. This paper mainly considers the user cost and the operating cost of the vehicle, as follows:

$$C = C_u + C_o \tag{5}$$

In the formula, C is the total cost; C_u is the user cost and C_o is the operating cost.

1) The waiting time cost of the user is :

$$C_u = a_w\left(\frac{\sum_{i=1}^{n} b^+}{f} + \frac{\sum_{i=1}^{n} b^-}{f}\right) \tag{6}$$

In the formula: f is the departure frequency,

b^+ is the waiting time for passengers who get on the bus,

b^- is the waiting time for passengers who get off the bus

The operating cost of the vehicle includes the driver's wage and welfare costs, the fuel consumption and maintenance costs of the daily operation of the bus.

$$C_o = f * P * (c_0 + c_1 * N + c_p) \tag{7}$$

In the formula, P is the length of the route (km);

c_0 is the initial fixed input cost (yua /km);

c_1 is the variable cost (yuan/km * seat);

c_p id the cost the drivers derive per kilometer (yuan /km).

According to questionnaire survey above, the basic data of cost analysis of Line 2 shows in Table 6. Calculate the user cost and operating cost in different departure frequencies, and look for the optimal departure frequency in a reasonable range to determine the minimum total cost. User cost and total operating cost are show in Table 7.

As what you see in Figure 6, along with the increase in departure frequency, the total cost decreases firstly and then rises with greater falling range and slow rising trend. The total cost is in a lower value when the departure frequency is among 12 and 15 veh/h.

Table 6. Basic data of cost analysis of Line 2.

parameter	value
a_w/(yuan/h)	6.74
c_0/(yuan/h)	1.50
c_1/(yuan /km*seat)	0.08
L/km	6.0
N	16.0

(Data collection is based on questionnaire survey).

Table 7. Table of user cost and total operating cost in different departure frequencies.

f	h/min	C_u	C_o	C
5	12	1.4	0.4	1.8
6	10	1.24	0.5	1.74
8	7.5	1.08	0.6	1.68
10	6	0.98	0.65	1.63
12	5	0.91	0.70	1.61
15	4	0.86	0.74	1.60
20	3	0.83	0.79	1.62
25	2.4	0.82	0.82	1.64
30	2	0.8	0.85	1.65

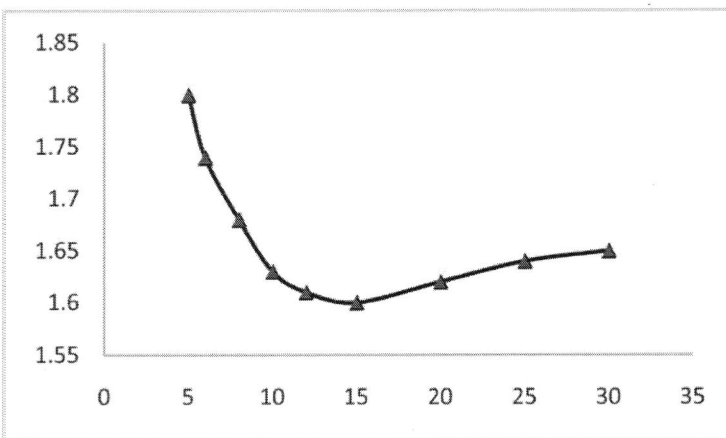

Figure 6. Total cost graph.

It is suggested that the departure time in the peak hours is 4 - 5 min, and the departure time in the off-peak hours period can extend by 6 - 8 min properly.

The operation mode adopts the form of bi-directional departure to increase the frequency of the campus bus; use the peak and the flat peak elastic design to increase the departure frequency in the peak and off-peak period. The average stopping time for each station is 30 s, and the time that vehicles travel along the route is:

Table 8. Operating schedule for new Line 2.

Time interval	Route	Departure interval/ min	Unidirectional departure frequency	Main function
7:30-8:15	residential area in east area —teaching building	3	15	attending class
8:15-9:40	route around campus	5	18	all travels
9:40-10:15	residential area in east area —teaching building	2	15	attending and finishing class

10:15-11:40	route around campus	5	20	all travels
11:40-12:30	residential area in east area —teaching building	2	15	finishing class
13:50-14:30	residential area in east area —teaching building	3	10	attending class
14:30-16:30	route around campus	5	30	all travel

Run time (min) = distance travelled (km) * 2 (min/km) + number of stations in the route * 0.5 (min/ station) [6.

In theory, the run time is 26min. Consider with the combination of cost, demand and safety. The operation of the campus bus is calculated according to 8 hours. The operating schedule is as Table 8 shows.

CONCLUSION

The reasonable design of campus bus routes can meet the travel needs of campus personnel, and at the same time it can realize the green, safe and harmonious campus traffic. This paper studies the planning and design method of the university campus bus, and conducts travel investigation among teachers and students in campus, analyzes the travel mode, travel distance, travel frequency, travel destination, travel time and people willingness feature. The school personnel travel has summarized as following characteristics: mainly by walking; travel and attractions are concentrated in the teaching area, dormitory area, office area and sports area. This paper also has built the travel model of campus traffic and allocated specific traffic volume to main roads of campus, and then put forward principles that suit campus traffic optimization of H University, including the principle of designing route and stations. The designed routes meet various requirements. At last, the optimized route is proposed to optimize the rate of coverage and operation cost, find the departure frequency after optimization, and propose campus bus design and planning program.

CONFLICTS OF INTEREST

The authors declare no conflicts of interest.

REFERENCES

1. Lin, Z. (2007) Research on Optimized Strategies of Chinese Campus Transportation under the Background of Mobilization. Shanghai Tongji University, 3, 281-293.
2. Xu, J.B. (2012) The University Campus Transportation Plan Theory and Design Method Research. Beijing University of Technology, 29, 1884-1894.
3. Liu, Z.Q. (2008) The Study on Bus Stop Time Modeling. City Public Transport, 218, 602-613.
4. Mao, B.L. (2009) Integrated Transportation Policy and Practice. Traffic Demand Management, 52, 1-14.
5. Xiong, P. (20012) Methodology Research of Campus Bus Planning and Designing—A Case of Shenzhen University. Journal of Transportation Engineering and Information, 31, 1-19.
6. Lu, F.Q., Chen, X.W. and Chen, S.P. (2009) Frequency Optimization of Bus Rapid Transit Based on Cost Analysis. Journal of Southeast University, 39, 1-24

CHAPTER 7

Route Optimization of Electric Vehicle considering Soft Time Windows and Two Ways of Power Replenishment

Ming Meng[1] **and Yun Ma**[1]

[1]Department of Economics and Management, North China Electric Power University, Baoding, Hebei 071003, China

ABSTRACT

Under the background of severe air pollution and energy shortage, electric vehicles (EVs) are promising vehicles to support green supply chain and clean production. In the world, the renewal of EVs has become a general trend. Therefore, the concern about EVs is a hot issue at present, but EVs have the characteristics of limited driving distance and long charging time. When the EVs are used in logistics transportation, these characteristics have a significant impact on the vehicle routing problems. Therefore, based on the research experience of traditional vehicle routing optimization, combining with the characteristics of EVs, this paper presents an optimal problem of

Citation: Ming Meng, Yun Ma, "Route Optimization of Electric Vehicle considering Soft Time Windows and Two Ways of Power Replenishment", Advances in Operations Research, vol. 2020,Article ID 5612872, 10 pages, 2020. https://doi.org/10.1155/2020/5612872.

electric vehicle routes with time windows based on two charging methods and it also designs a mathematical model which was caused by early and late arrival as the objective function to minimize the transportation cost, vehicle use cost, power supply cost, and penalty cost. The model is solved using an ant colony algorithm. Finally, the ant colony algorithm is tested and analyzed with an example.

INTRODUCTION

The research on vehicle routing began in the 1950s, Dantzig and Ramser firstly proposed the concept of vehicle routing problem (VRP), which refers to the purpose of distributing or collecting goods between distribution centres and a certain number of customers with different needs through the design of vehicle routing, and finally reached the goals, such as the shortest distance, the least time, and the least cost [1]. The importance of transportation in the logistics system distribution activities is undeniable, but in recent years, the large use of fossil fuel vehicles has resulted in the rapid consumption of oil resources and excessive emissions of greenhouse gases, so considering the balance and optimization of monetary costs and the environmental problems of fossil fuel vehicles, many vehicle routing problem models for fuel or emissions have risen, such as the fuel consumption rate of the VRP was considered by Xiao et al. [2], fuel consumption and carbon emission of the VRP considered by Zhang et al. [3], a time-dependent VRP model of minimizing fuel consumption of Norouzi et al. [4], the green vehicle routing problem (GVRP) model of Poonthalir and Nadarajan [5], and the vehicle scheduling problem of minimizing carbon emission of Wang et al. [6].

Because of the environmental protection characteristics of EVs, in recent years, with the rapid growth of the market share, EVs have been introduced into the market as personal and commercial alternative energy vehicles. In 2018, the number of electric vehicles worldwide exceeded 5.1 million, an increase of 2 million from 2017, and the growth of new vehicles almost doubled (IEA, 2019). In China, electric vehicles are growing at the rate of more than 50% per year (IEA, 2018). In 2018, the number of electric vehicles in China ranked first in the world [7]. Compared with traditional fuel vehicles, the main advantages of electric vehicles are zero greenhouse gas emissions, high efficiency, and low operating noise [8], which helps logistics companies get more and more social and environmental customer support. And it gets a green image [9]. Fernandez and Casals use sustainable analysis and practical estimation methods that take into account the life cycle carbon emissions of electric vehicles to analyze the contribution of electric vehicles to reducing greenhouse gas emissions [10, 11]. Wu et al. used the life cycle assessment method to estimate that the total life cycle greenhouse gas emission reduction potential of battery electric vehicles will gradually reach 13.4% in 2020 [12]. In addition to environmental benefits, EVs also have economic benefits. Compared with traditional fossil fuel powered vehicles, EVs consume 10% to 15% of the fuel cost of traditional vehicles at the same distance [13]. Therefore, the focus on EVs has become a hot issue at present. However, due to the characteristics of EVs such as low endurance mileage and long charging time, it becomes a significant and challenging task to study its VRP.

Electric vehicles refer to cars that use electric engines to provide energy through their own chemical batteries. The electrical energy used by EVs can be converted into many kinds of clean energy, which is the main force of environmental protection vehicles in the future with a high energy utilization rate and being clean and pollution-free. As a distribution vehicle, electric vehicles need to be charged much time for long-distance distribution, and the power replenishment time is longer than the refueling time. Therefore, it is necessary to consider the charging problems that may occur in the charging process. Based on traditional fuel vehicle route optimization, the introduction of the charging stations is the primary issue of electric vehicle upgrading combining with the characteristics of EVs. Electric distribution companies apply some coordination methods to control the charging load. That can affect the charging duration. Yang et al. studied the charging scheduling of electric vehicles on the highway [14]. Dogan and Alci optimized the charging schedule of electric vehicles considering the cost of battery degradation [15]. Dogan et al. based on a heuristic algorithm for charge and discharge coordination optimization [16]. Aravinthan and Jewell proposed a two-step method for scheduling EV charging, which limits the impact on EV charging on distribution assets [17].

At present, fast charging is the most common charging strategy. Schucking et al. proposed five charging strategies and believed that DC fast charging is essential [18]. In addition to fast charging on the issue of charging strategy, battery switching is also a relatively popular charging strategy. Adler and Mirchandani used real-time highway data to find the optimal battery exchange strategy [19]. Yang and Sun studied the location routing problem of the battery switching station of EVs with large capacity and optimized the routing plan and the selection of battery switching stations [20]. Dai et al. regarded the battery exchange strategy of EVs as the background and it provided reference for the determination of decision variables such as the number of backup batteries in AC power station and the charging selection of EVs [21]. Margaritis et al. analyzed the advantages and disadvantages of battery exchange strategy to the government, users, and enterprises based on the EU [22]. This study combines the two charging strategies of fast charging and battery switching. The fast charging time is related to the remaining power, fixed battery replacement time, and the power replenishment method with less time is selected.

In the large-scale application of EVs in modern logistics, in addition to considering the plan to charge or replace batteries on the way, the delivery time is also extremely important for logistics companies, so we need to consider some important practical factors, such as customer time windows. In the actual distribution activities, more and more logistics enterprises begin to pay attention to the timeliness of package delivery. For customers, "punctuality" is one of the important factors affecting customer experience, so the vehicle routing problem with time windows (VRPTW) has also become an important part of the research. By adding time windows constraints to the basic VRP model and Solomon built VRPTW model, in which time window is a hard time window that must be observed [23]. Qureshi further expanded the concept of the hard time window, extended the problem to the category of the soft time windows, and determined the strictness of time windows by setting penalty function [24]. The soft time windows problem is widely used. Goeke considered the problem between time windows and EVs is pickup and delivery [25]. Keskin and Catay studied partial charging strategies for electric vehicles with time windows [26]. Desaulniers et al. studied the effective route optimization of battery electric commercial vehicle fleets, and they considered four variants of the route problem of electric vehicle with time windows [9]. Goeke studied the pickup and delivery of EVs with time windows (PDPTW-EV). In the PDPTW-EV, access location is limited by time windows [25]. Many scholars introduced charging stations and time windows for discussion [27, 28]. This study combines the two hot charging strategies of fast charging and battery switching. The fast charging time is related to the remaining power, fixed the switching time, and the power replenishment method with less time is selected. This paper not only considers the problem of charging stations, but also uses the broken line time windows to limit the distribution time based on the soft time windows.

In summary, compared with other similar studies, the main contributions of this paper are as follows: (1) in order to calculate the cost of logistics enterprises more accurately and save cost, this paper considers the cost of transportation as much as possible, that is to say, fixed cost, transportation cost, charging cost, and time penalty cost. And it establishes a mixed-integer linear programming model (MILP) with the goal of minimizing the total cost; (2) in order to save charging time, two commonly used charging methods are considered, and one with shorter charging time is selected; and (3) considering the customer's time tolerance, the broken line soft time windows is adopted.

The rest of this paper is organized as follows: Section 2 describes the problem and identifications of the main assumptions. In Section 3, the MILP model is established and the description of the method of solving the model is provided. In Section 4, the test results of an example are given and the sensitivity analysis is carried out. Finally, the paper summarizes in Section 5.

PROBLEM DESCRIPTION

This problem can be abstracted as that an enterprise uses EVs to provide distribution services for *n* customers with time windows after the distribution centre is fully charged. Each customer's demand, service duration, good service time windows, and customer tolerance level are needed to be known. Finally, reasonable planning of the vehicle distribution route is necessary so that the total cost of distribution is small.

The soft time windows can relax the constraints of time windows, optimize resource allocation, and reduce energy consumption and road congestion, so the time windows studied in this paper mainly are soft time windows. As shown in Figure 1, in the traditional soft time windows, whether the vehicle arrives before or after *l*, the customer is allowed to serve, but they are required to pay the corresponding penalty fee, which is generally a simple linear relationship with the degree of time deviation. However, for the deviation of the best service time windows, customers have the difference between tolerance and intolerance. Therefore, based on the traditional soft time windows, considering the tolerance range of customers, this paper proposes a broken line time window.

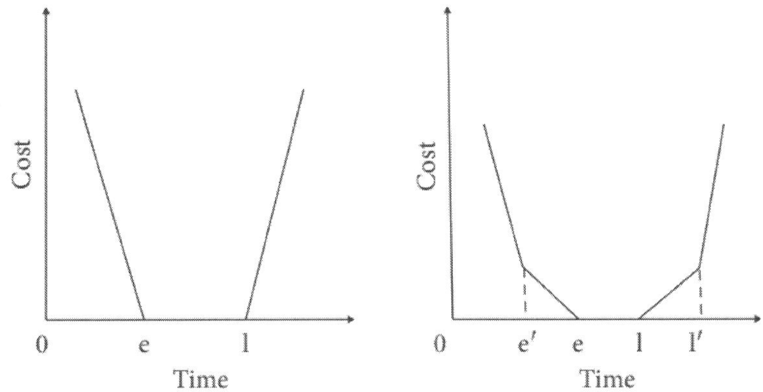

Figure 1. Penalty cost function under traditional soft time windows and polygonal line soft time windows.

According to the customer tolerance level (∂) and service duration (tf), the tolerance time window can be obtained on the basis of the optimal time window $[e',l']$, where $e' = e - \partial \times tf, l' = l + \partial \times tf$ If the vehicle provides services in the best time window $[e,l]$, there will be no penalty cost to be paid. If the vehicle provides services in the interval $[e',e][e',e]$ or $[l,l']$, there will be only less penalty cost to be paid. If the vehicle starts service earlier than e' or later than l', there will be more penalty cost to be paid. Compared with the traditional soft time windows, the broken line soft time windows take the actual feelings of customers into consideration, which will be conducive to better coordination between enterprises and customers, reasonable allocation, and optimization of resources.

In view of the above considerations, the problems and basic assumptions to be solved are as follows: (1) each vehicle can meet the needs of multiple customer points, and each customer point can only be served by one vehicle; after the completion of the distribution service, the vehicle must drive back to the distribution centre; (2) all vehicles are the same type, and the total transportation volume shall not exceed the capacity limit of EVs; (3) each customer's location coordinates, demand quantity, and service duration are known, and there are optimal service time windows and tolerance time windows; (4) the time window penalty coefficient of each customer node is the same; (5) EVs can only be charged or their battery can be replaced in the distribution centre or power station; (6) each customer must be visited and can only be visited once; (7) the road is smooth, without considering traffic congestion and other special situations; (8) it is assumed that the transportation cost generated by the vehicle is linearly related to the route length, and the use cost of each vehicle is fixed; and (9) the objective function of this problem is to minimize the total distribution cost.

MATERIALS AND METHOD

In this part, according to the electric vehicle routing problem with time windows (EVRPTW), we establish a MILP model and determine the constraints.

Defining Variables

The parameters and decision variables used to describe the MILP model are shown in the following:

N: set of all nodes n in networks.

V: set of all nodes v in networks.

C: set of all nodes c in networks.

D: network D = N ∪C ∪ o consisting of sets of nodes, $D = \{0, 1, 2, \ldots, |N| + |C|\}$.

n: set of customer nodes, where n ∈ N.

v: set of EV nodes, where v ∈ V.

c: set of charging station nodes, where c ∈ C.

o: set of distribution centre nodes.

C_0: fixed cost per vehicle.

C_1: electric vehicle unit distance transportation cost.

EC: total electricity supplementary cost.

C_2: charging cost per unit time.

C_3: single battery replacement cost.

i, j: index of nodes, i = 0, 1, 2, . . . , n.

d_{ij}: distance from the node i to the node j, i, j ∈ D.

q_n: demand of the customer node n, and n ∈ N.

Q: rated load capacity of electric vehicles.

P: battery capacity of electric vehicles.

g: charge coefficient.

P^1_{dv}: the residual power of EVs when v reaches the node D.

P^2_{dv}: residual electricity of the electric vehicle V leaving node D.

p: battery capacity of electric vehicles.

h: power consumption coefficient.

g: charge coefficient.

tf_i: if i represents the customer point, tf_i represents the service time of electric vehicle at the node i; if_i represents the charging stations, then tfi represents the electric vehicle's power replenishment time, i ∈ N ∪M

tc_i: single battery change time, and i ∈ C.

tw_i : waiting time of electric vehicles at the customer node i.

t_i^v: the time point when the vehicle v arrives at the customer node i, where $t_0^v = 0$

$t_i'^v$: the start time of the v-car service customer node i.

t_{ij}: time of electric vehicles from i to j.

speed: driving speed of electric vehicles.

e_i: the lower limit of the best service time windows of the customer node i.

l_i: the upper limit of the best service window of the customer node i.

e_i': the lower limit of tolerance time windows of the customer node i.

l_i': upper limits of tolerance time windows of the customer node i.

ρ_m: unit time penalty cost of vehicles violating the time windows, m = (1, 2, 3, 4).

x_{ijv}: if the vehicle v is from i to j, then

$x_{ijv} = 1$, otherwise $x_{ijv} = 0$.

x_{ov}: if the vehicle v returns to the distribution centre after delivering a customer group, then $x_{ov} = 1$, otherwise $x_{ov} = 0$. y_{iv}: the task of the customer node i is completed by the vehicle v, then $y_{iv} = 1$, otherwise $y_{iv} = 0$

Model and Method

The objective function is to minimize the comprehensive cost, including transportation cost, vehicle use cost, electricity replenishment cost, and time window penalty cost. The formula of the MILP model is as follows:

$$F_{\min} = C_0 \sum_{v\in V} x_{ov} + C_1 \sum_{v\in V} \sum_{i\in D} \sum_{j\in D, i\neq j} x_{ijv} d_{ij} + \sum_{v\in V} \sum_{i\in C} EC_i(tf_i) + \sum_{v\in V} \sum_{i\in N} pu_i(t_i^v). \tag{1}$$

Among them,

$$\mathrm{EC}_i(tf_i) = \begin{cases} C_2 \times tf_i, & tf_i \leq tc_i, \\ C_3, & tf_i > tc_i, \end{cases} \tag{2}$$

$$\mathrm{pu}_i\left(t_i^{'v}\right) = \begin{cases} \rho_1\left(e_i' - t_i^{'v}\right) + \rho_2\left(e_i - e_i'\right), & t_i^{'v} \leq e_i', \\ \rho_2\left(e_i - t_i^{'v}\right), & e_i' < t_i^{'v} \leq e_i, \\ 0, & e_i < t_i^{'v} \leq l_i, \\ \rho_3\left(t_i^{'v} - l_i\right), & l_i < t_i^{'v} \leq l_i', \\ \rho_3\left(l_i' - l_i\right) + \rho_4\left(t_i^{'v} - l_i'\right), & l_i < t_i^{'v}, \end{cases} \tag{3}$$

subject to

$$\sum_{i\in D, k\neq i} x_{ikv} = \sum_{j\in D, k\neq j} x_{kjv}, \quad k \in D, \tag{4}$$

$$\sum_{v\in V} y_{nv} = 1, \quad n \in N, \tag{5}$$

$$\sum_{n\in N} y_{nv} q_n \leq Q, \tag{6}$$

$$\sum_{v\in V} \sum_{i\in D, i\neq 0} x_{oiv} \leq |V|, \tag{7}$$

$$\sum_{i\in N} \sum_{j\in D, i\neq j} x_{ijv} \leq |N|, \tag{8}$$

$$t_0^v = 0, \tag{9}$$

$$t_{ij} = \frac{d_{ij}}{\text{speed}}, \quad i, j \in D, \tag{10}$$

$$tf_i = \min\left[tc_i, \left(\frac{P - p_{iv}^1}{g}\right)\right], \tag{11}$$

$$t_i^v = t_i^{'v} + tf_i + tw_i, \quad i \in N, \tag{12}$$

$$t_i^{'v} \geq t_i^v, \quad i \in N, v \in V, \tag{13}$$

$$t_j' = \sum_{i \in D} \sum_{j \in D, i \neq j} x_{ijv}(t_i'' + t_{ij}), \tag{14}$$

$$p_{iv}^1 = p_{iv}^2, \quad i \in N, \tag{15}$$

$$p_{ov}^1 = 100, \tag{16}$$

$$p_{jv}^1 = p_{iv}^2 - x_{ijv} \times \frac{d_{ij}}{h}, \quad i, j \in D, i \neq j, v \in V, \tag{17}$$

$$p_{iv}^1 \geq 0, \quad i \in D, v \in V, \tag{18}$$

$$x_{ijv}, x_{ov}, y_{iv} \in \{0, 1\}, \quad i, j, o \in D, v \in V. \tag{19}$$

In the model given above, Constraint (4) ensures that the vehicle starts from the distribution centre returns to the distribution centre and focuses on distribution centre. Constraint (5) ensures that each customer is served only once by one vehicle. Constraint (6) is a restraint on vehicle loading. Constraint (7) indicates that the number of EVs serving the customer is less than or equal to the total number of vehicles owned by the distribution centre. Constraint (8) requires that the number of customers served by each vehicle is less than or equal to the total number of customers. Constraint (9) requires that when the electric vehicle starts from the distribution centre, the time is 0. Constraint (10) indicates that the travel time of electric vehicles from the point i to the point j is the ratio of the distance between two points and the travel speed. Constraint (11) indicates that when the electric vehicle passes through the charging stations, the charge replenishment time is the minimum of the battery replacement time and fast charging time. Constraint (12) indicates that the time the electric vehicle leaves the customer node i is the sum of arriving at the customer node, waiting time at the customer node i, and servicing time. Constraint (13) shows the relationship between the starting service time and the arrival time of vehicles at the customer point. Constraint (14) indicates that the time the electric vehicle arrives at the node j is the accumulation of the previous time. Constraint (15) means that neither power consumption nor power replenishment occurs at the customer node i. Constraint (16) indicates that when the electric vehicle starts from the distribution centre, the state of charge is 100. Constraint (17) indicates that the remaining power to node j is equal to the remaining power to leave node i minus the power consumed on the way. Constraint (18) indicates that the state of charge of electric vehicles is nonnegative at any position. Constraint (19) indicates that the decision variable is constrained to 0-1.

In this paper, the ant colony algorithm is used to solve the approximate optimal solution of the NP hard problem. The ant colony algorithm has the characteristics of distributed computing, positive information feedback, and heuristic search. In essence, it is a heuristic global optimization algorithm in the evolutionary algorithm. The algorithm imitates the social behavior of ants in order to find the shortest route from nest to food source. In the ant colony algorithm, each ant performs four basic activities in the process of route construction: (1) select the next customer based on the probable function of the distance from the current location to the customer and the route strength on the arc; (2) save the taboo list of the customers in the current route; (3) update the residual capacity of the vehicle; and (4) update the track intensity on the access arc, and use the method of local search to improve the quality of the solution. Finally, the taboo lists are deleted, and a new iteration is started. When the ant colony algorithm solves the MILP model, the specific steps are as follows:

Step 1: importing data and setting basic parameters.

Step 2: calculating the distance between customer nodes and the cost and time of distance between customer points.

Step 3: initializing and iterating in order to find the best route.

Step 4: terminating the algorithm and reporting the best solution.

Figure 2 shows the traversing process of a single ant in the iterative search for the best route.

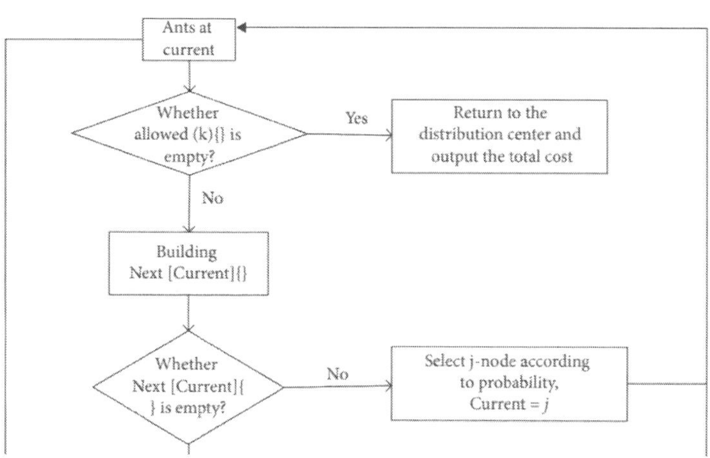

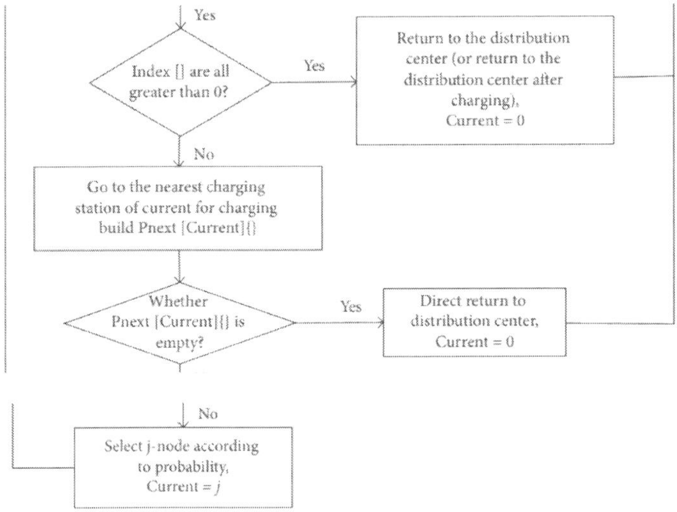

Figure 2. Single ant traversal process to find the best route during iterative search.

The meanings of the mathematical symbols involved in Figure 2 are explained as follows:

Current: the current location of ants.

Allowed(k){}: the collection of client nodes that have not been accessed, initially including all client nodes.

Next[i] {}: the set of optional customer nodes when the electric vehicle is in *i*-node.

P_{next}[i] {}: the collection of optional customer nodes for EV charging closest to *i*-node.

CALCULATION EXPERIMENT AND COST ANALYSIS

In order to verify the proposed MILP model, the calculation experiments are carried out based on the known benchmark instances, and the ant colony algorithm is used to solve the model.

Test Examples

The experimental data are from Solomon's VRPTW standard problem set, and the data number is R101 [29], which is characterized by uniform distribution of customer points and narrow time windows. In this case, a vehicle is generally only responsible for the distribution of several customer points, and the vehicle route cost is greatly affected by the time windows, so the data selection of the case is reasonable. In order to draw a clear road map, this paper only selects the first 26 data, including 1 distribution centre and 25 customer nodes.1 is the distribution centre, 2–26 is the customer node, 27–31 is the charging stations node. The specific parameters of the example are shown in Table 1, and the node information is shown in Table 2.

Table 1. Temperature and wildlife count in the three areas covered by the test examples.

Name	Parameter
Maximum loading capacity of vehicles	80 pieces
Use cost of vehicles	1000 RMB/vehicle
Unit transportation cost	2 RMB/km
Speed of vehicles	40 km/h
Full charge at the charging stations	60 kWh
Power consumption per unit distance	1 kWh
Cost per battery replacement	30 RMB
Battery replacement time once	31 min
Service time per customer	10 min
Penalty coefficient of time windows	(1, 0.5, 1.5, 2)
The customer tolerance level	0.5
Maximum number of iterations	200

Table 2 ode information.

Number	X-coordinate	Y-coordinate	Demand	Ready time	Due date
1	35	35	0	0	230
2	41	49	10	161	171
3	35	17	7	50	60
4	55	45	13	116	126
5	55	20	19	149	159
6	15	30	26	34	44
7	25	30	3	99	109
8	20	50	5	81	91
9	10	43	9	95	105
10	55	60	16	97	107
11	30	60	16	124	134
12	20	65	12	67	77
13	50	35	19	63	73
14	30	25	23	159	169
15	15	10	20	32	42
16	30	5	8	16	71
17	10	20	19	75	85
18	5	30	2	157	167
19	20	40	12	87	97
20	15	60	17	76	86
21	45	65	9	126	136
22	45	20	11	62	72
23	45	10	18	97	107
24	55	5	29	68	78
25	65	35	3	153	163
26	65	20	6	172	182
27	10	32	0	0	230
28	27	47	0	0	230
29	40	30	0	0	230
30	50	50	0	0	230
31	60	10	0	0	230

In this paper, according to the actual situation, we made assumptions about the required data in Table 1, and for this part of the program design, we reserved a data change area, which is malleable.

The MILP model is solved by MATLAB programming. In order to reduce the influence of random factors as much as possible, this paper repeatedly tests the example 10 times to get the total cost optimal solution (*C*), the number of vehicles (*N*), route length (*L*), and penalty cost (*P*) when the optimal solution is reached.

As can be seen from Table 3, the optimal solution of this example is 7283.08, including the route length when the optimal solution is reached at 270.61, and Table 4 shows that the optimal allocation scheme includes 4 routes; the route map is shown in Figure 3.

Table 3. Related results of the test examples.

Operation times	*C*	*N*	*L*	P
1	7404.18	4	272.08	2372.11
2	7613.71	4	278.04	2544.15
3	7764.36	3	262.93	3814.29
4	7586.94	4	280.75	2669.48
5	7283.08	4	270.61	2238.96
6	7567.74	3	258.58	3629.46
7	7590.47	3	269.97	3617.05
8	7491.76	4	269.34	2314.07
9	7603.67	3	271.03	3724.88
10	7466.96	3	263.06	3578.25
Average	7537.29	3.5	269.64	3050.27

Table 4. Sequence of routes when the test examples reaches the optimal solution.

Vehicle	Route number
1	1–14–22–23–24–27–9–1
2	1–7–6–18–17–15–16–28–19–20–1
3	1–3–5–26–25–4–13–28–8–1
4	1–2–21–10–11–12–1

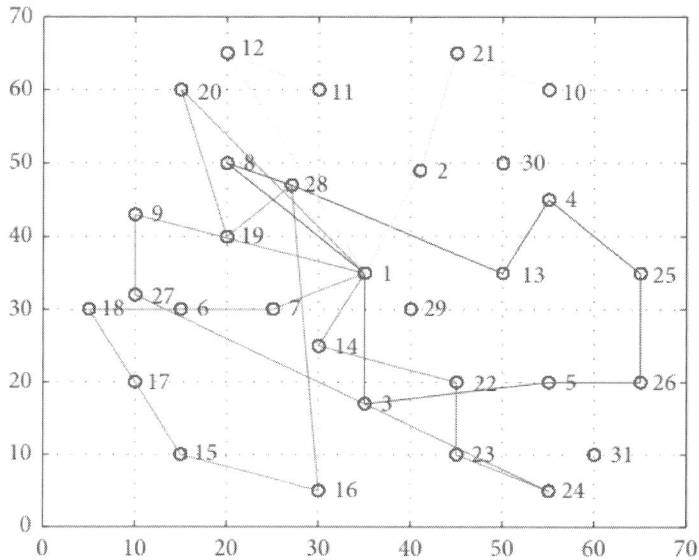

Figure 3. Corresponding route trajectory when the example reaches the optimal solution.

Cost Analysis

Based on the calculation experiment of the Solomon benchmark example, the sensitivity analysis of the cost affecting the electric vehicle route planning is carried out.

Use Cost and Transportation Cost

First, adjust the use cost from 1000 for each vehicle to 5000 for each vehicle, keep the other parameters unchanged, repeat the test for 10 times, and see Table 5 for the results when the optimal solution is reached.

Table 5. Impact of adjusting use cost on results.

Operation times	*N*	*L*	*P*
1	2	281.25	6036.13
2	2	275.01	6646.81
3	2	268.50	6809.43
4	2	281.49	6021.61
5	2	280.76	6722.07
6	2	268.62	6450.82
7	2	272.84	6257.53
8	2	274.15	5855.41
9	2	285.66	6344.63
10	2	280.63	6025.55
Average	2	276.89	6317.00

It can be seen from Table 5 that after adjusting the use cost of vehicles from 1000 to 5000, after 10 times of operation, compared with Table 3, the number of vehicles is reduced from 4 to 2, which is 50%. It can be seen that the higher the use cost of vehicles is, the smaller the number of vehicles used is. When the use cost of vehicles is much higher than other costs, the vehicle route distribution scheme with the minimum number of vehicles will be preferred. At the same time, the route length is only increased by 3% compared with the result in Table 3, while the time window penalty is increased by 52%. This is understandable, because the proportion of vehicle use cost in the objective function increases, so the algorithm will tend to find the solution with the minimum vehicle use.

Keep the cost of using vehicles at 5000 per vehicle, and adjust the unit transportation cost from 2 to 10, while the penalty factor of time windows remains unchanged. Run the program 10 times, and see Table 6 for relevant results when the optimal solution is reached.

Table 6. Impact of adjusting use cost and unit transportation cost on results.

Operation times	*N*	*L*	*P*
1	2	281.25	6524.18
2	2	275.01	6646.81
3	2	272.41	6992.54
4	2	281.49	6021.61
5	2	281.23	6679.20

6	2	268.62	6593.43
7	2	272.84	6257.53
8	2	274.15	6323.09
9	2	285.38	6344.63
10	2	280.63	6025.55
Average	2	277.30	6440.86

It can be seen from Table 6 that after adjusting vehicle use cost and unit transportation cost at the same time, the number of vehicles obtained is significantly reduced compared with Table 3. It can be seen that when vehicle transportation cost and use cost are much higher than other costs, the route distribution scheme with the least number of vehicles will be preferred. The route length in Table 6 is similar to that in Table 3, but the penalty cost of time windows is increased by 53%. By increasing the factors of vehicle use cost and unit freight in the objective function, the proportion of time window penalty cost is greatly reduced. Therefore, the length of the route and the number of vehicles are given priority in the result, so the time window penalty is greatly increased. This shows the effectiveness of the algorithm again in this paper.

Time Window Penalty Cost

Adjust the penalty coefficient of the time windows from $\rho_1 = 1$, $\rho_2 = 0.5$, $\rho_3 = 1.5$, and $\rho_4 = 2$ to $\rho_1 = 5$, $\rho_2 = 2.5$, $\rho_3 = 7.5$, and $\rho_4 = 10$, keep other parameters unchanged, repeat the test 10 times, and see Table 7 for the results when the optimal solution is reached.

Table 7. Impact of adjusting time window penalty costs on results.

Operation times	*N*	*L*
1	8	279.77
2	7	234.70
3	7	291.22
4	7	245.15

5	6	284.25
6	6	271.80
7	8	265.79
8	6	259.40
9	7	289.76
10	7	267.66
Average	6.9	268.95

After the penalty cost coefficient increases 5 times, the calculation results are shown in Table 7. The route length is reduced by 0.3% compared with the result in Table 3, but the number of optimal vehicles has increased from 4 to 7, indicated an increase of 75%, and we can see that the number of vehicles has increased significantly. This is understandable, for the change of the penalty cost coefficient increases the penalty cost in the objective function, so the algorithm will tend to find more solutions for vehicles to serve customers. The increase in the number of vehicles will inevitably lead to a decrease in the average driving route length, which proves the effectiveness of the algorithm again. Therefore, the higher the cost of fines is, the more vehicles are required. This has led logistics companies to prepare more vehicles under strict time window conditions, making the customer's total demand unchanged or even increased.

Electricity Replenishment Cost

The battery capacity is 60 kWh, the charging time needs 60 min, and the power change time is 31 min. The shortest power supply mode is selected. Therefore, when the supplementary power is not more than half, quick charging is preferred. Otherwise, choose the battery replacement strategy. The charging cost per unit time is 1 RMB/min, and the cost of single battery replacement is 30 RMB. Now, the cost of power supply is increasing. The charging cost per unit time is 50 RMB/min, and the cost of single battery replacement is 1500 RMB. The results are shown in Table 8, and the vehicle route, when the optimal solution is reached, is shown in Figure 4.

Table 8. Impact of adjusted electricity supplement costs on results.

Operation times	N	L	
1	4	284.15	3492.10
2	5	273.77	2099.19
3	5	272.84	2372.20
4	5	268.24	2601.64
5	5	244.32	2244.60
6	5	251.66	2644.77
7	4	271.03	3498.75
8	5	266.05	2281.85
9	4	281.59	3563.48
10	5	265.55	2552.16
Average	4.7	267.92	2735.07

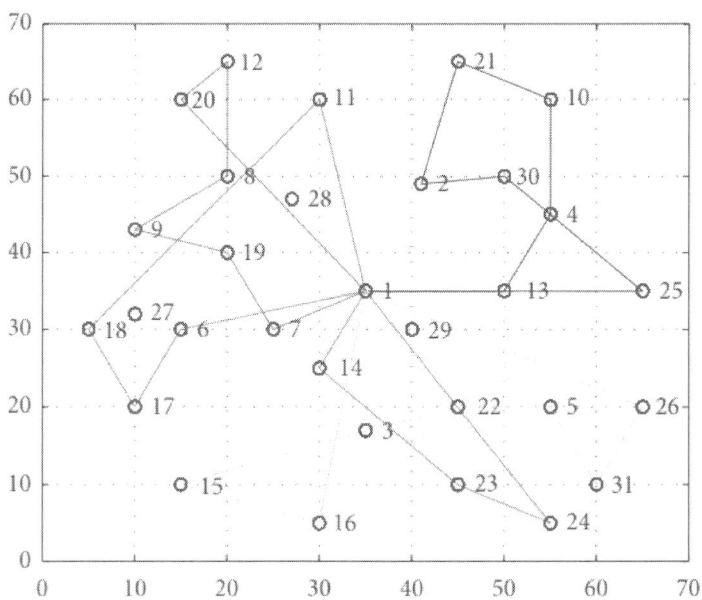

Figure 4. Vehicle route at high electricity replenishment cost.

It can be seen from Table 8 that after the unit charge cost is adjusted from 1 to 50 and the single battery change cost is adjusted from 30 to 1500, after ten runs, compared with Table 3, the number of vehicles increased from 4 to 5 vehicles, indicated an increase of 25%. At the same time, it can be seen that the number of charging stations passed in Figure 3 is 3; however, the number of charging stations passed in Figure 4 is 2. Increased costs have drived companies to choose more vehicles to avoid charging. In addition, compared with Table 3, the route length is reduced by 0.6%, together with the penalty cost is reduced by 10.3%. And the increase in the number of vehicles serving customers leads to a reduction in the possibility of reducing time, penalties, and costs. This proves the effectiveness of the algorithm. Therefore, when the charging facilities are not perfect, together with the power replenishment cost is high, the logistics companies tend to use more vehicles to avoid charging as much as possible.

CONCLUSIONS

At present, most of the literature studies use a charging strategy, and the customer tolerance is not considered while studying the customer time windows. In this paper, the charging strategy of combining two charging ways is adopted, the broken line soft time windows is used considering the customer tolerance, and the MILP model of the EVRPTW is finally established. And through the calculation experiment of the Solomon benchmark example, the cost of affecting the route optimization of electric vehicle is analyzed, the availability as well as effectiveness of data and algorithm are proved, and the management opinions on the practical application of MILP model are put forward.

(1)The higher the use cost and freight of vehicles are, the less the number of vehicles used. When the use cost and freight of vehicles are much higher than other costs, the vehicle route distribution scheme with the least number of vehicles is preferred.

(2)The higher the penalty cost of time windows is, the higher the number of vehicles used for vehicle distribution is. In the case of strict time windows, logistics companies need to prepare more vehicles than the total demand of customers.

(3) Logistics companies tend to use more EVs to avoid charging as much as possible when charging facilities are not perfect and the cost of power supply is high.

The model can help the logistics enterprises provide some suggestions for the route optimization, help to promote the application of EVs in the field of logistics, and improve energy efficiency, energy conservation, and emission reduction. The proposed MILP model is helpful to reduce the logistics cost. However, the traditional shortcomings of VRP still exist in the EVRPTW model. For example, it is still a NP-hard, and it is difficult to solve large-scale problems. The focus of future research may be to developing more effective heuristic algorithm to solve large-scale problems, consider the real-time changes of traffic conditions, and expand the direction of the dynamic vehicle route.

DATA AVAILABILITY

The data used to support the findings of this study are included within the article.

CONFLICTS OF INTEREST

The authors declare that there are no conflicts of interest regarding the publication of this paper.

ACKNOWLEDGMENTS

This research was funded by the National Natural Science Foundation of China, grant no. 71471061, and the Fundamental Research Funds for the Central Universities, grant no. 2017MS171.

REFERENCES

1. G. B. Dantzig and J. H. Ramser, "The truck dispatching problem," *Management Science*, vol. 6, no. 1, pp. 80–91, 1959.
2. Y. Xiao, Q. Zhao, I. Kaku, and Y. Xu, "Development of a fuel consumption optimization model for the capacitated vehicle routing problem," *Computers & Operations Research*, vol. 39, no. 7, pp. 1419–1431, 2012.
3. J. Zhang, Y. Zhao, W. Xue, and J. Li, "Vehicle routing problem with fuel consumption and carbon emission," *International Journal of Production Economics*, vol. 170, pp. 234–242, 2015.
4. N. Norouzi, M. Sadegh-Amalnick, and R. Tavakkoli-Moghaddam, "Modified particle swarm optimization in a time-dependent vehicle routing problem: minimizing fuel consumption," *Optimization Letters*, vol. 11, no. 1, pp. 121–134, 2017.
5. G. Poonthalir and R. Nadarajan, "A fuel efficient green vehicle routing problem with varying speed constraint (F-GVRP)," *Expert Systems with Applications*, vol. 100, pp. 131–144, 2018.
6. J. Wang, S. Yao, J. Sheng, and H. Yang, "Minimizing total carbon emissions in an integrated machine scheduling and vehicle routing problem," *Journal of Cleaner Production*, vol. 229, pp. 1004–1017, 2019.
7. S.-C. Ma, Y. Fan, J.-F. Guo, J.-H. Xu, and J. Zhu, "Analysing online behaviour to determine Chinese consumers' preferences for electric vehicles," *Journal of Cleaner Production*, vol. 229, pp. 244–255, 2019.
8. B. G. Pollet, I. Staffell, and J. L. Shang, "Current status of hybrid, battery and fuel cell electric vehicles: from electrochemistry to market prospects," *Electrochimica Acta*, vol. 84, pp. 235–249, 2012.
9. G. Desaulniers, F. Errico, S. Irnich, and M. Schneider, "Exact algorithms for electric vehicle-routing problems with windows," *Operations Research*, vol. 64, no. 6, pp. 1177–1565, 2016.
10. R. A. Fernandez, "A more realistic approach to electric vehicle contribution to greenhouse gas emissions in the city," *Journal of Cleaner Production*, vol. 172, pp. 949–959, 2018.
11. L. C. Casals, E. Martinez-Laserna, B. A. Garcia, and N. Nieto, "Sustainability analysis of the electric vehicle use in Europe for CO_2 emissions reduction," *Journal of Cleaner Production*, vol. 127, pp. 425–437, 2016.

12. Z. Wu, M. Wang, J. Zheng, X. Sun, M. Zhao, and X. Wang, "Life cycle greenhouse gas emission reduction potential of battery electric vehicle," *Journal of Cleaner Production*, vol. 190, pp. 462–470, 2018. Y. Xiao, X. Zuo, I. Kaku, S. Zhou, and X. Pan, "Development of energy consumption optimization model for the electric vehicle routing problem with time windows," *Journal of Cleaner Production*, vol. 225, pp. 647–663, 2019.
13. S.-N. Yang, W.-S. Cheng, Y.-C. Hsu, C.-H. Gan, and Y.-B. Lin, "Charge scheduling of electric vehicles in highways," *Mathematical and Computer Modelling*, vol. 57, no. 11-12, pp. 2873–2882, 2013.
14. A. Dogan and M. Alci, "Heuristic optimization of EV charging schedule considering battery degradation cost," *Elektronika ir Elektrotechnika*, vol. 24, no. 6, pp. 15–20, 2018.
15. A. Dogan, S. Bahceci, F. Daldaban, and M. Alci, "Optimization of charge/discharge coordination to satisfy network requirements using heuristic algorithms in vehicle-to-grid concept," *Advances in Electrical and Computer Engineering*, vol. 18, no. 1, pp. 121–130, 2018.
16. V. Aravinthan and W. Jewell, "Controlled electric vehicle charging for mitigating impacts on distribution assets," *IEEE Transactions on Smart Grid*, vol. 6, no. 2, pp. 999–1009, 2015.
17. M. Schücking, P. Jochem, W. Fichtner, O. Wollersheim, and K. Stella, "Charging strategies for economic operations of electric vehicles in commercial applications," *Transportation Research Part D: Transport and Environment*, vol. 51, pp. 173–189, 2017.
18. J. D. Adler and P. B. Mirchandani, "Online routing and battery reservations for electric vehicles with swappable batteries," *Transportation Research Part B: Methodological*, vol. 70, pp. 285–302, 2014.
19. J. Yang and H. Sun, "Battery swap station location-routing problem with capacitated electric vehicles," *Computers & Operations Research*, vol. 55, pp. 217–232, 2015.
20. Q. Dai, T. Cai, S. Duan, and F. Zhao, "Stochastic modeling and forecasting of load demand for electric bus battery-swap station," *IEEE Transactions on Power Delivery*, vol. 29, no. 4, pp. 1909–1917, 2014.
21. D. Margaritis, A. Anagnostopoulou, A. Tromaras, and M. Boile, "Electric commercial vehicles: practical perspectives and future research directions," *Research in Transportation Business & Management*, vol. 18, pp. 4–10, 2016.

22. M. M. Solomon and J. Desrosiers, "Survey paper-time window constrained routing and scheduling problems," *Transportation Science*, vol. 22, no. 1, pp. 1–13, 1988.
23. A. G. Qureshi, E. Taniguchi, and T. Yamada, "An exact solution approach for vehicle routing and scheduling problems with soft time windows," *Transportation Research Part E: Logistics and Transportation Review*, vol. 45, no. 6, pp. 960–977, 2009.
24. D. Goeke, "Granular tabu search for the pickup and delivery problem with time windows and electric vehicles," *European Journal of Operational Research*, vol. 278, no. 3, pp. 821–836, 2019.
25. M. Keskin and B. Çatay, "Partial recharge strategies for the electric vehicle routing problem with time windows," *Transportation Research Part C: Emerging Technologies*, vol. 65, pp. 111–127, 2016. M. Schneider, A. Stenger, and D. Goeke, "The electric vehicle-routing problem with time windows and recharging stations," *Transportation Science*, vol. 48, no. 4, pp. 500–520, 2014.
26. G. Hiermann, J. Puchinger, S. Ropke, and R. F. Hartl, "The electric fleet size and mix vehicle routing problem with time windows and recharging stations," *European Journal of Operational Research*, vol. 252, no. 3, pp. 995–1018, 2016.
27. Solomon benchmark problems, http://w.cba.neu.edu/~msolomon/r101.htm.

CHAPTER 8

Integrating Origin-Destination Survey and Stochastic User Equilibrium: A Case Study for Route Relocation

Deo Chimba, Daniel Emaasit and Boniphace Kutela

Department of Civil Engineering, Tennessee State University, Nashville, USA.

ABSTRACT

The paper analyses integrating origin-destination (O-D) survey results with stochastic user equilibrium (SUE) in traffic assignment. The two methods are widely used in transportation planning but their applications have not yet fully integrated. While O-D gives a generalized trip patterns, purpose and characteristics, SUE provides optimal trip distributions using the characteristics found in O-D survey. The paper utilized O-D and SUE in route relocation study for the town of Coamo in Puerto Rico. The O-D survey was used initially in studying possible trip distribution and

Citation: D. Chimba, D. Emaasit and B. Kutela, "Integrating Origin-Destination Survey and Stochastic User Equilibrium: A Case Study for Route Relocation," Journal of Transportation Technologies, Vol. 2 No. 4, 2012, pp. 297-304. doi: 10.4236/jtts.2012.24032.

assignment for the new route. Initial distribution and assignment of traffic to the existing roadway networks and the proposed route were allocated utilizing the O-D survey findings. The SUE was then used to optimize the assignments considering roadway characteristics such as number of lanes, capacity limits, free flow speed, signal spacing density, travel time and gasoline cost. The travel time was optimized through the Bureau of Public Roads (BPR) equation found in 2000 HCM. The optimal trips found from the SUE were then used to propose the final alignment of the new route. Traffic assignment from the SUE was slightly different from those initially assigned using O-D, indicating there was optimization. The assignment on new route was increased by 13.8% from the one assigned using O-D while assignment on the existing link was reduced by 22%.

Keywords:- Route Relocation; Origin-Destination; Stochastic User Equilibrium

INTRODUCTION AND BACKGROUND

Knowledge of the travel patterns for a defined jurisdiction or roadway network is an important aspect in transportation planning [1]. The patterns may include vehicle classifications, trip purposes, travel time, age differentiations, life styles and vehicle occupancy among others. The information can be used for different purposes including traffic impact studies, corridor and area planning, zoning, master plans, traffic projection and traffic assignments. There are different methodologies used in studying traffic patterns, one of them being Origin and Destination (O-D) survey. While some studies use O-D as a stand-alone approach in traffic pattern studies, some have combined the information from O-D report with other supporting traffic data to facilitate the findings and conclusions. For instance, O-D can be used in demand estimation using turning movement counts [2]. There are some studies whose objectives can be fulfilled with the O-D survey information only, but most of them will need supporting data or analysis in order to draw practical conclusions. Taking example of the O-D survey giving the percentage of trips from city A to city B, in rare cases the same survey will give the route assignment used by the interviewed travelers. In this case, while the percentage of trips from city A to city B will be obtained for planning purposes, supporting information related to the route assignments will be needed. In other words, the results from O-D study will need supporting analysis to make final recommendations. The O-D study gives the details of what type of trips in terms of purpose at origin and destination are made by the travelers. Through O-D, one can determine which among home-based, education, shopping, recreational or any other trip purpose are dominant in the area. In case of route relocation study, diversions, road expansion and other similar kind of projects, O-D survey becomes not a stand-alone but a supporting document [3].

Study Objectives

This study therefore combines the O-D survey with the traffic assignment analysis using stochastic user equilibrium in route relocation study. The objective is to evaluate which approaches (between utilizing O-D survey only, SUE only or integrated O-D and SUE) yield the optimal and desired results. The desired results in this case are defined in terms of attributes such as traffic volumes which eventually lead to choosing number of lanes and intersection configurations. The project required the formulation and evaluation of a number of alternatives and combination of alternatives for the selection of a new improved roadway connection between the PR-52 interchange and the central area of the town of Coamo in Puerto Rico, Figure 1. PR-52 is the 4-lane limited access highway running E-W just south of the town connecting to the other major cities to the west, south and east. The existing roadway system PR-153 and PR-545 are the main roadways connecting central Coamo and PR-52. PR-545 connects PR-52 to PR-14, which runs to the downtown area. Currently PR-153 is operating beyond the capacity while PR-545 is substandard. The proposed new alternative route (Route A in Figure 1) is expected to capture some traffic currently using PR-153 and used as a substitute for the trips using PR-545.

The Use of O-D Survey and SUE in Transportation Planning

The use of O-D survey in transportation planning has been applied widely under various scenarios [4-9]. One of the previous studies which align with the objectives of this paper is the one conducted by Yang and Zhou [10] who highlighted that the quality of an estimated O-D matrix depends much on the reliability of the input data, and the number and locations of traffic counting points in the road network. They then addressed the problem of how to determine the optimal number and locations of traffic counting points in a road network for a given prior O-D distribution pattern. Origin and destination surveys can also be used for public transportation studies. Hu et al. [11] proposed the origin-destination of public transportation to help optimize layout of bus stops, reduce the influences of origin-destination of public transportation and improve the traffic efficiency of bus stations. They based their study on the nagel-schreckenberg traffic flow models and used the two-lane aggressive lane-changing rule to examine the influence of the origin-destination of the public transportation on the urban bidirectional four-lane mixed traffic flow. As for the O-D studies, user equilibriums has also been applied in different transportation planning studies [12-14]. For instance, Hazelton [15] indicated that the behavioral foundation of Stochastic User Equilibrium is that each traveler attempts to minimize his or her perceived travel costs, where these costs are composed of a deterministic measured cost and a random term which can be interpreted as perceptual error.

Figure 1: Town of Coamo existing road network.

In his study, he presented Stochastic User Equilibrium as a probability distribution defined by the conditional route selection of each individual given the choices of all other travelers. He also investigated the limiting behavior as the travel demand becomes large. Some of the methodologies and procedures used in these previous studies which utilized O-D and SUE are replicated in this paper.

Study Data and Methodology

To achieve the project objectives, the existing traffic condition for the PR-52, PR-153, PR-545 and PR-14. The 7-days, 24-hours count along these roadways were used to develop existing traffic characteristics. Historical traffic data, population growth, economic trends, employment growths and number of registered vehicles for the past years was used to develop the growth rates for traffic projection to the year 2027. The origin-destination survey was then conducted to determine major areas where the traffic enter and leave the Coamo town. The O-D provided the percentage of trips to and from the town of Coamo from different cities and zones, breakdown of trips by purpose, vehicle occupancy and classifications. Apart from revealing which highways were currently the major collectors and distributors to and from Coamo, the O-D study survey was also used to determine the possible future traffic pattern changes. The projected ADT (annual daily traffic) was portioned along PR-545 and PR-153 and the proposed new roadway based on the O-D survey results. As shown in Figure 1, Route A was proposed as an alternative for the traffic currently using PR-545 and those, which will be diverted from PR-153. Stochastic User Equilibrium trip assignment was used to distribute and balance the trips to and from Coamo along PR-153 and the proposed Route-A. Stochastic User Equilibrium (SUE) was used due to its underlying principle, which considers a population of drivers with homogeneous characteristics and perceive the same set of network costs except for the variation allowed by the stochastic choice model considered. With the fixed origin-destination within the known roadway networks, stochastic equilibrium assumes that the drivers will react to changes in network conditions as a result of change in route characteristics stochastically.

FORMULATION OF ORIGIN-DESTINATION MATRIX AND FINDINGS

As introduced earlier, Town of Coamo was the epicenter of this study; hence, all trips surveyed were coded with respect to routes to and from this town. Let i denote town of Coamo, j denote the external cities, and n number of cities or separate routes to those cities, then

T_{ii}= Internal Trips within Coamo

T_{ij}=Trips from Coamo to other cities

T_{ji}= Trips from other cities to Coamo

The proportion trip among all surveyed vehicles is given by;

Proportion of internal trips to total survey, P_{ii};

$$P_{ii} = \frac{T_{ii}}{\sum_{i=1}^{1}\sum_{j=1}^{n} T_{ij} + \sum_{j=1}^{n}\sum_{i=1}^{1} T_{ji} + \sum_{i=1}^{1} T_{ii}} \tag{1}$$

Proportion of outbound external trips to total survey,P_{ij};

$$P_{ij} = \frac{T_{ij}}{\sum_{i=1}^{1}\sum_{j=1}^{n} T_{ij} + \sum_{j=1}^{n}\sum_{i=1}^{1} T_{ji} + \sum_{i=1}^{1} T_{ii}} \tag{2}$$

Proportion of inbound external trips to total survey, P_{ji};

$$P_{ji} = \frac{T_{ji}}{\sum_{i=1}^{1}\sum_{j=1}^{n} T_{ij} + \sum_{j=1}^{n}\sum_{i=1}^{1} T_{ji} + \sum_{i=1}^{1} T_{ii}} \tag{3}$$

Both trip proportions P_{ii},P_{ij} and P_{ji} are utilized in determination of the trips along the existing and proposed routs.

The study was there re i itiate y preparing an O-D questionnaire for designated locations along PR-52, PR153 and PR-14. Different considerations were taken into account for effective O-D results including avoidance of uncertainties. One of the uncertainties avoided was to choose the interview locations which could have brought conflicting responses. According to some previous studies, the O-D location should consider traffic flow coverage and minimize the expected uncertainties [16,17]. Interviews were conducted on March 15, 2007 on PR 52 at the entrance and exit ramps with PR 153 and on PR 14 at the intersection with PR 153. The findings from O-D survey were summarized as

- PR 153 was the main highway from PR 52, used by motorist to and from Coamo,
- The intersection of PR 14/PR 153 was the major intersection used by motorist to/from Coamo,
- The PR 14 link from PR153 to downtown was the main receiver and deliverer of traffic from/to PR 153,

- Home, work and personal based trips were the major trip purposes to and from Coamo.

Figure 2 shows some of the results found from the O-D survey. For all of the trips to and from the town, 22% originate or ended west of Coamo, 13% south of Coamo and 13% east of Coamo.

PROPOSED ALTERNATIVE ROUTE

The 2027 projected ADT(Annual Daily Traffic) was 33,800 vpd on PR-153 and 6200 vpd on PR-545. These ADT were taken as external trips. From the Origin Destination (O-D) matrix developed, it was found that of all external traffic to Coamo from PR-153, 46% originated from the west, 28% from the south and 26% from the east. This distribution led to traffic assignment with respect to proposed route with 4-lane section and PR-153, which remained as a 2-lane section.

Table 1 elaborates the trip assignment developed based on origin destination survey analysis. Furthermore, from the survey it was observed that, out of 26% of the traffic going to Coamo from the east using PR 153, 11% will be diverted to proposed Route A while 15% will continue using PR-153 to Coamo. For the 28% of the traffic originating from the South to Coamo, 21% will continue using PR-153 while 7% will be diverted to Route A. Traffic from the west which make up 46% of external trips, 37% was found that utilize Route A and 9% PR-153. All traffic currently using PR-545 to downtown Coamo were assumed to be diverted to Route A. After all of the analysis using the O-D survey, PR 153 was found that will remain with 15,400 vehicles per day vpd which according to HCM is Level of Service (LOS) D,that was a reduction of 18,400 vpd (54%) from originally 33,800 vpd projected. The proposed Route A was expected to receive 18,400 vpd diverted from PR-153 and 6200 currently using PR-545, a total of 24,600 vpd (LOS B) by the year 2027. These O-D survey results were implemented in the Stochastic User Equilibrium analysis to find the final traffic balance based on the characteristics of existing PR-153 and the proposed Route A.

THEORY OF USER EQUILIBRIUM (UE) STOCHASTIC USER EQUILIBRIUM (SUE)

User Equilibrium can be derived from Wardrop's firstprinciple which states that, under equilibrium conditions traffic arranges itself in congested networks in such a way that no individual trip maker can reduce his travel cost by switching routes or all used routes between an origin and destination pair have equal and minimum costs while all unused routes have greater or equal costs [14,18]. A SUE condition is derived from the user equilibrium (UE) assumption which can be written as a given O-D pair as:

$f_i\,(c_i - u) = 0$ for all i

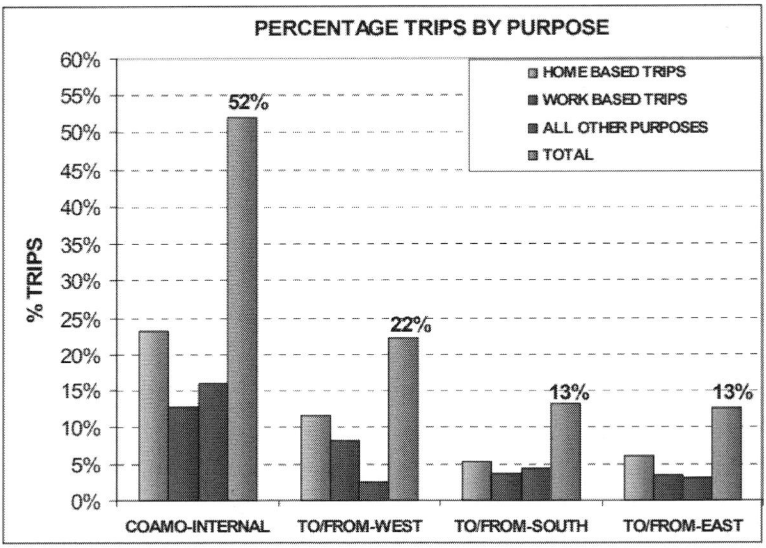

Figure 2: Percentage trips by location and purpose.

Table 1: TripAssignment along PR-153 and new route using O-D survey results.

	To/From Coamo		Through PR-153	Through New-Route
	%	ADT	ADT	ADT
To/From East	26%	8900	5300	3600
To/From South	28%	9300	7000	2300
To/From West	46%	15,600	3100	12,500
Pr-545				6200
Total	100%	33,800	15,400	24,100

$(c_i - u) > 0$ for all i

$\sum f_i = q$ and $f_i > 0$

where,

f_i is the flow on Route i
c_i is the travel cost on Route i
u is the minimum cost
and q is the combined flow.

Solution to the above conditionare obtained by solving an equivalent optimization program

$$\text{Min } z = \sum_i \int_0^{v_i} t_i(v)\,dv$$

The assumptions regarding UE is include the road user having perfect knowledge of the rout cost, travel time on a given link is a function of the flow on that link only and travel time functions being positive and increasing. The SUE extends the UE concept by considering the following assumptions and implications:

- Traveler has perfect knowledge of the network and travel cost,
- Traveler will choose the perceived least cost route,
- Different traveler perceives differently, thus introducing stochasticity.

The SUE route choice can then be analyzed using logit model, which treats all alternatives statistically independent.The formulation of SUE can be summarized asin Equation (4) [19]:

$$\text{Min } z(v) = -\sum q\left[\min c(v)\right] + \sum v_i t_i(v_i) - \sum_i \int_0^{v_i} t_i(v)\,dv \qquad (4)$$

where,

- The first term represent expected minimum route cost for all links,
- The second term represents expected total system travel time,
- The third term represents UE formulation,
- $z(v)$ = function for a given traffic volume v,
- $c(v)$ = route capacity for given traffic volume v,
- v_i = traffic volume for route i,
- t_i = travel time for route i,
- dv = change in traffic volume.

SUE utilizes multinomial models in stochastic assignment. Multinomial models use utilities which areindependent and identically distributed mainly with a Gumbel distribution. They response homogeneity across individuals and there is error variance-covariance homogeneity across individuals.

APPLICATION OF SUE COAMO ROUTE RELOCATION STUDY

ADT volumes along PR-153 and Route A shown in figure 3. The travel times were formulated using posted speed limits, link lengths and s density along each segment utilizing bureau of Public ads (BPR) [19].

$$t = t_0 \left[1 + a \left(\frac{V}{C} \right)^b \right] \tag{5}$$

$$t_0 = \frac{L}{S_0} \tag{6}$$

Combining Equations (5) and (6) gives;

$$t = \frac{L}{S_0} \left[1 + a \left(\frac{V}{C} \right)^b \right] \tag{7}$$

where,

t = link average travel time (hr),
t_0 = free flow link traversal time (hr),
L = link length (mi),
S_0 = link Free-Flow Speed (FFS),
V = link traffic volume in ADT,
C = link ADT capacity,

a and b = The BPR function parameters are from Exhibit C30-1 and C30-2 of 2000 HCM [20].

The initial traffic flow for each link V1 and V2 were based on the O-D survey results discussed earlier. As shown in Figure 3,Route A is approximately 5 miles by length while PR-153 isapproximately5.5 miles. The first section of Route A from PR-52 is an uninterrupted flow while the second section which connects to Pr-14 to downtown is an arterial with two signalized intersection. Therefore signal density for Route A was taken as 0.5 per mile. PR-153 is an arterial with 4 signalized intersection making signal density to be 0.73 per mile. Posted speed limit for Route A is expected to be 55 mph while for Pr-153 is 40 mph. Using exhibit C30-2 in 2000 HCM [18], "a" and "b" values for each link were estimated as; a = 0.31 and b = 3.64 for Route A and a = 0.38 and b = 5.0 for PR-153

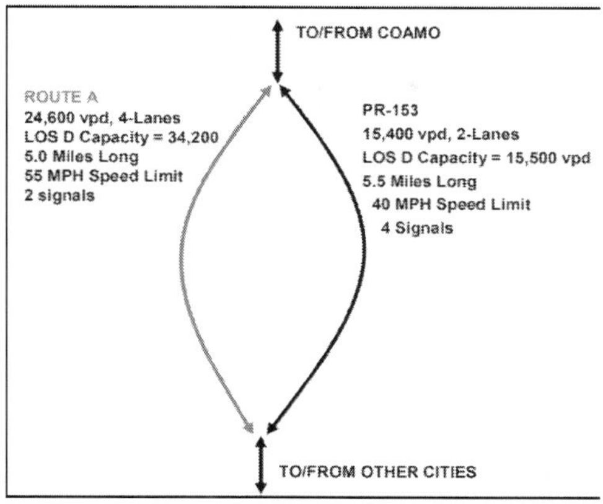

Figure 3: PR-153 and proposed Route A characteristics.

By inserting these values in Equation (7) above yields the following formulations which are also illustrated in Figure 4:

Route A;

$$t_1 = 0.091\left[1 + 0.31\left(\frac{V_1}{34200}\right)^{3.64}\right]$$

with Initial V_1 =24,600 vpd

PR-153;

$$t_2 = 0.138\left[1 + 0.38\left(\frac{V_2}{15500}\right)^{5.0}\right]$$

with Initial V_2 =15,400 vpd.

SUE Logit Utility Function

The generalized cost functions used for each route are set based on theory developed by Kato et al. [20] using logitutility function in Equations (8) and (9).

The logit utility function which is formulated as;

$$GC_1 = \theta_1 * t_1 + \theta_2 * \text{Cost} + \theta_3 * \left(\frac{v_1}{C_1}\right)^2 * t_1 \tag{8}$$

$$GC_2 = \theta_1 * t_2 + \theta_2 * \text{Cost} + \theta_3 * \left(\frac{v_2}{C_2}\right)^2 * t_2 \tag{9}$$

where

GC_1 = general Cost for Route 1,
GC_2 = general Cost for Route 2,
t_1 = travel time along Route A,
t_2 = travel time along PR-153,
θ_1 = In vehicle time constant coefficient, –0.094,
θ_2 = Total Cost constant coefficient, –0.002,
θ_3= Congestion Index constant Coefficient, –0.009,
Cost = in this study refers to gasoline consumption

cost,
v_i =traffic volume for route i,
C_i =capacity for route i,

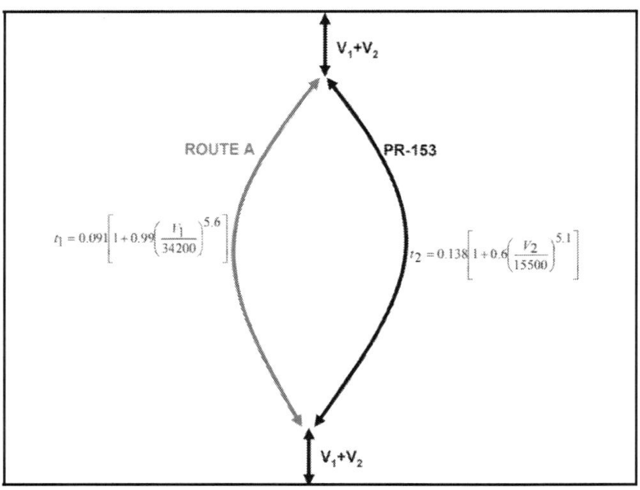

Figure 4: Link travel time and flow formulation along Route A and PR-153.

$$\left(\frac{v_i}{C_i}\right)^2 * t_i = \text{Congestion Index}$$

The value of coefficients , θ 2 and θ1 θ3 were adopted from previous researches by Kato et al.[6].

Link Cost Due to Gasoline Expenses

The "cost" which appears in the utility functions (8) and (9) refers to the situation where some travelers may choose the route based on the gasoline consumption. In this study, it is a cost with respect to gasoline consumption. The study compared Route A and PR-153 using the link length with respect to average fuel efficiency and price per gallon. The average fuel efficiency is taken 20(miles/ gallon, while gasoline price was taken as $4.15/ gallon (the gas price when the study was conducted).)

Therefore;

For Route A Link: length = 5.00 miles,

Gasoline cost = 4.15 ×5. 00/20 = $1.04 per link complete trip.

For PR-153 Link: length = 5.50 miles,

Gasoline cost = 4.15 × 5.50/20 = $1.14 per link complete trip,
Probability of the traveler choosing Route A over PR-153 is

$$P_A = \frac{e^{-GC_1}}{e^{-GC_1} + e^{-GC_2}} \tag{10}$$

Probability of the traveler choosing PR-153 over Route A is

$$P_{153} = \frac{e^{-GC_2}}{e^{-GC_1} + e^{-GC_2}} \tag{11}$$

To determine the optimal balance volumes which take into account the described road characteristics and costs, the following conditions must be fulfilled;

$$\begin{aligned} &(V_1+V_2)*\mathrm{P_A} - V_1 \approx 0.0 \\ &(V_1+V_2)*\mathrm{P_{153}} - V_2 \approx 0.0 \end{aligned} \tag{12}$$

The outputs from Equation (12) gives the optimal traffic volume assignment and travel time by taking into account all variable and utility costs. The equation was programmed in Matlab for optimization and yielded;

GC_1	GC_2	V_1 (vph)	V_2 (vph)
−0.95	−0.405	28,000	12,000

Therefore, as a result of Stochastic User Equilibrium (SUE), the originally traffic assigned to Route A and PR-153 using O-D study percentage were optimized to the final assignments as follows:

	Traffic Assignment		
	From O-D Survey	From SUE	Deviation (%)
Route A	24,600	28,000	13.8%
PR-153	15,400	12,000	22%

CONCLUSIONS

The paper integrated the findings from origin-destination (O-D) survey and stochastic user equilibrium approach in route relocation study. The new route is proposed in the town of Coamo, Puerto Rico. The proposed new route is expected to capture diverted traffic currently using existing routes. The O-D provided the existing traffic pattern and characteristics with respect to trip purposes, and the percentages for internal and external trips. Percentages of trips from major cities surrounding the town were determined from the O-D survey. The initial trip assignments on the new and existing routes were based on the findings from O-D survey.

Stochastic User Equilibrium (SUE) was then applied to the network using the initial traffic volumes assigned from O-D survey findings. Included in the SUE was travel time on the links which is controlled by the free flow speed, maximum capacity, length of the link and signal spacing density. Apart from link travel time, the utility function of SUE contained other link measures of effectiveness such as time spent in the vehicle (in-vehicle time coefficient), congestion index and cost due to gasoline consumption (cost coefficient). The gasoline cost considered vehicle fuel efficient of 20 miles/gallon, gasoline price of $4.15/gallon and length of the link. All these link characteristics were used to optimize the traveler choice of the route.

Traffic assignment from the SUE was slightly different from those initially assigned using O-D, indicating there was optimization. The assignment on new route was increased by 13.8% from the one assigned using O-D while assignment on the existing link was reduced by 22%. The final optimized volumes were within capacity limits for each link indicating successful optimization. The final traffic assignment from SUE was used in the new route design. The findings from this study showed the possible benefit of integrating O-D with other trip assignment optimization approaches. By integrating O-D survey with optimization algorithms like UE or SUE can result in a well balanced links which take into account all possible constrains.

REFERENCES

1. S. Jiang, B. Bai, G. Hong and J. O'Leary, "Understanding Travel Expenditure Patterns: A Study of Japanese Pleasure Travelers to the United States by Income Level," Journal of Tourism Management. Vol. 25, No. 3, 2004, pp. 331-341. doi:10.1016/S0261-5177(03)00141-9
2. H. Alibabai and H. S. Mahmassani, "Dynamic OriginDestination Demand Estimation Using Turning Movement Counts," The 87th Annual Meeting of the Transportation Research Board Proceedings, Washington DC., 2008.
3. H. Spiess, "A Maximum Likelihood Model for Estimating Origin-Destination Matrices," Transportation Research, Vol. 21, No. 5, 1987, pp. 395-412. doi:10.1016/0191-2615(87)90037-3
4. M. Hazelton, "Some Comments on Origin-Destination Matrix Estimation," Transportation Research Part A: Policy and Practice, Vol. 37, No. 10, 2003, pp. 811-822. doi:10.1016/S0965-8564(03)00044-2
5. S. Han, "Dynamic Traffic Modeling and Dynamic Stochastic User Equilibrium Assignment for General Road Networks," Transportation Research Part B: Methodological, Vol. 37, No. 3, 2003, pp. 225-249. doi:10.1016/S0191-2615(02)00009-7
6. M. Hazelton, "Estimation of Origin-Destination Matrices from Link Flows on Uncongested Networks," Transportation Research, Vol. 34, No. 7, 2000, pp. 549-566. doi:10.1016/S0191-2615(99)00037-5
7. K. Jeornsten and S. Wallace, "Overcoming the (Apparent) Problem of Inconsistency in Origin-Destination Matrix Estimations," Transportation Science, Vol. 27, No. 4, 1993, pp. 374-380. doi:10.1287/trsc.27.4.374
8. H. Lo., N. Zhang and W. Lam, "Estimation of an Origin-Destination Matrix with Random Link Choice Proportions: A Statistical Approach," Transportation Research, Vol. 30, No. 4, 1996, pp. 309-324. doi:10.1016/0191-2615(95)00036-4
9. M. Hazelton and M. Gordon, "Estimation of Origin-Destination Trip Matrices from Link Counts," Proceedings of the 2002 European Transport Conference, London, 2002.
10. H. Yang and J. Zhou, "Optimal Traffic Counting Locations for Origin-Destination Matrix Estimation," Transportation Research Part B: Methodological, Vol. 32, No. 2, 1998, pp. 109-126. doi:10.1016/S0191-2615(97)00016-7

11. J. Hu, L. Yang, L. Kong and Y. Yang, "Urban Mixed Traffic Flow Considering the Influence by Origin-Destination of Public Transportation," Journal of Transportation Systems Engineering and Information Technology, Vol. 11, No. 1, 2011, pp. 102-107. doi:10.1016/S1570-6672(10)60107-9

12. S. Clark and P. Watling, "Sensitivity Analysis of the Probit-Based Stochastic User Equilibrium Assignment Model," Transportation Research Part B: Methodological, Vol. 36, No. 7, 2002, pp 617-635. doi:10.1016/S0191-2615(01)00021-2

13. Y. Lim and B. Heydecker, "Dynamic Departure Time and Stochastic User Equilibrium Assignment," Transportation Research Part B: Methodological, Vol. 39, No. 2, 2005, pp. 97-118. doi:10.1016/j.trb.2003.08.003

14. A. Nielsen, D. Frederiksen and N. Simonsen, "Stochastic User Equilibrium Traffic Assignment with Turn-Delays in Intersections," International Transactions in Operational Research, Vol. 5, No. 6, 1998, pp. 555-568. doi:10.1111/j.1475-3995.1998.tb00137.x

15. M. Hazelton, "Some Remarks on Stochastic User Equilibrium," Transportation Research Part B: Methodological, Vol. 32, No. 2, 1998, pp. 101-108. doi:10.1016/S0191-2615(97)00015-5

16. X. Fei and H. S. Mahmassani, "A Two-Stage Stochastic Model for the Sensor Location Problem in a Large-Scale Network," The 87th Annual Meeting of the Transportation Research Board Proceedings, Washington DC, 2008.

17. K. N. Paul and H. K. L. William, "Evaluation of Count Location Selection Methods for Estimation of O-D Matrices," Journal of Transportation Engineering, Vol. 124, No. 4, 1998, pp. 376-383. doi:10.1061/(ASCE)0733-947X(1998)124:4(376)

18. Highway Capacity Manual, "Transportation Research Board," National Research Council, Washington DC, 2000.

19. M. G. H. Bell and H. W. K. Lam, "Ploss, G. and Inaudi, D. Stochastic User Equilibrium Assignment and Iterative Balancing," Proceedings of the 12th International Symposium on Transportation and Traffic Theory, Berkeley, 1993.

20. S. Kato, Y. Kaneko and M. Inoue, "Empirical Analysis on Relationship Between Types of Travel Demand Techniques and Estimated User's Benefit Stemming from Transportation Investment," Journal of the Eastern Asia Society for Transportation Studies, Vol. 6, 2005, pp. 3937-3947.

CHAPTER 9

Research of UAV Flight Planning Parameters

Jing He[1], Yongshu Li[1] and Keke Zhang[2]

[1]Southwest Jiaotong University, GIS Engineering Center, Chengdu, China; [2]Engineering Design Co., Ltd., CNPC, Chengdu, China

ABSTRACT

UAV remote sensing as a digital aerial photography, not only has some basic photogrammetry features, but also has some other features. In this paper, aim at the characteristics of UAV remote sensing, begin with image data acquisition, the various parameter setting in the route planning were introduced, some of the principle was analyzed, the design of control points was described, and some of the considerations when laid control points were summarized.

Keywords: Unmanned Aerial Vehicles Image; Flight Planning

Citation: J. He, Y. Li and K. Zhang, "Research of UAV Flight Planning Parameters," Positioning, Vol. 3 No. 4, 2012, pp. 43-45. doi: 10.4236/pos.2012.34006.

INTRODUCTION

Unmanned Aerial Vehicles remote sensing technology is an important means of spatial data access, which has advantages of long endurance, real-time image transmission, high-risk areas detection, the low cost, the flexible etc, and complement the technology of satellite remote sensing and manned aerial vehicles remote sensing. Unmanned Aerial Vehicles remote sensing technology countries has been widely used [1,2] in foreign. However, low-altitude digital aerial photography is defined by "Specifications for low-altitude digital aerial photography" issued and implemented by State Bureau of Surveying and Mapping [3] in 2010:

1) Use the light small aircraft, don't rely on the airport take off and landing;
2) The low relative flight height, commonly below 2000 m;
3) Small format digital camera with above 20 million pixels as the sensor for aerial photogrammetry.

The widely use of domestic UAV aerial photography system marks that the UAV aerial photography system comes to a new age in our country.

However, with the aerial photography system equipment to production unit, it has some phenomenon of field data acquisition quality not qualified. Consequently it leads to the increased difficulties of processing and even is unable to deal with the problem. This paper introduces the low number of aerial photography UAV air route planning, and analysis some specific parameter settings in detail.

UAV IMAGES ACQUISITION

General Situation of Test Area

Before the flight, it is necessary to collect local meteorological information and the weather conditions. Generally it is good to choose the sunny day without wind. According to the topographic map data and the remote sensing image map information, we learn the terrain information of the test area. If there are mountains, it needs to assess whether the height of the mountains influences flight safety. If there is snow, it needs to access whether the snow influences image quality.

The experimental area located in the hilly region is about 20 km^2, and is located in urban and rural area. The height of the highest hills in the test area is about 100 m and has no snow and the elevation is between 490 m and 590 m.

Flight Parameter Analysis and Settings

The flight parameter including ground sampled distance (GSD), longitudinal overlap degree (p_x), side overlap degree (q_y) and some line design related parameters.

UAV fly in the air so fast that adjacent image exposure interval time is short. To ensure that the focal length of the image and other parameters are the same, the camera of UAV takes pictures in the way of the fixed-focus. Focus mode sets in infinity to ensure that it can get the deepest of field range, make the object in the farthest distance and focus distance clear, regardless of short comings of scope of the depth of field.

Ground resolution is the minimum distance to distinguish the two goals in the image, but it doesn't mean the minimal size to recognize the ground object in the image [4]. For example: a goal with the size of 0.3, in the ground for 0.3 m GSD, is just a pixel, no matter how many times the image zooms in, it is still just a pixel. So, it needs to have several pixels to identify a target in the image. Generally, the smallest size of recognizing target from the image should be the 5 - 10 times of GSD. GSD depends on the focal length, the flight height and pixel size. The pixel size can be written as the following equations (1)-(2):

$$\mathrm{GSD} = \frac{H}{f}\mu \tag{1}$$

$$\mu = \frac{W}{S_W} = \frac{H}{S_H} \tag{2}$$

where H is flight height (m), f is focal length (mm), GSD is ground sampled distance (m), μ is pixel size (μm), W is the width of CCD (mm), H is the height of CDD(mm), S_W is the number of pixels for W, S_H is the number of pixels for H.

Due to the use of single spell camera, it can satisfy the largest mapping precision scale for 1:2000 [5]. Figure 1 analysis the geometric relations between the flight height and ground width, where W is ground width, H is flight height. If the camera angle of view is 53°tan26.5 $= \frac{AO}{SO} = \frac{AO}{H} = \frac{1}{2}$, the flight height H is the ground width of the mages covered. When H increases, the ground width of the mages covered is getting bigger.

Usually, the focal length f, μ of the digital camera is fixed. According to equation (1), when H increases, GSD decreases, and from the analysis of the figure 1, it is known that the single image coverage area also increases. Table 1 gives ground coverage area with single digital camera image and the corresponding ground resolution with different flight height for navigation of the view angle of 53°.

From Table 1 we can see, in the area of certain high, the higher the flight is, the less image pictures we can get. Therefore, after determining the mapping scale, based onthe principle of the shortened mapping cycle, reducing cost, improving the comprehensive effective of surveying and mapping ,we choose what ground resolution we need. In principle at the beginning, the resolution should not be too high, once we choose the resolution for the ground; we can determine the flight height. Table 2 gives the requirement between the mapping scale and the GSD.

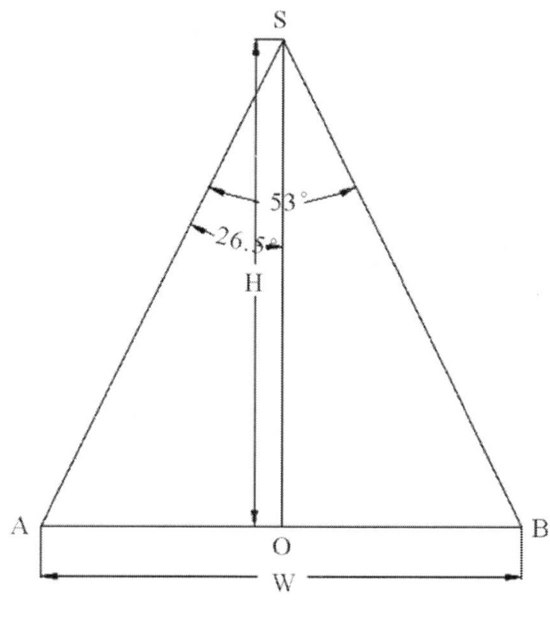

Figure 1: The geometric relations between the flight height and ground width.

Table 1: Coverage area with single digital camera image

Flight height (m)	100	200	300	400	500	600
Picture format width (m)	100	200	300	400	500	600
Picture format length (m)	150	300	450	600	750	900
Coverage area (km^2)	0.015	0.06	0.135	0.24	0.375	0.54
GSD (m)	0.027	0.053	0.08	0.107	0.133	0.16

After we determine the GSD and the flight height, we need to set longitudinal overlap degree and side overlap degree, table 3 gives the requirements for the overlap degrees. Because the flight attitude in the air is not stable, the preset overlap degree should be larger. At the same time overlap degree is affected by the weather conditions. The default overlapping degree keeps better effect without wind while the default overlapping degree would be not very good because of the dual action of wind and the shaking and the default overlapping degree will need to raise some. In the experiment the longitudinal overlap degree is 65%, side overlap degree 30%.

The above parameters settings did not consider some special terrain of the surveyed area, such as mountains, the lowlands. Whether the image for these places meets the requirements should do some calculation.

In the terrain of the test, overlap degree of the highest point and the GSD can be calculated as the following equation.

1) Calculation datum elevation

$$h_{base} = \frac{h_{max} + h_{min}}{2} \tag{3}$$

where h_{max} is the elevation of the highest point, h_{min} is the elevation of the lowest point.

2) Calculation the smallest GSD

$$GSD_{min} = \frac{h_{base} - h_{min} + H}{f} \times \mu \tag{4}$$

where H is the related flight height in design, according to Equation (1) the smallest GSD is the lowest point.

3) Calculation the smallest overlap degree The smallest overlap degree includes longitudinal overlap degree, side overlap degree. It is expressed as the following equation.

Table 2: GSD

The mapping scale	GSD	The mapping scale	GSD	The mapping scale	GSD
1:500	≤0.05	1:1000	0.08 - 0.1	1:2000	0.15 - 0.2

Table 3: Requirement of overlap degree

Longitudinal overlap degree	**60% - 80%, not be less than the minimum 53%**	**Side overlap degree**	**15% - 60%, not be less than the minimum 8%**

$$p_{x_{\min}} = \frac{p_x - \frac{h_{\max} - h_{\text{base}}}{H}}{1 - \frac{h_{\max} - h_{\text{base}}}{H}}$$

$$q_{y_{\min}} = \frac{q_y - \frac{h_{\max} - h_{\text{base}}}{H}}{1 - \frac{h_{\max} - h_{\text{base}}}{H}} \quad (5)$$

where p_x is the default longitudinal overlap degree, q_y is the default side overlap degree. $p_{x_{\min}}$ is the smallest default longitudinal overlap degree. $q_{y_{\min}}$ is the smallest default side overlap degree. According to Equation (5) we can learn that the smallest overlap degree is the highest point in survey area.

When the smallest GSD can not meet the requirement for mapping, or the overlapping degree of the highest point is less than the requirement, GSD should be changed, and the relative flight light, GSD of the lowest point, the overlap degree of the highest point should be calculated once more till it meets the requirements.

Field Control Points Measurement

The outside directional of an image or stereo can be determined corresponding relation between image and the object space through the geometry transform. To confirm the transform parameter, we need to measure some control point in the image. In order to recognize the control points, the control points logo should be designed carefully, should have the appropriate size and shape, and the contrast between the foreground and background should be high.

In order to ensure that the location of the control points in unmanned aerial vehicle (UAV) flight and measurement process is not moved, we need to mark the location of the four corners for the control points. After flight, when we use RTK to measure the control points. We should check the previous mark and see whether the mark points have been moved. If it has been moved, we should mark notes for remarks in the field measurement. All images control points are measured by RTK. Measurement precision and requirements for the control points is carried out according to "Specifications for aerophotogrammetric office operation of 1:500 1:1000 1:2000 topographic maps" GB/T7930 2008 [6]. All of the control points need to be numbered when measured, and take pictures in the field in order to easily search for prick point in office operation.

In experiment, we come to a conclusion that the marks should not be put on the roadside and the place with too many people, because these control points can be damaged, and even moved easily. We should put it to a relatively empty and hidden place so as to easily find the control points in the UAV image, and also effectively guarantee the safety of the control points marks.

CONCLUSION

Jing He, Yongshu Li, Keke Zhang With development of the social informatization level ceaselessly, requirements for the speed of obtaining the information become higher and higher. UAVS low-altitude digital aerial photography with the characteristics of its mobile, flexible and efficient just meet the needs of society. In the background that the country promotes the low-altitude photogrammetry UAV development, this paper begins with the data acquisition for low-altitude photography measurements, analyzes flight route planning parameters deeply, optimizes some related parameters combined with requirement the latest promulgated and implemented by the low-altitude digital aerial photography measurement standard, and summarizes some practical attention in operation.

REFERENCES

1. P. van Blyenburgh, "UAVs: An Overview," Air & Space Europe, Vol. 1, No. 5-6, 1999, pp. 43-47.
2. J. Sun, Z.-J. Lin and H.-X. Cui, "Low-Altitude UAV Remote Sensing Monitor System," Remote Sensing Information, Vol. 1, No. 1, 2003, pp. 49-50.
3. State Bureau of Surveying and Mapping, "Specifications for Low-Altitude Digital Aerial Photography," Surveying and Mapping Press, Beijing, 2010.
4. D.-Z. Gui, "Study on Construction of 3D Building Based on Wide-Angle and Combine Camera Images from UAV," China University of Mining and Technology, Xuzhou and Beijing, 2010.
5. Z.-J. Lin, "UAV Borne Low Altitude Photogrammetry System," Science of Surveying and Mapping, Vol. 36, No. 1, 2011, pp. 5-9.
6. China National Standardization Management Committee, "Specifications for Aerophotogrammetric Office Operation of 1:500 1:1000 1:2000 Topographic Maps," China Standards Press, Beijing, 2008.

Section 3: Operational Research Applications in Logistics

CHAPTER 10

Optimization of Urban Rail Transportation in Emerging Countries Using Operational Research Techniques

M. C. Agarana, T. A. Anake and H. I. Okagbue

Department of Mathematics, College of Science and Technology, Covenant University, Ota, Nigeria.

ABSTRACT

Rail systems are gradually becoming the most desirable form of transit infrastructure around the world, partly because they are becoming more environmentally friendly compared with airplanes and automobiles. This paper examines the place of emerging countries in this move of implementing modern rail system that will eventually enhance the realization of a low-carbon society. Network model, transportation model and linear programming algorithms are used to model the present urban rail transport system in Nigeria, as an emerging country, in order to optimize it. Operational research methods, including simplex method and MODI, with the aids of computer software (excel solver and LIP solver) were adopted

Citation: Agarana, M. , Anake, T. and Okagbue, H. (2016) Optimization of Urban Rail Transportation in Emerging Countries Using Operational Research Techniques. Applied Mathematics, 7, 1116-1123. doi: 10.4236/am.2016.710099.

to solve the resulting models. The results showed that optimization of rail transport system will not only reduce carbon emission but also bring about economic development which is required for the eradication of prevalent poverty in these emerging countries.

Keywords:- Rail System, Emerging Countries, Operational Research, Network Model, Transportation Model, Linear Programming Algorithm

INTRODUCTION

Emerging countries are developing countries which have achieved individual capacity and are on the path to becoming industrialized countries [1] . The list of emerging countries according to the World Bank index (May 2010) [2] includes: Afghanistan, Albania, Algeria, American Samoa, Angola, Argentina, Armenia, Azerbaijan, Bangladesh, Belarus, Belize, Benin, Bhutan, Bolivia, Bosnia and Herzegovina, Botswana, Brazil, Bulgaria, Burkina Faso, Burundi, Cambodia, Cameroon, Cape Verde, Central African Republic, Chad, Chile, China, Colombia, Comoros, Congo, Dem. Rep., Congo, Rep., Costa Rica, Côte d'Ivoire, Croatia, Cuba, Djibouti, Dominica, Dominican Republic, Ecuador, Egypt, Arab Rep., El Salvador, Eritrea, Ethiopia, Fiji, Gabon, The Gambia, Georgia, Ghana, Grenada, Guatemala, Guinea, Guinea-Bissau, Guyana, Haiti, Honduras, India, Indonesia, Iran, Islam, Rep., Iraq, Jamaica, Jordan, Kazakhstan, Kenya, Kiribati, Korea, Dem. Rep. , Kyrgyz Republic, Lao PDR, Latvia, Lebanon, Lesotho, Liberia, Libya, Lithuania, Macedonia, FYR, Madagascar, Malawi, Malaysia, Mali, Marshall Islands, Mauritania, Mauritius, Mayotte, Mexico, Micronesia, Fed. Sts., Moldova, Mongolia, Montenegro, Morocco, Mozambique, Myanmar, Namibia, Nepal, Nicaragua, Niger, Nigeria, Pakistan, Palau, Panama, Papua New Guinea, Paraguay, Peru, Philippines, Poland, Romania, Russian Federation, Rwanda, Samoa, São Tomé and Principe, Senegal, Serbia, Seychelles, Sierra Leone, Solomon Islands, Somalia, South Africa, Sri Lanka, St. Kitts and Nevis, St Lucia, St. Vincent and the Grenadines, Sudan, Suriname, Swaziland, Syrian Arab Rep., Tajikistan, Tanzania, Thailand, Timor-Leste, Togo, Tonga, Tunisia, Turkey, Turkmenistan, Uganda, Ukraine, Uruguay, Uzbekistan, Vanuatu, Venezuela, RB, Vietnam, West Bank and Gaza, Yemen, Rep, Zambia, Zimbabwe, and few other.

A lot of these emerging countries now have urban rail transport projects. This is largely due to the increase in commercial transportation activities brought by development. Nigeria's largest city, Lagos, for instance is constructing a light rail system under a Public Private Partnership (PPP). The first two lines of the urban rail transport project are estimated to cost $1.4 billion [2] . The development objective of the intermodal and Rail Development Project for Tanzania is to deliver a reliable open access infrastructure on the Dar es Salam-Isaka rail segment. Total project cost include funding from World Bank and non-bank sources in US$ millions [3] . The developing world is embarking on a massive infrastructure boom, and China wants to build it—and bankroll it— a $12 billion contract to build a railway stretching to more than 1400 kilometers along the coast of Nigeria. It was China's biggest overseas contract [4] . Kenya Railways Corporation is developing a new standard railway line for passengers and Congo transportation between Mombasa, the largest port in East Africa and Nairobi, the capital city [5] . With a population of 17 million, Cairo is one of the most densely populated cities in Africa. Its railway system operated by Egyptian National Railway (ENR) transports nearly 500 million passengers and 12 million tons of freight each year. Construction of line 3 began in 2007 to accommodate the ever-growing population in Cairo [6] . The Oran tramway project is part of Enterprise du Metro d'Alger's (EMA) plan to build a tram network to improve public transport in Algeria, which was announced in 2006. Light rail development in Algeria is currently targeting the completion of three new systems by 2020. Oran tramway, being one of them [7] . Rail is an effective way to deliver mass transit of large number of people. It provides fast, safe and comfortable transportation from suburbs into city centers or city centers to another city center. The cost of rail projects, both the upfront land acquisition and construction, as well as the ongoing operations and maintenance, is considerable. Thus, governments and institutions must carefully plan, structure and implement rail and road projects to ensure that they deliver value for money, same time and make sure that transport policy objectives are met optimally [8] [9] . Application of operational research technique to urban rail transportation projects in these emerging countries will help to maximize value for money and meeting policy objectives. Linear programming and other transporting models are operational research tools used to optimize the urban rail transportation projects of the developing countries in this paper [10] .

Linear programming model is a planning technique that uses mathematical model in maximizing or minimizing appropriate measure to optimize the value of some objective after identifying some constraints [10] [11] . It may be used for solving broad range of problems in different sectors of an economy including health and transportation sectors [11] .

The urban rail transit transportation system, in most of the emerging countries, consists of three parts; the carrying tool, infrastructure and operation management [12] .

i) carrying tool is a general term of various types of equipment required by the railway and urban transit transport; it consists of safety monitoring, facilities of maintenance and protection.

ii) Infrastructure is the basic device of rail transit transportation, and it mainly consists of public works engineering system, tractive power supply system, and communication signal system. The public works engineering consists of lines, stations and bridges and tunnels.

The tractive power supply system consists of substation, overhead line system and electric power supply. Also the communication signal consists of two parts namely: the communication and signal parts.

iii) Operation management mainly refers to the operation organization and service management system organizing various transportation resources scientifically and reasonably, and provides high quality transportation services for passengers and owners of cargo based on the requirements on the transport of travellers and goods. This involves mainly: The passenger flow demand forecasting, demand rules analysis, Passenger transport path, transportation organization model, train operation plan, train graph, motor train unit application, train and station organization. The major constraints of rail transport market in sub-Sahara Africa are as follows:

1) The railway network is limited.

2). The technical condition of railway infrastructure is very poor. This have manifested in form of

(a). Aging tracks-insufficient, rail wear, deteriorating earthworks.

(b). Poor conditions of most structures.

(c). Lack of technical expertise and poorly motivated workforce.

(d). Obsolete signals and telecommunication, and lack of spare parts

(e). Lack of investment in the sector and corruption.

3) (i). Africa operates a very limited volume of traffic compared with the rest of the world: With a length of around 54,000 km, the network of rail of Africa represents only 5 percent of the worldwide network [13] . Railway has a competitive advantage

to other mode of transportation on large volumes of traffic. Sub-Sahara Africa operates less than 1 percent of the world volume of freight and passenger rail traffic [13] .

3 (ii). The inter-country trade in sub-Saharan Africa is very limited: The percentage of inter-African export exchanges is still low as most Sub-Sahara African countries are still economically dependent on overseas markets, both for export and import. The railway pattern is still mainly oriented on exporting the resources to other continents, and less on regional trade [13] .

FORMULATION OF PROBLEM

In order to solve the associated problems of urban rail transportation, especially in the emerging countries, we optimize the three parts that makes up the urban rail transportation system, namely; carrying tool, infrastructure and operation management. These three parts constitutes our decision variables as follows:

Decision Variables

Let

x_1 represent the carrying tool

x_2 represent the infrastructure

x_3 represent the Operation management

Carrying Tool (x_1)

Let

x_{11} = safety monitoring

x_{12} = facilities of maintenance and protection

Infrastructure (x_2)

Let

x_{21} = Public works engineering system

x_{22} = tractive power supply system

x_{23} = communication signal system

Operation Management (x_3)

Let x_{31} = Passenger flow demand forecasting

x_{32} = demand rules analysis

x_{33} = Passenger transport path

x_{34} = transportation organization model

x_{35} = train operation plan

x_{36} = train graph

x_{37} = motor train unit application

x_{38} = train and station organization

Contributions

From the data gathered, the percentage of success (contribution) of each of the decision variables are given in Table 1 [12] [13] .

Therefore, the contributions c_{1j} as regards the carrying tool are as follows:

$c_{11} = 10$, $c_{12} = 15$

The contributions c_{2j} as regards the infrastructure are as follows:

$c_{21} = 5$, $c_{22} = 10$, $c_{23} = 5$

The contributions c_{3i} as regards the operation management are as follows:

$c_{31} = 6$, $c_{32} = 5$, $c_{33} = 5$, $c_{34} = 10$, $c_{35} = 10$, $c_{36} = 4$, $c_{37} = 7$, $c_{38} = 8$

The performances of the decision variables towards optimization of urban rail transportation in emerging countries, therefore, are respectively:

$$\sum_{j=1}^{2}\sum_{i=1}^{2} c_{1j}x_{1i} = c_{11}x_{11} + c_{12}x_{12} \tag{1}$$

$$\sum_{j=1}^{3}\sum_{i=1}^{3} c_{2j}x_{2i} = c_{21}x_{21} + c_{22}x_{22} + c_{23}x_{23} \tag{2}$$

$$\sum_{8}^{j=1}\sum_{8}^{i=1} c^{3j}x^{3i} = c^{31}x^{31} + c^{35}x^{35} + c^{33}x^{33} + c^{34}x^{34} + c^{32}x^{32} + c^{36}x^{36} + c^{37}x^{37} + c^{38}x^{38} \tag{3}$$

Objective

The objective is to maximize total performances of the decision variables. That is:

Maximize

$$Z = \sum_{j=1}^{2}\sum_{i=1}^{2} c_{1j} x_{1i} + \sum_{j=1}^{3}\sum_{i=1}^{3} c_{2j} x_{2i} + \sum_{j=1}^{8}\sum_{i=1}^{8} c_{3j} x_{3i} \quad (4)$$

$$= c_{11}x_{11} + c_{12}x_{12} + c_{21}x_{21} + c_{22}x_{22} + c_{23}x_{23} + c_{31}x_{31} + c_{32}x_{32} + c_{33}x_{33}$$
$$+ c_{34}x_{34} + c_{35}x_{35} + c_{36}x_{36} + c_{37}x_{37} + c_{38}x_{38} \quad (5)$$

Resource Utilization

Let a_{ij} be the amount of resource j used to achieve one unit of decision variable x_i. From the data gathered and with some simulations, we have the following:

Table 1. Decision variables and their corresponding contribution.

Decision variable	Sub Decision variable	Description	Percentage (weight) of success/ contribution
X_1	X_{11} X_{12}	safety monitoring facilities of maintenance and protection	10 15
X_2	X_{21} X_{22} X_{23}	Public works engineering system tractive power supply system communication signal system	5 10 5
X_3	X_{31} X_{32} X_{33} X_{34} X_{35} X_{36} X_{37} X_{38}	Passenger flow demand forecasting Demand rules analysis Passenger transport path Transportation organization model Train operation plan Train graph Motor train unit application Train and station organization	6 5 5 10 10 4 7 8

a_{11} = unit cost of achieving safety monitoring

a_{12} = unit cost for acquiring facility of maintenance and production

a_{22} = unit cost for tractive power supply system

a_{23} = unit cost of installing communication signal system
a_{31} = unit cost to achieve passenger flow demand forecast
a_{32} = unit cost to carry out demand rules analysis
a_{33} = unit cost to achieve the transport path
a_{34} = unit cost of formulating transport organization model
a_{35} = unit cost of building train operations plan
a_{36} = unit cost of making train graph available
a_{37} = unit cost of motor train unit application
a_{38} = unit cost of proper train and station organization
The amount of available resources used to achieve a unit of the decision variables are represented in Table 2 as follows [12] [13] .

Constraints

$$a_{11}x_{11} + a_{12}x_{12} \leq 25\% \text{ of T}$$

$$a_{21}x_{21} + a_{22}x_{22} + a_{23}x_{23} \leq 20\% \text{ of T}$$

$$a_{31}x_{31} + a_{32}x_{32} + a_{33}x_{33} + a_{34}x_{34} + a_{35}x_{35} + a_{36}x_{36} + a_{37}x_{37} + a_{38}x_{38} \leq 55\% \text{ of T}$$

$$a_{11}x_{11} + a_{12}x_{12} + a_{21}x_{21} + a_{22}x_{22} + a_{23}x_{23} + a_{31}x_{31} + a_{32}x_{32} + a_{33}x_{33}$$
$$+ a_{34}x_{34} + a_{35}x_{35} + a_{36}x_{36} + a_{37}x_{37} + a_{38}x_{38} \leq T$$

$$a_{ij} \geq 0; i = 1,2,3,; j = 1,2,\cdots,8$$

where T is the total amount available for rail transport in sub-Sahara African countries, which is put at \$20 million. in a year, on the average, for the purpose of this paper.

The Model

Maximize

$$Z = 10x_{11} + 15x_{12} + 5x_{21} + 10x_{22} + 5x_{23} + 6x_{31} + 5x_{32} + 5x_{33}$$
$$+ 10x_{34} + 10x_{35} + 4x_{36} + 7x_{37} + 8x_{38}$$

Table 2. Values of resources used to produce a unit of different decision variables.

Resources	Values
a_{11}	25
a_{12}	50
a_{22}	25
a_{23}	25
a_{31}	5
a_{32}	5
a_{33}	5
a_{34}	5
a_{35}	10
a_{36}	5
a_{37}	7.5
a_{38}	12.5

Subject to

$$25x_{11} + 50x_{12} \leq \$5 \text{ million}$$

$$25x_{21} + 25x_{22} + 25x_{23} \leq \$4 \text{ million}$$

$$5x_{31} + 5x_{32} + 5x_{33} + 5x_{34} + 10x_{35} + 5x_{36} + 7.5x_{37} + 12.5x_{38} \leq \$11 \text{ million}$$

$$25x_{11} + 50x_{12} + 25x_{21} + 25x_{22} + 25x_{23} + 5x_{31} + 5x_{32} + 5x_{33}$$
$$+5x_{34} + 10x_{35} + 5x_{36} + 7.5x_{37} + 12.5x_{38} \leq \$20 \text{ million}$$

$$x_{ij} \geq 0; i = 1,2,3; j = 1,2,\cdots,8$$

MODEL SOLUTION

The model is first of all written in its standard form before applying simplex method to solve it.

Standardized Model

Maximize

$$Z = 10x_{11} + 15x_{12} + 5x_{21} + 10x_{22} + 5x_{23} + 6x_{31} + 5x_{32} + 5x_{33}$$
$$+ 10x_{34} + 10x_{35} + 4x_{36} + 7x_{37} + 8x_{38}$$

Subject to:

$$25x_{11} + 50x_{12} + s_1 \leq 5,000,000$$

$$25x_{21} + 25x_{22} + 25x_{23} + s_2 \leq 4,000,000$$

$$5x_{31} + 5x_{32} + 5x_{33} + 5x_{34} + 10x_{35} + 5x_{36} + 7.5x_{37} + 12.5x_{38} + s_3 \leq 11,000,000$$

$$25x_{11} + 50x_{12} + 25x_{21} + 25x_{22} + 25x_{23} + 5x_{31} + 5x_{32} + 5x_{33} + 5x_{34} + 10x_{35}$$
$$+5x_{36} + 7.5x_{37} + 12.5x_{38} + s_4 \leq 20,000,000$$

$$x_{ij} \geq 0; i = 1,2,3; j = 1,2,\cdots,8$$

Simplex Method of Solution

In this research work, we used the simplex method algorithm to solve the standardized linear programming model. The initial tableau was formed from the standardized linear programming model, after which linear programming solver (LIPS) software was adopted to arrive at the following results, with optimum solution equals 25,600,000.

Table 3 shows the values of the different decision variables using LIPS [14] .

RESULTS DISCUSSION

The Results obtained in table show that three of the decision variables are very significant, namely: x_{11}, x_{22} and x_{34}. That is, Safety monitoring, Tractive Power Supply, and Transportation organization model. In any of the emerging countries, these three factors, classified under carrying tools, Infrastructure and operational management respectfully, are very critical in optimizing the urban rail transport in emerging countries. The result revealed that if 200,000 USD is set aside to take care of the safety monitoring, 160,000 USD is set aside for tractive power supply, and 2,200,000 USD is set aside to take care of transportation organization model for urban rail transportation, then the optimum value of the objective, in terms of USD is 25,600,000. Substituting the values of x_{11}, x_{22} and x_{34} into the objective function gives us the optimum value. This value is greater than the average amount available for investment in Rail transportation, which was put at 20,000,000 USD on the average, by 5,600,000 USD. This is an advantage. It is important to note that x_{34} representing transportation organization model, has the highest value, which suggests its importance among the decision variables. Finally, the factors capable of hindering the achievement of the optimization of rail transportation in the emerging countries, such as corruption and unnecessary bottle necks should be discouraged.

Table 3. Optimal feasible values of the decision variables.

Variables	Represented by	Value
X_{11}	X_1	200,000
X_{12}	X_2	0
X_{21}	X_3	0
X_{22}	X_4	160,000
X_{23}	X_5	0
X_{31}	X_6	0
X_{32}	X_7	0
X_{33}	X_8	0
X_{34}	X_9	2,200,000
X_{35}	X_{10}	0
X_{36}	X_{11}	0
X_{37}	X_{12}	0
X_{38}	X_{13}	0
S_1	X_{14}	0
S_2	X_{15}	0
S_3	X_{16}	0
S_4	X_{17}	0

CONCLUSION

In this research paper, optimization of urban rail transportation was considered. The problem was modeled as a linear programming problem, which was solved using simplex method with the aid of computer software—Linear programming solver (LIPS). It is shown that all the three categories of factors leading to optimization of urban rail transportation in emerging countries are important in the process of maximization. Specifically, a good transportation organization model must be built in order to attain optimality of urban rail transportation in these countries.

CONFLICTS OF INTEREST

The authors declare no conflicts of interest.

REFERENCES

1. Business Dictionary.com
2. Lagos Rail Mass Transit System, Nigeria, Railway-Technology.com. www.railway.technology.com/projects/Lagosrailmsstransit
3. Tanzania Intermodal and Rail Develop ment Project, The World bank, Project. www.Worldbank.org/projects/p127241/intermodal-rail-development-project?lang=en
4. Schaen, J.W. China Wants to Build the World a Railway. CNBC. www.cnbc.com/2014/11/20/china-wants-to-build-the-world-a-railway.html
5. Mombasa—Nairobi Standard Gauge Railway Project, Railway-Technology.com. www.railway.technology.com/projects/region/africa
6. Cairo Metro, Egypt, Railway-Technology.com. www.railway.technology.com/projects/cairo-metro
7. Oran Trainway, Algeria, Railway-Technology.com. www.railway.technology.com/projects/oran-trainway/
8. Agarana, M.C., Owoloko, E.A. and Kolawole, A.A. (2016) Enhancing the Movement of People and Goods in a Potential World Class University Using Transportation Model. Global Journal of Pure and Applied Mathematics, 12, 281- 294.
9. Rama Murthy, P. (2007) Operations Research. 2nd Edition, New Age International Publication.
10. Agarana, M.C., Anake, T.A. and Adeleke, O.J. (2014) Application of Linear Programming Model to Unsecured Loans and Bad Debt Risk Control in Banks. International management Information Technology and Engineering, 2, 93-102.
11. Agarana, M.C. and Olokunde, T.O. (2015) Optimization of Health Care Pathways in Covenant University Health Centre Using Linear Programming Model. Far East Journal of Applied Mathematics, 91, 215-218. http://dx.doi.org/10.17654/FJAMJun2015_215_228
12. Xin, X. and Jing, Z.C. (2015) Current Situation and Existing Problems of Urban Rail Transit Industry Development in China. Eastern Academic Forum, 805-811.
13. Varile Nicolae Olievschi (2013) Rail Transport: Framework for Improving Railway Sector Performance in SubSaharan Africa. SSTAP Africa Transport Policy Program, Working Paper No. 94.

CHAPTER 11

A Review on Strategic, Tactical and Operational Decision Planning in Reverse Logistics of Green Supply Chain Network Design

Farahanim Misni[1,2] and Lai Soon Lee[1,3]

[1]Department of Mathematics, Faculty of Science, Universiti Putra Malaysia, Serdang, Selangor, Malaysia.

[2]Faculty of Sciences & Technology, Universiti Malaysia Pahang, Gambang, Kuantan, Malaysia.

[3]Laboratory of Computational Statistics and Operations Research, Institute for Mathematical Research, Universiti Putra Malaysia, Serdang, Selangor, Malaysia.

ABSTRACT

The environmental impact has been a critical issue in supply chain network research. Handling it efficiently will help the firm reduce pollution and

Citation: Misni, F. and Lee, L. (2017) A Review on Strategic, Tactical and Operational Decision Planning in Reverse Logistics of Green Supply Chain Network Design. Journal of Computer and Communications, 5, 83-104. doi: 10.4236/jcc.2017.58007.

save cost. In order to improve the environmental effect on this supply chain network, recovery activity becomes a significant part in the reverse logistics network, due to complexity and heterogeneity. In a reverse logistics network, there are many decision problems to be taken into consideration such as facility location, capacity allocation, production planning and vehicle routing. Each of these problems can be categorized into three types of decision planning: strategic, tactical or operational, based on the time planning horizon. In this paper, we review the literature of supply chain network in an effort to achieve green through implementing reverse logistic practices, and classify the defined problems by the group of decision planning. In general, most of the literatures are focused on the strategic and tactical decision planning, only a few are operational-based. The objective of this paper is to analyze the decision planning being done in this area of research and provide future insights on how to design the reverse logistics network with different decision planning which will reflect the real life scenario.

Keywords:- Reverse Logistics, Supply Chain, Strategic, Tactical, Operational

INTRODUCTION

Supply chain is one of the active application areas in operational research and management science. The success of a company depends on the efficiency of design, manufacture and product distribution in a network. The supply chain is designed to meet consumers' demands and satisfactions in a timely manner and as efficiently and profitably as possible. Dealing with the demands that need to match with the supply, reliability, and consistently in high quality at acceptable cost are the major problems in supply chain network design (SCND) [1] .

The SCND is a network connecting facilities from different levels. The top level corresponding to the suppliers or plants while the bottom level corresponding to the consumers or demand zones. The paths joining between these facilities represent sequences of supply chain network activities to ensure the product or service is delivered. The network activities are referring to the economic activities which associated with the flow of products or services through the early raw material stage to the end user. The suppliers will provide the raw materials to the manufacturer where the production takes place. Then the product will be stored at distribution centres prior to despatch, which is dependent on a customer's specific delivery instructions.

The SCND can have different stages or phases depending on the number of facilities being connected (see, Figure 1).The origin for all activities is the predefined deterministic demand of the consumers, where the products are manufactured or transported only if requested by the subsequent members of the supply chain. However, it can be a stochastic demand with uncertainty. Supply and demand are rarely concentrated in the same place, therefore transportation is vital to ensure the reliable and affordable flow. The supply chain can be categorized as: lean, agile, green, closed- loop, sustainable and risk supply chain. Figure 2 shows the definition for each type of the SCND.

In a lean supply chain, elimination of waste or non-adding elements will reduce the cost. Increasing the efficiency through reduction of waste and cost of inventories by carrying small batches of production are two major features in lean manufacturing. Flexibility is the key characteristic of agile supply chain. From inside supply chain, flexibility means non-fixed configurations and structures. While from the point of consumer perspective, the supply chain must deliver products and services at the beginning of the usual short profit widows in less time [2] .

In sustainable supply chain practices, all parties involved will conserve the natural resources for the next generation to avoid the overconsumption of resources [3] . It is a combination of economic and environmental issues, but it al so adds social aspect into the consideration.

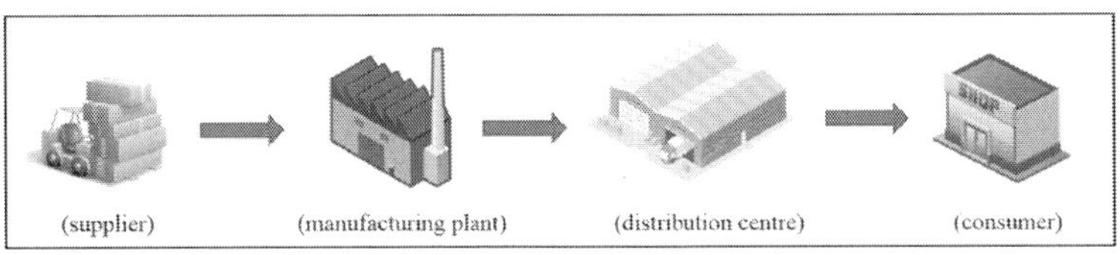

Figure 1. Supply chain network design.

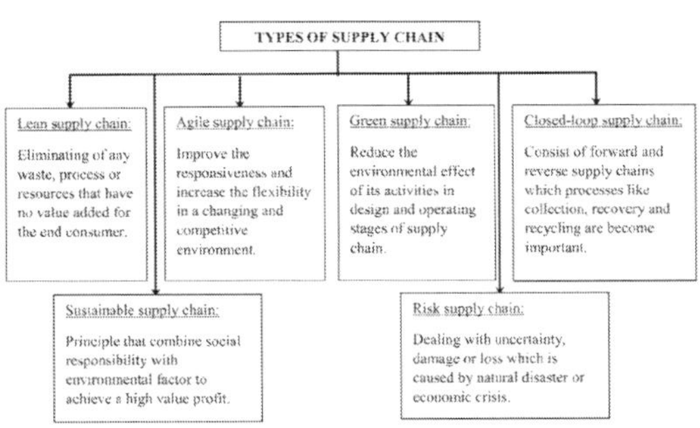

Figure 2. Type of supply chain network.

Predicting and dealing with the various type of risk such as delay, inaccurate forecast, failures in procurement, disruption, system breakdowns and problem with capacity and inventory are risk management concerned in supply chain [3] . Hence, designing a robust network with efficient product flow management can handle the risks in the supply chain.

In a closed-loop supply chain, it combines both forward and reverse flows. The forward supply chain includes the activities of procurement, design, manufacture and distribution to consumer while reverse supply chain is related to process of inspecting, sorting, disassembly for the purposes of reuse, reprocessing and redesign.

The increases of waste disposal and limited space of landfill show that the environmental problem becomes a critical issue among the industries. The hazardous risk through the contamination and emission will affect the human health. Moreover, the waste disposal are costly since the location of landfills are commonly far away to be sent. Therefore, industries have started to integrate environmental factors throughout the organizations by implementing green supply chain. They integrate the green technologies into their product designs, production and distribution process. The green distribution involves efficient route planning, fuel reduction and promotes the eco-friendly products as well [4] . Green supply chain enhances the productivity and environmental performance for sustainable development to achieve a competitive advantage. In order to respond to the consumer for the environmental based products, the green aspect in supply chain has now become more significant among the industries. Besides, the environmental legislation forces the practitioners to set the environmental effect as one of the main objectives. The trade-offs between the cost and environmental objectives must be conducted to optimize both dimensions [5] . Due to the importance of the study, the main focus of this review is on the green supply chain network design (GSCND).

The remainder of this paper is organized as follows: Section 2 provides an introduction on the GSCND and the green processes involved. The network design, facilities and recovery process in reverse logistics are describe in Section 3. Section 4 review the strategic, tactical and operational planning in reverse logistics and Section 5 present the conclusions and future directions.

GREEN SUPPLY CHAIN NETWORK DESIGN

The GSCND can be defined as "greening" one or some of the processes in SCND such as product design, material sourcing and selection, manufacturing operations and logistic of final products to the end consumers by considering the environmental effect with the support of the firm and enforced by the government regulations [6] . By implementing GSCND practices, it can improve the environmental performance of product and process.

Although GSCND has many benefits, there are also some challenges to be concerned. There must be a high level of uncertainty about the market position which means that green is not warranty for better income. To be green, there are many elements need to be tackled. Green product, green procurement, green manufacturing, green consumer, end-of-life management and green logistics are the key elements in GSCND [7] . Each of them has their own characteristics and decisions to be made.

Green product relates to the design and material of the product. The manufacturer needs to design the product to be environmental friendly and scored high in the life cycle assessment (LCA). LCA is the evaluation of the product that could affect the environmental and related resources through all phases of its life. It should be designed purposely for the continuity of the life-cycle process. In the process of making a product, green material minimize the content and types, packaging, energy consumption during usage and product development as well as the disposal at product ends-of-life. It is about material sourcing and selection. To be green, the firm needs to choose and use less polluting materials that can be recycled.

Green procurement is a supplier selection with the risk analysis. Hazardous and harmful substances accumulated throughout the process seek green suppliers to provide cleaning materials and components [7] . In a production process, the manufacturer implement green manufacturing by reducing energy consumption, production waste and the use of clean technologies. Remanufacturing is one of the processes that protecting the environment from pollution. It is the manufacturing of the recycling product. It offers a potential advantage such as increased profitability through reduction in the new raw material usage. Remanufactured parts can be used for assembly of new product or resold if the entire products are upgraded [8] .

The end-of-life management contributed to the green supply chain. The recovery activities such as recycling, remanufacturing, refurbishing, repairing and reusing are the main areas in life-cycle process. The returned products can be reprocessed to bring quality back to a reusable level, obtain recoverable components and materials. It is much cheaper to use returned component and recycle material than using a new part.

Logistics associated with various products according to its size, weight, volume as well as the quality and safety. Implementation of green logistics is supported by consumer pressure, different type of logistics activity and the development of innovation. Green logistics is the integration of the environmental aspect into logistics activities, and it is significant in every decision making process across the logistics network. This practice becomes one of the critical issues among practitioners which are associated with climate change, air pollution, noise and other environmental effects. Green logistics approaches depending on the type of companies' service provided to the consumer and the size of the companies. Logistics can be "green" by considering the alternative mode of transportation, the energy usage during logistics activities and material usage along the route of the shipping [5] . It relates to the CO_2 and greenhouse gas emission. Reverse logistics can be defined as one of the green logistics which is the focus on this review study.

REVERSE LOGISTICS

Reverse logistics can be defined as the process of handling and collecting all the returned end-of-life equipment, materials, products or components from end user back to the point of origin. It is also for all operations related to the recovery process of any possible used products and materials. The process of collecting sometime faces difficulty depending on the willingness of the consumer and the strategic location of collection centre.

Reverse logistics can also be related to the process of planning the reverse flow network, implementing the process of recovery and controlling the related costs for the purpose of creating value or disposal [9] . It is known as all activities associated with a product or service after the point of sale and set up the goal to optimize the aftermarket activity development. In addition, it can save in inventory carrying transportation and waste disposal costs as well as improving consumer service [10] .

Reverse logistics can be classify as a closed loop (see Figure 3) if it integrated with the forward logistics. When only the reverse flow is being considered, it is an open-loop supply chain. In an open-loop, the flow of materials and products are delivered starting from the point of consumer to the secondary user through the collection centre and processing centre.

Facilities in Reverse Logistics Network

In a reverse flow, there are four facilities or centers that need to be considered in the network: collection, disassembly, processing and disposal [11] .

Collection centre is the place where all the returned products or used products being collected. There are two stages involved in collection centre which are the pick-up activities and the transportation of the returned products from one facility to another.

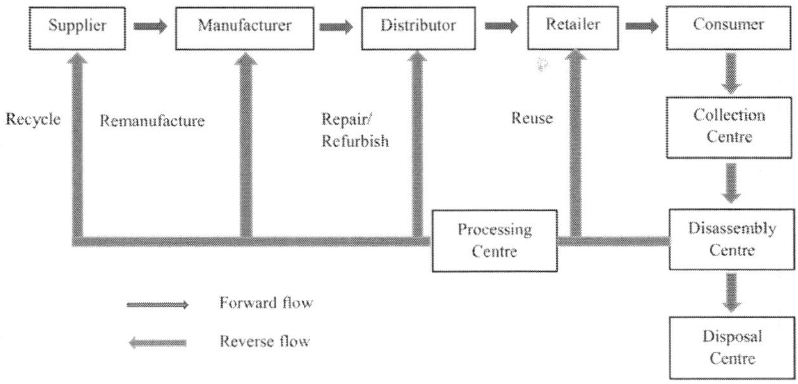

Figure 3. Closed-loop flow [11] .

Usually, the products will be self-collected by the company. However, due to some reasons, the company may outsource to third party for collection or the products are returned by the consumers. The choice of collection method is depends on the cost, flexibility, location, complexity of the product, and the reason for returning it.

After all the returned products have been collected, the next process is to sort them according to the quality level of the products at the disassembly centre. Disassembly centre will first inspect and separate each product. During the sorting process, the conditions of the returned products are inspected concurrently. It determines either the returned products need to be recovered or sent to the disposal centre. It is important to ensure the right recovery process for the next stage. Inspection and separation process may include the disassembly, shredding, testing and sorting steps. In addition, the company should prepared guidelines for the returned products that meet the criteria, to be considered as an inventory and the intricacies of the sorting process.

Processing centre transformed a returned product into a usable product again. This is the place where all the treatment options are performed such as recycling, remanufacturing, repairing and repackaging of the returned products [12] . The activities such as cleaning and replacement may be involved too. The choices of recovery process are depended on the condition and the quality level of the returned products. Upon completion, the recovered products will be sent back to the facilities based on the type of recovery processes. For example, if the product is recycled, then it will be sent to the supplier. If a returned product cannot be recovered, it will be disposed at the disposal centre. The disposal centre is the exit of the reverse logistics system. A company would prefer to send the product to the disposal centre when there is no any beneficial value to the product.

Recovery Process in Reverse Logistics

Due to limited landfill capacity, waste disposal become a major concern among industries. At the same time, consumers are expecting that company will produce a green product to reduce the environmental pollution. At the processing centre, the recovery decision of the returned products is very crucial as it will be sent back to the facility after the product is recovered. Besides, product recovery may be attractive from an economical perspective as well [9] .

Repair

Repairing can be defined as recovering the defected, broken or damaged in some parts of the product. It could be a failed working device, or deficiency of an item. The purpose of repairing is to ensure the returned product is functioning [13] . Repairing require lower capital investment and more to skill-based [6] . After the product has been repaired, it will be sent to the distributor.

Refurbish

Refurbishing is similar to the repairing but the process will upgrade the quality level of the product and usually less complicated process than a new production. It is more skill-based in order to make them almost as good as new [6] . Compared to the repairing process, refurbishing needs more attention since it involved disassembly process, inspection and replacement of critical parts. Com- monly, it is closely related to the electronic or electric product that required testing before the product can be returned. Refurbished product is normally returned under warranty or unused consumer returns. It will be sent to the distributor after conducting a proper inspection.

Reuse

The reused products are sent to the retailer for the same function or for others activity. Reuse is the simplest process as it does not require any high technology tools. Hence, it can minimize the cost compared to other recovery processes.

Remanufacture

Remanufacturing is the process of rebuilding or reconstructing the product by using the combination of reused, repair and import of new parts. It requires the replacement of obsolete parts. It is more specific and challenging. To remanufacture it to a new product, it involved the process of disassembly, detailed inspection of all parts and finally assembled them properly. Remanufacturing process is quite different with other recovery processes in its completeness where the remanufactured product usually has the same quality as the new product. The remanufactured product will be sent back to the manufacturer to be sold at the same price [14] .

Recycle

Recycling is a process of converting waste material to a new raw material. The recycled products will be sent to the supplier and enter the supply chain again as a raw material for new products [4] . The purpose of recycling is to reuse the material from the returned product that is cannot be remanufactured or repaired. In order to reduce the cost of raw material, recycling could be the best alternative. This process can reduce the environmental impact and it is a good alternative to save the material from been disposed.

STRATEGIC, TACTICAL AND OPERATIONAL DECISIONS

The main area in reverse logistics are logistics distribution, production planning and inventory capacity. Reverse logistics, as part of supply chain management, provides useful quantitative models for network design in the management of end-of-life products [15] . Location-allocation problem is particularly relevant in a reverse logistics context that affects both costs and consumer service levels. Inventory also may become significant, especially in case of high valuable goods (e.g. electronic equipment) [16] . There are many others decision can be listed and each of it has a different level of planning horizon, which can be categorized into the following three levels:

i) strategic decision

ii) tactical decision

iii) operational decision

Decision planning can be integrated either "horizontal integration" or "vertical integration". Horizontal integration means by integration the same decision level of optimization problems while vertical integration combines two different levels of decision [17] . For example, strategic-tactical, strategic-operational or tactical-operational.

Strategic Decision Planning

Strategic decision planning is a long term decision planning between 2 to 5 year timeline. At this strategic level, the decisions made are high level because there are difficult to change. It is associated with the recovery process, especially during the design of the product. Coordination of supply chain network, capacity planning and designing of the remanufacturing systems with environmental consideration also fall under this category [11] .

Product design plays an important role in a profitable prospect of product return for recycling and remanufacturing. The company should investigate the effect of product design, especially in a closed-loop network. However, during the process design, the impact of the product design has always been neglected. In order to be more practical, company should plan the appropriate design for their products to continue the recovery process of the returned products in reverse logistics flow. A company, who is actively practicing the remanufacturing in their network, should competence in defining the quality level of the reliability components. Original equipment manufacturer (OEM) could decide to permanently seal or weld a product, as opposed to using fasteners, during its initial assembly but this makes disassembly of the product (during recycling or remanufacturing) much more difficult and costly. Thus, product design decisions have a significant impact on a firm's strategic decisions regarding remanufacturing [18] .

Before a company starts the production operation, they must develop the network for used products recovery [19] . The coordination of reverse logistics network design should be efficient in order to collect the used products in a cost-effective manner and gives profitable to the remanufacture. Reverse logistics network consist of network design of collection activity and its physical transportation to support the system. One of the critical components in the network design is the decision on capacity and location of each facility. The facilities should be placed in a strategic location to maximize the flexibility and responsiveness of consumers. Recently, some industries prefer to have a hybrid system where two activities are concurrently working in one location as they share the same component and processing. By implementing this, the operating and transportation cost can be reduced. However the complexity to handle hybrid system will be increased. In this case, the sufficient space in the facilities must be well organized.

The challenging part faced by company is to decide on how to build the reverse logistic network such that it meets the objective and constraints. Besides, a company needs to set up a reverse logistics network based on the requirement of environmental legislation to ensure the proper disposal of its products [18] . The role of environmental legislations could affect the company's remanufacturing decision in a long term. Based on [12] , return policy is also one of the strategic planning in reverse logistics. One study has been found where the decision is to obtain optimal policies for price and return policy with the aim of maximizing the profit [20] .

As mentioned earlier, the researchers pointed out that at the strategic level decision; the design of the recovery network has to be decided first. Therefore, some researchers proposed the network design for reverse logistics such as: design of collection network for take-back of white goods for remanufacturing [21] , design of generic reverse logistics model considering distribution and recovery network [22] , design of integrate traditional and reverse supply chain model [23] , model of the reliable network for facility location [24] , model of a reverse SCND problem considering specified returns [25] , planning a forward/reverse logistics considering capacity expansion and dynamic transportation link [26] and develop optimal model for cross-stage reverse logistics [27] . Among these, only [24] deal with the multi-objective and uncertainty. The objectives of their study are to minimize the total cost and expected transportation cost after failure of facility. They used fuzzy multi-objective programming together with robust optimization and queuing theory to solve the problems.

Most of the studies in this review deal with the decision on how to locate the facilities, the optimal number of facilities to be opened, the capacity of production for each facility and the products flow between facilities. This can be found in the following multi-objective problems. Monte-Carlo simulation was chosen to solve the cost and environmental risk minimization under uncertainty [28] . The heuristic (scatter search) and dual simplex method are used for multi-objective total cost and total tardiness of cycle time minimization [19] . The bi-objective of minimizing total cost and at the same time maximizing the consumer responsiveness been solved by using multi-objective memetic algorithm [17] . Corley method which is the hybrid of weighted-sum and ε -constraint methods, has been used in order to maximize the on-time delivery from supplier as well as maximizing the total profit [29] and multi-objective particle swarm optimization is used to minimize the total cost and CO_2 emission [30] .

The facility location-allocation problems with single objective have been found in many literatures [8] [9] [15] [31] [32] [33] [34] . Among them, [32] proposed forward and reverse logistics to extend the single product unlimited capacity to multi-product capacitated. Reference [35] focused on the optimal capacity in the production of manufactured and remanufactured products for forward and reverse flow supply chain while [36] proposed mathematical model for reverse logistics handling product returns. They use GAMS software to solve the problems.

Since some problems in location-allocation are considered large scale problem and exact methods are not appropriate to obtain the optimal solution for this size, the researchers decided to use heuristic and/or metaheuristic approaches. Jenkins' heuristic is used by [9] to solve piecewise linear in generic facility location for product recovery. Reference [37] solved the location and material flows for take back and the recovery of end-of-life vehicle (ELV) by using heuristic method based on greedy algorithm. Genetic algorithm is widely used in facility location problem such as reverse logistics problem involving product returns [33] , dynamic modeling approach for 3rd party logistic provider (3PLs) [38] , location-allocation problem of repair facilities for 3PLs [39] , and optimization of location of collection points based on RFID [40] . Priority-based encoding simulated annealing with special neighbourhood search mechanism has been used to solve multi-stage reverse logistics problem [41] . Others metaheuristic methods such as ant colony optimization for nonlinear dimension factor called SCAnt- NL Design [42] , hybrid genetic algorithm with particles warm optimization [34] and hybrid discrete artificial bee colony [43] are also been studied.

Dynamic location-allocation model with forward/reverse under uncertainty also has been proposed. This problem has been solved by using integrated sample average approximation and simulated annealing [44] , scenario-based stochastic approach using stochastic programming model [45] , and multi objective differential evolution to minimize both total cost and CO_2 emission cost for fuzzy programming [46] .

Tactical Decision Planning

The tactical decision problem is a medium term decision. It takes 1 to 2 year timeline and generally it comes after the strategic decision planning. The decisions are related to the production planning and inventory management as well as procurement and integrated management of product returns with the overall organization. Return forecasting, product return handling and aggregate production planning are related to the product returns activities [11] .

A company that decides to remanufacture their products has some issues of tactical decision. They need to plan on the number of used products that should be recovered and decide on the capacity limit of the recovery products for each time period. The decision is more complex compared to the traditional forward supply chain. In a forward supply chain, the components or materials are purchased from the supplier with the exact amount. The number of materials is predetermined since the target unit of the final products to be produced is decided earlier. In contrast with the reverse flow, the production planning is necessary to be organized systematically since there are many unpredictable quality levels as well as a capacity constraint in product return of the remanufacturing process. The variation of quality level may due to the different ways of the product maintenance conducted by the owner. Some products required additional costs to remanufacture and not all the products can be recovered. The functionality of production planning is to determine the optimum quantity of each unit for each period to be processed. This decision is crucial in order to obtain the cost-saving effectively. Different demands and limitation on the number of processed unit in each period would influence the successes of the whole planning [18] .

Since the returned products will have different levels of quality, the disposition decision must be made. The highest quality level of returned products are the least expensive to remanufacture. The disposition decision in the reverse supply chain allocates returned products to an appropriate processing option. This decision is important in order to reduce pollution during the recovery processes.

In inventory planning, there are two main areas. First is the management of uncertainty of product returns and secondly is the management of inventory of new components when both regular procurement and recovered components can satisfy the demands. Aggregate planning concerned with the amount of inventory to be held in stock and backlogged for each period. This planning is considered as a tactical decision planning. A safety stock planning approach in reverse logistics is provided in [47] . An optimal inventory policy on changes of manufacturing, remanufacturing and disposal rate in reverse logistics is proposed by [48] . Both studies aim to reduce the holding cost in their objective. Other researchers proposed inexact reverse logistics in planning the inventory control, waste disposal and distribution [49] . The collection of returnable combines with recovery modules has been solved by [50] .

The general framework of remanufacturing system has been developed by [51] . They applied the framework with some numerical examples in making a decision on the quantity of parts needed to remanufacture and number of parts needed to purchase from the subcontractor. This followed by other researchers where the decision is to determine the optimal capacity of manufacturing and remanufacturing product units per period [14] . In the study, a variable neighborhood descent heuristic algorithm is applied for solving a multi-product dynamic lot-sizing problem.

Besides the inventory control, the hazardous waste treatment and transportation problem are also included under tactical planning. About the problems of waste treatment and management, [52] formulated a discrete-time linear analytical model to determine the time-varying and hazardous waste in reverse logistics. An activity-based model is presented by [53] to allocate the discarded product, disassembly activity and fraction of waste of electrical and electronic equipment (WEEE) treatment. An incremental risk-induced penalty is used by [54] to estimate the risk for multi-objective linear programming. The study is conducted to minimize both reverse logistics operational cost and risk-induced penalties.

Multi-objective can be found in transportation model proposed by [5] . They aim to determine the dependencies of transport cost, emission and fuel consumption. The optimization model is calculated by computer program LOMP. Others researchers use priority-based encoding genetic algorithm (priGA) and heuristic approach for transportation flow problem [10] .

Tactical planning can be integrated with strategic decision planning. Usually, the researchers will combine the decision of location and capacity planning of facilities with the optimal inventory capacity to be held or purchased. The generalized algebraic modeling system (GAMS) is used by [55] to solve 0 - 1 mixed integer programming. Reference [56] determined the location of third checking sites and inventory policy for B2C (business to consumer) e-market and modeled it as a bi-level programming problem while [16] used technique of differential evolution and incorporated the

MINLP with queuing model for inventory. Other software are also been used such as XpressSP software for multi-stage stochastic demand [57] and CPLEX to tackle product returns in multi-period reverse logistics [58] . Differential evolution also been proposed by [59] to deal with the stochastic delay due to uncertainty.

The facility location-allocation problem can be combined with disposition decision. A bi-level optimization model was coded in GAMS 21.2 with CPLEX 7.5 [6] , and a two-stage stochastic mixed-integer solved by ILOG CPLEX 10.1, has been used to solve these problems [60] . They considered the multi-objective in their problem which is minimizing the operating cost and minimizing the obnoxious effect caused by the reverse network. Meanwhile, the decision of disassembly and processing centers to be opened with the transportation strategy has been solved by priGA [61] .

Operational Decision Planning

Operational decision planning is a short term planning which is on a day-to-day decision. Vehicle planning and scheduling are under operational planning decision. The majority of remanufacturing shops are low volume job-shop type operations [18] . The remanufacturing operations become more challenging as the uncertainty in the supply and the large variance along the possible routing are increased compared to the traditional forward production. The issues under the scheduling planning are the decision of optimal disassembly sequences, the parts should be released to the floor, the rules for shop floor and the route of the products [18] .

Operational decisions come as consequences of the strategic and tactical decisions. Dynamic control of product recovery operations and dynamic pricing of products are included in operational planning decision [11] . Some other interesting operational issues are warehousing operations problems such as order batching and picking, routing pickers in warehouse are also considered in this operational planning [62] .

Operational planning found in this review study is mostly related to vehicle routing problem. Vehicle routing and scheduling problem may have issues on mixed pick-up and deliveries and time windows for delivery. A heuristic construction procedure is used by [63] where the initial solution obtained by a local search to solve the vehicle routing problem with simultaneous delivery and pickup. The same vehicle routing problem as in [63] is solved by [64] using tabu search to calculate the transportation route for recycling. An integrated ant colony and tabu search algorithm is also proposed by [65] for the same problem. Both papers have the same objective which is to minimize the total distance and cost by determining the route of vehicle.

Recently, many researchers integrated the operational decision planning with strategic or tactical decision planning. The decision on multi-stop vehicle route with return policy and optimal quantity of return material to be picked up been solved by heuristic approach [66] . Location and critical path of WEEE intermediate storage points are determined using CPLEX 9.1 [67] . A framework is proposed by [68] that separated into operational and strategic flexible measure for 3PLs performance outcome and solved it by novel neighborhood rough set approach. A vehicle routing problem and location-allocation problem is solved by [69] by implementing greedy adjustment in a novel discrete artificial bee colony algorithm. The artificial immune system and particle swarm optimization are used by [4] to obtain the vehicle routes and quantities to be produced at each facility.

The integration of tactical and operational decision planning which is defined by the service area of each depot and scheduling of collection route has been found [70] . The study solved the combination of three objectives: minimize distance, CO_2 emission and working hours by the augmented ε -constraint method. The paper balances the economic factor with environmental issues and social aspect. The approximation to the Pareto front is obtained to solve the multi-objective problem. To the best of our knowledge, only one paper integrated all three planning: facility-inventory-routing problem in their study [71] . They used a hybrid ant colony optimization algorithm to minimize the total cost in reverse logistics. To summarize, all the collected articles published since 1999 that related to the reverse logistic problems based on the strategic, tactical and operational planning are listed in Table 1 below.

Table 2 below shows the distribution of decision planning of the selected articles in the reverse logistics. It clearly shows that most of the studies are in the strategic decision planning. Compared to the strategic and tactical decision planning, operational decision planning is the lowest. Before the operational decision can been made, the strategic and tactical planning should be decided first. Since the strategic and tactical decision should be determined first, it becomes a constraint for operational decision [42] . Besides, among these three decisions planning, operational give less effect to the entire performance of supply chain since it is a short term planning. Due to this, most of the researchers are more focusing on the strategic and tactical decisions planning as compared to the operational. Only one paper has been found that proposed the combination of all three different decision planning.

Table 1. Classification of strategic (S), tactical (T) and operational (O) planning based on the decision problems.

Authors	Decision Problems	S	T	O
[55]	· Location of distribution/remanufacturing facilities. · Optimal inventories quantities of remanufactured products. · Transportation of cores between facilities.	✓	✓	
[63]	· Vehicle routing with simultaneous delivery and pick-up			✓
[9]	· Location of plant and distribution warehouse and allocation of goods flows between facilities.	✓		
[47]	· Production and inventory for safety stock planning.		✓	
[31]	· Number of location for storage and treatment facilities · Optimal physical flow of end-of-life products.	✓		
[52]	· Time-varying collection and treatment amount of hazardous waste		✓	
[48]	· Optimal inventory policies in reverse logistics.		✓	
[20]	· Optimal price and return policies.	✓		
[53]	· Optimal allocation of WEEE treatment in reverse logistics system.		✓	
[28]	· Location and allocation at various facilities. · Optimal waste quantity travelling.	✓		
[51]	· Quantity parts to be processed and purchased from subcontractor.		✓	

[9]	· Transportation model (dependencies of cost, emission and fuel consumption)		✓	
[33]	· Location of collection and repair facilities for 3PLs under capacity limit.	✓		
[32]	· Warehouse location-allocation model.	✓		
[66]	· Multi-stop vehicle route design with return policy. · Optimal quantity of return material to be pick up along delivery route.	✓		✓
[21]	· Collection network design for take-back of white goods for remanufacturing.	✓		
[34]	· Capacitated location model of warehouse and repair centre. · Allocation of customer within capacity limitations.	✓		
[16]	· Facility location-allocation problem. · Assign optimal flow of returned goods including inventory in reverse logistics.	✓	P	
[22]	· Design generic reverse logistics model considering distribution and recovery network.	✓		
[54]	· Time-varying amount of physical flow for treatment hazardous waste.		✓	
[56]	· Location and inventory policy in B2C firms.	✓	✓	
[36]	· Location facilities problem. · Optimal production and transportation quantities of manufacture and remanufacture products.	✓		
[19]	· Location-allocation for installing repair facilities and collection sites. · Transportation flows between collection sites and facility sites.	✓		

[37]	· Location of collection centres and dismantlers. · Optimal material flows of end-of-life vehicle (ELV).	✓		
[35]	· Location-allocation of repair facilities for 3PLs.	✓		
[6]	· Location-allocation and capacity decision for facilities. · Disposition decision for various grade of returned products.	✓	✓	
[15]	· Facility location of end-of-life vehicle (ELV) collection network.	✓		
[64]	· Capacitated vehicle routing problem.			✓
[40]	· Dynamic location and allocation model in reverse logistics.	✓		
[38]	· Location problem for collection points. · Optimal quantity of products to be shipped.	✓		
[10]	· Transportation problem of recycling process.		✓	
[39]	· Optimal location and capacities of various facilities. · Allocation of material flows between facilities.	✓		
[67]	· Location of WEEE intermediate storage points. · Critical path of WEEE considering intermediate storage points.	✓		✓
[57]	· Facility location, production of each facilities and transported quantity between locations. · Optimal inventory to be held at each period.	✓	✓	
[60]	· Location problem with choice of technology.	✓		

[17]	· Facility location and capacity of distribution, production/recovery and disposal centres. · Optimal product flows between facilities.	✓		
[41]	· Facility location of hybrid distribution/collection, production/recovery and disposal centres. · Optimal quantity flows between network facilities.	✓		
[42]	· Number and location of collection/ inspection centres. · Optimal quantity of flow between facilities.	✓		
[8]	· Location of collection and return centre. · Allocation of product flow between facilities.	✓		
[23]	· Design integrated traditional and reverse supply chain model.	✓		
[49]	· Production, inventory and transportation planning for the waste management.		✓	
[58]	· Location and capacity decision of inspection centres and remanufacturing facilities. · Optimal inventory to be held or purchased at remanufacturing plants.	✓	✓	
[25]	· Model a reverse supply chain network design problem considering specified returns.	✓		
[50]	· Optimize quantity of new and recovered products.		✓	
[59]	· Location and flow allocation of products for each facilities integrated with queuing relationship.	✓	✓	
[24]	· Design and model the reliable network for facility location.	✓		

[68]	• Flexible measure of performance outcomes for 3PLs in reverse logistics framework.	✓		✓
[26]	• Design and planning forward and reverse logistics.	✓		
[27]	• Develop optimal cross-stage reverse logistics model.	✓		
[69]	• Location-allocation problem of reverse logistics. • Vehicle routing problem.	✓		✓
[70]	• Service area establishment for each depot. • Scheduling of collection routes for each vehicles.		✓	✓
[61]	• Subset of disassembly centres and processing centres to be opened. • Transportation strategy that satisfy demand by manufacturing centres and recycling centres.	✓	✓	
[65]	• Vehicle routing problem with simultaneous delivery and pick-up.			✓
[46]	• Optimal location and products assign for each facilities.	✓		
[14]	• Optimal number of manufacturing and remanufacturing product per period.		✓	
[44]	• Design and planning the facility location and allocation.	P		
[43]	• Facility location and capacities.	P		
[29]	• Facility location problem. • Optimal flow of products between facilities.	✓		
[71]	• Location-inventory-routing problem.	✓	✓	✓

[4]	· Optimal vehicle routing. · Optimal quantities of production to be produced and obtained from the supplier.	✓		✓
[30]	· Facility location, capacity expansion and transportation quantity between facilities.	✓		
[45]	· Location-allocation problem in reverse logistics.	✓		

Table 2. The number of publication of different decision planning in reverse logistic.

Decision planning	Number of Publication
Strategic	33
Tactical	11
Operational	3
Strategic and Tactical	8
Strategic and Operational	5
Tactical and Operational	1
Strategic, Tactical and Operational	1

CONCLUSIONS AND FUTURE DIRECTIONS

This paper reviewed the strategic, tactical and operational decision planning in the reverse logistics supply chain network design problem. The review shows the proportion of each decision planning and the trends of the reverse logistics problem studied between 1999 and 2017.

From the literature review, the future work in reverse logistics can be identified. Integration of the three different decisions planning (strategic, tactical and operational) should be further explored. The combination of these will give significant impact to the research of the reverse logistics network design problem in terms of the cost reduction and environmental impact. Operational decision planning requires more attention since only a handful of researches are conducted. The possibilities of integrating two different decisions planning should also be considered.Instead of concentrating only to one flow which is the reverse flow, the network should be integrated with the forward flow to form a closed-loop. The integration of the forward and reverse flows will increase the problem complexity. In order to obtain a realistic reverse logistics problem implemented in the industry, all stages of closed loop supply chain must be considered.Most of the previous studies focused on the uncertainty in demand and supply but very few studies investigated the uncertainty product returns. The main issue in reverse logistics is the process of the recovery which relates to product return. Researchers should investigate the uncertain quality and quantity of the recovery and returned products respectively. The model will be set as a stochastic model and can be solved using fuzzy or stochastic algorithm. To be more practical, the researchers should consider multi-objectives that combine the economic factor with the environmental impact as well as the social aspect. Since the objective of the total cost reduction and profit maximization are commonly used in the supply chain, other objectives such as maximizing the consumer satisfaction and responsiveness and minimizing the environmental impact caused by the pollution from the transportation and cleaning process should also be highlighted.A more realistic network design problem should also consider multi-period, multi-product and multi-facilities problems. During the setup of the network facility location, the network with multi collection and disassembly centers should be taken into account. The recovery operations should be well planned too. These decisions will give significant impact to the cost and the reduction of the gas emission from the production process. In multi-period problem, the impact of transportation lead time and time horizon for demand are worth to explore.When taking the environment issue into consideration, the main factor is due to the transportation. Researchers should consider different transportation decision variable such as the mode of transportation, the fuel consumption of transportation and freight weight capacity of transportation. These decisions will bring insights on how transportation is needed in generating large enough volume of recycling and remanufacturing process.

CONFLICTS OF INTEREST

The authors declare no conflicts of interest.

REFERENCES

1. MacCarthy, B.L. and Ramanathan, R. (2007) Recent Developments in Operations Strategy, Supply Chain Design and Production Systems. International Journal of Production Economics, 110, 1-4. https://doi.org/10.1016/j.ijpe.2007.02.010
2. Lu, D. (2011) Fundamentals of Supply Chain Management. Ventus Publishing ApS, Frederiksberg.
3. Farahani, R.Z., Rezapour, S., Drezner, T. and Fallah, S. (2014) Competitive Supply Chain Network Design: An Overview of Classifications, Models, Solution Techniques and Applications. Omega, 45, 92-118. https://doi.org/10.1016/j.omega.2013.08.006
4. Kumar, V.N.S.A., Kumar, V., Garza-Reyes, J.A. and Simpson, M. (2016) Resolving Forward-Reverse Logistics Multi-Period Model Using Evolutionary Algorithms. International Journal of Production Economics, 183, 458-469. https://doi.org/10.1016/j.ijpe.2016.04.026
5. Logozar, K., Radonjic, G. and Bastic, M. (2006) Incorporation of Reverse Logistics Model into In-Plant Recycling Process : A Case of Aluminium Industry. Resources, Conservation and Recycling, 49, 49-67. https://doi.org/10.1016/j.resconrec.2006.03.008
6. Srivastava, S.K. (2008) Network Design for Reverse Logistics. Omega, 36, 535-548. https://doi.org/10.1016/j.omega.2006.11.012
7. Wang, H.F. and Gupta, S.M. (2011) Green Supply Chain Management—A Product Life Cycle Approach. McGraw-Hill Education, New York.
8. Sasikumar, P., Kannan, G. and Haq, A.N. (2010) A Multi-Echelon Reverse Logistics Network Design for Product Recovery—A Case of Truck Tire Remanufacturing. International Journal of Advance Manufacturing Technology, 49, 1223-1234. https://doi.org/10.1007/s00170-009-2470-4
9. Fleischmann, M., Beullens, P., Bloemhof-ruwaard, J.M. and Wassenhove, L.N.V. (2001) The Impact of Product Recovery on Logistics Network Design. Production and Operations Management, 10, 156-173. https://doi.org/10.1111/j.1937-5956.2001.tb00076.x
10. Lee, J., Gen, M. and Rhee, K. (2009) Network Model and Optimization of Reverse Logistics by Hybrid Genetic Algorithm. Computers & Industrial Engineering, 56, 951-964. https://doi.org/10.1016/j.cie.2008.09.021

11. Sheriff, K.M.M., Gunasekaran, A. and Nachiappan, S. (2012) Reverse Logistics Network Design : A Review on Strategic Perspective. International Journal of Logistics Systems and Management, 12, 171–194. https://doi.org/10.1504/IJLSM.2012.047220
12. Lambert, S., Riopel, D. and Abdul-Kader, W. (2011) A Reverse Logistics Decisions Conceptual Framework. Computers & Industrial Engineering, 61, 561-581. https://doi.org/10.1016/j.cie.2011.04.012
13. Srivastava, S.K. (2007) Green Supply-Chain Management: A State-of-the-Art Literature Review. International Journal of Management Reviews, 9, 53-80. https://doi.org/10.1111/j.1468-2370.2007.00202.x
14. Sifaleras, A. and Konstantaras, I. (2015) Variable Neighborhood Descent Heuristic for Solving Reverse Logistics Multi-Item Dynamic Lot-Sizing Problems. Computers & Operations Research, 78, 385-392. https://doi.org/10.1016/j.cor.2015.10.004
15. Cruz-Rivera, R. and Ertel, J. (2009) Reverse Logistics Network Design for the Collection of End-of-Life Vehicles in Mexico. European Journal of Operational Research, 196, 930-939. https://doi.org/10.1016/j.ejor.2008.04.041
16. Lieckens, K. and Vandaele, N. (2007) Reverse Logistics Network Design with Stochastic Lead Times. Computers & Operations Research, 34, 395-416. https://doi.org/10.1016/j.cor.2005.03.006
17. Pishvaee, M.S., Farahani, R.Z. and Dullaert, W. (2010) A Memetic Algorithm for Bi-Objective Integrated Forward/Reverse Logistics Network Design. Computers & Operations Research, 37, 1100-1112. https://doi.org/10.1016/j.cor.2009.09.018
18. Ferguson, M. (2010) Strategic and Tactical Aspects of Closed-Loop Supply Chains. Foundation and Trends in Technology, Information and Operations Management, 3, 101-200. https://doi.org/10.1561/0200000019
19. Du, F. and Evans, G.W. (2008) A Bi-Objective Reverse Logistics Network Analysis for Post-Sale Service. Computers & Operations Research, 35, 2617-2634. https://doi.org/10.1016/j.cor.2006.12.020
20. Mukhopadhyay, S.K. and Setoputro, R. (2004) Reverse Logistics in e-Business: Optimal Price and Return Policy. International Journal of Physical Distribution & Logistics Management, 34, 70-89. https://doi.org/10.1108/09600030410515691

21. Kara, S.Ã., Rugrungruang, F. and Kaebernick, H. (2007) Simulation Modelling of Reverse Logistics Networks. International Journal of Production Economics, 106, 61-69. https://doi.org/10.1016/j.ijpe.2006.04.009
22. Salema, M.I.G., Barbosa-povoa, A.P. and Novais, A.Q. (2007) An Optimization Model for the Design of a Capacitated Multi-Product Reverse Logistics Network with Uncertainty. European Journal of Operational Research, 179, 1063-1077. https://doi.org/10.1016/j.ejor.2005.05.032
23. Banyai, T. (2011) Optimisation of a Multi-Product Green Supply Chain Model with Harmony Search. DAAAM International Scientific Book, 15-30. https://doi.org/10.2507/daaam.scibook.2011.02
24. Vahdani, B., Tavakkoli-moghaddam, R., Modarres, M. and Baboli, A. (2012) Reliable Design of a Forward/Reverse Logistics Network under Uncertainty : A Robust-M/M/c Queuing Model. Transportation Research Part E: Logistics and Tranportation Review, 48, 1152-1168. https://doi.org/10.1016/j.tre.2012.06.002
25. Che, Z.-H., Chiang, T.-A. and Kuo, Y.-C. (2012) Multi-Echelon Reverse Supply Chain Network Design with Specified Returns using Particle Swarm Optimization. International Journal of Innovative Computing, Information and Control, 8, 6719- 6731. http://www.ijicic.org/ijicic-10-09080.pdf
26. Cardoso, S.R., Barbosa-póvoa, A.P.F.D. and Relvas, S. (2013) Design and Planning of Supply Chains with Integration of Reverse Logistics Activities under Demand Uncertainty. European Journal of Operational Research, 226, 436-451. https://doi.org/10.1016/j.ejor.2012.11.035
27. Chiang, T., Che, Z.H. and Cui, Z. (2014) Designing a Multistage Supply Chain in Cross-Stage Reverse Logistics Environments : Application of Particle Swarm Optimization Algorithms. The Scientific World Journal, 2014, Article ID: 595902. https://doi.org/10.1155/2014/595902
28. Ahluwalia, P.K. and Nema, A.K. (2006) Multi-Objective Reverse Logistics Model for Integrated Computer Waste Management. Waste Management & Research, 24, 514- 527. https://doi.org/10.1177/0734242X06067252
29. Amin, S.H. and Baki, F. (2017) A Facility Location Model for Global Closed-Loop Supply Chain Network Design. Applied Mathematical Modelling, 41, 316-330. https://doi.org/10.1016/j.apm.2016.08.030

30. Chen, Y., Wang, L., Wang, A. and Chen, T. (2017) A Particle Swarm Approach for Optimizing a Multi-Stage Closed Loop Supply Chain for the Solar Cell Industry. Robotics and Computer Integrated Manufacturing, 43, 111-123. https://doi.org/10.1016/j.rcim.2015.10.006
31. Shih, L. (2001) Reverse Logistics System Planning for Recycling Electrical Appliances and Computers in Taiwan. Resources, Conservation and Recycling, 32, 55-72. https://doi.org/10.1016/S0921-3449(00)00098-7
32. Salema, M., Povoa, A. and Novais, A. (2006) A Warehouse-Based Design Model for Reverse Logistics. Journal of Operational Research Society, 57, 615-629. https://doi.org/10.1057/palgrave.jors.2602035
33. Min, H., Jeung, H. and Seong, C. (2006) A Genetic Algorithm Approach to Developing the Multi-Echelon Reverse Logistics Network for Product Returns. Omega, 34, 56-69. https://doi.org/10.1016/j.omega.2004.07.025
34. Soleimani, H. and Kannan, G. (2015) A Hybrid Particle Swarm Optimization and Genetic Algorithm for Closed-Loop Supply Chain Network Design in Large-Scale Networks. Applied Mathematical Modelling, 39, 3990-4012. https://doi.org/10.1016/j.apm.2014.12.016
35. Demirel, N.Ö. and Gökçen, H. (2008) A Mixed Integer Programming Model for Remanufacturing in Reverse Logistics Environment. International Journal of Advance Manufacturing Technology, 39, 1197-1206. https://doi.org/10.1007/s00170-007-1290-7
36. Mutha, A. and Pokharel, S. (2009) Strategic Network Design for Reverse Logistics and Remanufacturing using New and Old Product Modules. Computers & Industrial Engineering, 56, 334-346. https://doi.org/10.1016/j.cie.2008.06.006
37. Mansour, S. and Zarei, M. (2008) A Multi-Period Reverse Logistics Optimisation Model for End-of-Life Vehicles Recovery Based on EU Directive. International Journal of Computer Integrated Manufacturing, 21, 37-41. https://doi.org/10.1080/09511920701685325
38. Ko, H.J. and Evans, G.W. (2007) A Genetic Algorithm-Based Heuristic for the Dynamic Integrated Forward/Reverse Logistics Network for 3PLs. Computers & Operations Research, 34, 346-366. https://doi.org/10.1016/j.cor.2005.03.004
39. Min, H. and Ko, H. (2008) The Dynamic Design of a Reverse Logistics Network from the Perspective of Third-Party Logistics Service Providers. International Jour-nal of Production Economics, 113, 176-192. https://doi.org/10.1016/j.ijpe.2007.01.017

40. Lee, C.K.M. and Chan, T.M. (2009) Development of RFID-Based Reverse Logistics System. Expert Systems with Applications, 36, 9299-9307. https://doi.org/10.1016/j.eswa.2008.12.002
41. Pishvaee, M.S., Kianfar, K. and Karimi, B. (2010) Reverse Logistics Network Design using Simulated Annealing. International Journal of Advance Manufacturing Technology, 47, 269-281. https://doi.org/10.1007/s00170-009-2194-5
42. Viera, P.F., Viera, S.M., Gomes, M.I., Sousa, J.M.C. and Barbosa-póvoa, A.P. (2015) Designing Closed-Loop Supply Chains with Nonlinear Dimensioning Factors Using Ant Colony Optimization. Soft Computing, 19, 2245-2264. https://doi.org/10.1007/s00500-014-1405-7
43. Li, J., Wang, J., Pan, Q., Duan, P., Sang, H., Gao, K. and Xue, Y. (2017) A Hybrid Artificial Bee Colony for Optimizing a Reverse Logistics Network System. Soft Computing, 1-18. https://doi.org/10.1007/s00500-017-2539-1
44. Lee, D. and Dong, M. (2009) Dynamic Network Design for Reverse Logistics Operations under Uncertainty. Transportation Research Part E: Logistics and Tranportation Review, 45, 61-71. https://doi.org/10.1016/j.tre.2008.08.002
45. Pishvaee, M.S., Jolai, F. and Razmi, J. (2009) A Stochastic Optimization Model for Integrated Forward/Reverse Logistics Network Design. Journal of Manufacturing Systems, 28, 107-114. https://doi.org/10.1016/j.jmsy.2010.05.001
46. Saffar, M.M., G, H.S. and Razmi, J. (2015) A New Multi Objective Optimization Model for Designing a Green Supply Chain Network under Uncertainty. International Journal of Industrial Engineering Computations, 6, 15-32. https://doi.org/10.5267/j.ijiec.2014.10.001
47. Minner, S. (2001) Strategic Safety Stocks in Reverse Logistics Supply Chains. International Journal of Production Economics, 71, 417-428. https://doi.org/10.1016/S0925-5273(00)00138-9
48. Dobos, I. (2003) Optimal Production—Inventory Strategies for a HMMS-Type Reverse Logistics System. International Journal of Production Economics, 81-82, 351- 360. https://doi.org/10.1016/S0925-5273(02)00277-3
49. Mei, Y., He, G. and He, L. (2011) An Inexact Reverse Logistics Model for Municipal Solid Waste Management Systems. Journal of

Environmental Management, 92, 522- 530. https://doi.org/10.1016/j.jenvman.2010.09.011

50. Das, K. and Chowdhury, A.H. (2012) Designing a Reverse Logistics Network for Optimal Collection, Recovery and Quality-Based Product-Mix Planning. International Journal of Production Economics, 135, 209-221. https://doi.org/10.1016/j.ijpe.2011.07.010
51. Kim, K., Song, I., Kim, J. and Jeong, B. (2006) Supply Planning Model for Remanufacturing System in Reverse Logistics Environment. Computers & Industrial Engineering, 51, 279-287. https://doi.org/10.1016/j.cie.2006.02.008
52. Hu, T. and Sheu, J. (2002) A Reverse Logistics Cost Minimization Model for the Treatment of Hazardous Wastes. Transportation Research Part E: Logistics and Transportation Review, 38, 457-473. https://doi.org/10.1016/S1366-5545(02)00020-0
53. Walther, G. and Spengler, T. (2005) Impact of WEEE-Directive on Reverse Logistics in Germany. International Journal of Physical Distribution & Logistics Management, 35, 337-361. https://doi.org/10.1108/09600030510607337
54. Sheu, J. (2007) A Coordinated Reverse Logistics System for Regional Management of Multi-Source Hazardous Wastes. Computers & Operations Research, 34, 1442- 1462. https://doi.org/10.1016/j.cor.2005.06.009
55. Jayaraman, V., Guide Jr, V.D.R. and Srivastava, R. (1999) A Closed-Loop Logistics Model for Remanufacturing. Journal of Operational Research Society, 50, 497-508. https://doi.org/10.1057/palgrave.jors.2600716
56. Wang, Z., Yao, D. and Huang, P. (2007) A New Location-Inventory Policy with Reverse Logistics Applied to B2C e-Markets of China. International Journal of Production Economics, 107, 350-363. https://doi.org/10.1016/j.ijpe.2006.09.012
57. El-Sayed, M., Afia, N. and El-Kharbotly, A. (2010) A Stochastic Model for ForwardReverse Logistics Network Design under Risk. Computers & Industrial Engineering, 58, 423-431. https://doi.org/10.1016/j.cie.2008.09.040
58. Alumur, S.A., Nickel, S., Saldanha-da-gama, F. and Verter, V. (2012) Multi-Period Reverse Logistics Network Design. European Journal of Operational Research, 220, 67-78. https://doi.org/10.1016/j.ejor.2011.12.045

59. Lieckens, K. and Vandaele, N. (2012) Multi-Level Reverse Logistics Network Design under Uncertainty. International Journal of Production Research, 50, 23-40. https://doi.org/10.1080/00207543.2011.571442
60. Fonseca, M.C., García-sánchez, Á., Ortega-mier, M. and Saldanha-da-gama, F. (2010) A Stochastic Bi-Objective Location Model for Strategic Reverse Logistics. Top, 18, 158-184. https://doi.org/10.1007/s11750-009-0107-2
61. Roghanian, E. and Pazhoheshfar, P. (2014) An Optimization Model for Reverse Logistics Network under Stochastic Environment by Using Genetic Algorithm. Journal of Manufacturing Systems, 33, 348-356. https://doi.org/10.1016/j.jmsy.2014.02.007
62. Dekker, R., Fleischmann, M., Inderfurth, K. and van Wassenhove, L.N. (2013) Reverse Logistics: Quantitative Models for Closed-Loop Supply Chains. Springer Science & Business Media, Berlin.
63. Dethloff, J. (2001) Vehicle Routing and Reverse Logistics : The Vehicle Routing Problem with Simultaneous Delivery and Pick-Up. OR Spectrum, 23, 79-96. https://doi.org/10.1007/PL00013346
64. Kim, H., Yang, J. and Lee, K. (2009) Vehicle Routing in Reverse Logistics for Recycling End-of-Life Consumer Electronic Goods in South Korea. Transportation Research Part D: Transport and Environment, 14, 291-299. https://doi.org/10.1016/j.trd.2009.03.001
65. Zhang, T. and Chaovalitwongse, W.A. (2014) Integrated Ant Colony and Tabu Search Approach for Time Dependent Vehicle Routing Problems with Simultaneous Pickup and Delivery. Journal of Combinatorial Optimization, 28, 288-309. https://doi.org/10.1007/s10878-014-9741-1
66. Alshamrani, A., Mathur, K. and Ballou, R.H. (2007) Reverse Logistics : Simultaneous Design of Delivery Routes and Returns Strategies. Computers & Operations Research, 34, 595-619. https://doi.org/10.1016/j.cor.2005.03.015
67. Achillas, C., Vlachokostas, C., Aidonis, D., Moussiopoulos, N., Iakovou, E. and Banias, G. (2010) Optimising Reverse Logistics Network to Support Policy-Making in the Case of Electrical and Electronic Equipment. Waste Management, 30, 2592- 2600. https://doi.org/10.1016/j.wasman.2010.06.022

68. Bai, C. and Sarkis, J. (2013) Flexibility in Reverse Logistics : A Framework and Evaluation Approach. Journal of Cleaner Production, 47, 306-318. https://doi.org/10.1016/j.jclepro.2013.01.005
69. Guo, K.U.N. and Zhang, Q. (2014) A Discrete Artificial Bee Colony Algorithm. International Journal of Information Technology & Decision Making, 13, 1-19.
70. Ramos, T.R.P., Gomes, M.I. and Barbosa-póvoa, A.P. (2014) Planning a Sustainable Reverse Logistics System : Balancing Costs with Environmental and Social Concerns. Omega, 48, 60-74. https://doi.org/10.1016/j.omega.2013.11.006
71. Deng, S., Li, Y., Guo, H. and Liu, B. (2016) Solving a Closed-Loop Location-Inventory-Routing Problem with Mixed Quality Defects Returns in e-Commerce by Hybrid Ant Colony Optimization Algorithm. Discrete Dynamics in Nature and Society, 2016, Article ID: 6467812. https://doi.org/10.1155/2016/6467812

CHAPTER 12

Research on Agri-Food Cold Chain Logistics Management System: Connotation, Structure and Operational Mechanism

Jiuyi An[1], Lei Wang[2], Xiaohua Lv[1]

[1]Beijing Wuzi University, Beijing, China.

[2]Beijing Benz Automotive Co. Ltd., Beijing, China.

ABSTRACT

Along with the demand of development, this paper will define agri-food cold chain logistics management system. It is a general term of management system, organization system, standard system, running mechanism, etc. According to the activities and characters of agri-food cold chain logistics, the content of agri-food cold chain logistics management system includes three levels of management activities. There are macro management and part of the industry management of the government-led agricultural cold chain

Citation: Jiuyi An, Lei Wang, Xiaohua Lv (2015) Research on Agri-Food Cold Chain Logistics Management System: Connotation, Structure and Operational Mechanism. Journal of Service Science and Management,08,894-902. doi: 10.4236/jssm.2015.86090.

logistics activity, industry management of industry association as the main body, and micro enterprise operation management of enterprise as the main body. In addition, this paper will draw lessons from the city of agricultural products cold chain logistics management experience in the international, the main behavior of the logistics market, strengthen monitoring and early warning, strengthen the whole process, and make market access strictly and other key links have to be closely surrounded. Through improving agricultural products cold chain logistics management laws and regulations system, standard system, management system, testing system, certification system, technology supporting system, information service system, and establishing emergency mechanism and other agricultural products cold chain logistics security support system, government, industry, consumers, media, education and scientific research institutions and other relevant parties can take many aspects, many angles, many levels mutually supporting measures, in order to establish and improve the agricultural products cold chain logistics safety control system, and establish the whole process control system from farm to table, to ensure the safety of agricultural products cold chain logistics.

Keywords:- Fresh Agricultural Products, Management System of Cold Chain Logistics, Operational Mechanism

INTRODUCTION

In recent years, along with the structure adjustment of agriculture and the improvement of level of consumption, the production and circulation of fresh agricultural products increase year by year; the whole society puts forward higher requirements on safety and quality of fresh agricultural products, so as to drive the rapid development of cold chain logistics. The state attaches great importance to the development of cold chain logistics. In recent years the No. 1 document of the Central Committee had stressed the need to accelerate the construction of cold chain logistics system of agricultural products, and promote the circulation of agricultural products. Some national standards, industry standards and local standards of cold chain logistics have been issued and carried out. "Food Safety Law" and other important laws and regulations are gradually perfect. But our management of cold chain logistics of agricultural products is still in its early stage, and the systematic cold-chain logistics management system has not yet formed. However, China's cold chain logistics management of agricultural products is still in its initial stage; the system of the cold chain logistics management system has not yet formed. The research on the connotation, structure and operation mechanism of the cold chain logistics management system of fresh agricultural products has certain theoretical and practical significance.

THE CONNOTATION AND CHARACTERISTICS OF MANAGEMENT SYSTEM OF COLD CHAIN LOGISTICS OF AGRICULTURAL PRODUCTS

The definition of agri-food cold chain logistics management system is a special supply chain system. It collect the meat, poultry, aquatic products, vegetables, fruits, eggs, fresh agricultural products from origin of harvest, slaughter or fishing, then let the products process, storage, transportation, distribution, retail always in a suitable environment of low and control temperature, its aim is to extremely ensure the quality of product and quality safety, reduce loss and prevent the pollution that in a special supply chain system [1] .

Agri-food cold chain logistics is a chain network system which "from farm to table", the system is made of the producers (farmers/production base) of agricultural products, the purchasing and processing enterprise of agricultural products, the distribution enterprise of agricultural products, the subject of cold chain logistics activity of agricultural products. The work of cold chain logistics of agricultural products includes products production, processing, and distribution, those are happen in the process of transportation, storage, loading and unloading, handling, packaging, distribution, information processing, etc.According to the sorted of the cold chain logistics activities of agricultural products, this paper will define agricultural products' cold chain logistics management system as a general designation of management system, organization system, standard system, running mechanism, and so on, which in order to perform the industry regulatory responsibilities, build industry healthy and orderly market environment, ensure the quality of product quality and safety. Above those are according to the requirements of the market economy system efficient allocation of resources, and aim to adapt to the development of modern circulation of agricultural products demand. Combine the property of cold chain logistics activity of agricultural products with the definition of the connotation, the category of agricultural products' cold chain logistics management includes three levels management activities, they are macro management of government-led cold chain logistics activity of agricultural products and some industry management, industry management that the industry association act as the main body and the enterprising micro operation management that enterprises act as the main body.

The basic connotation of cold chain logistics management system of agricultural products includes the following meanings:

1) Operation mechanism is the soul of the management system. If has no Decision-making, implementation, feedback and other continuous operating activities, the cold chain logistics management system of agricultural products cannot work and have none value. The cold chain logistics management system of agricultural products rely on legal and administrative means to regulate and control the main body of market, through the government enforcement regulation or economic means to guide the market to form industry self-discipline mechanism and control mechanism of the enterprises. Through perfect the laws and standards to establish the quality inspection and supervision mechanism.
2) The organizational structure of cold chain logistics management system of agricultural products is the forms of management system. The organizational structure is the carrier of management function, if has no organization, management system is impossible. The existing cold chain logistics system of agricultural products builds organizational structure according to departments build the government organizational structure, industry association in accordance with the link of cold chain logistics and the product categories.
3) The core content of cold chain logistics management of agricultural products is the division of management responsibilities in the management system and the configuration of the management functions. The division of management responsibilities decides the configuration of the management functions, and the configuration of the management functions affects the work of system. The current management system takes governmental agricultural departments at all levels, health department, quality checking department, food and drug quality supervision department, the ministry of commerce and industry and so on as the main body, takes the industry associations and a variety of chamber of commerce that the government authorized them parts industry management, market entity as the architecture of auxiliary system.

4) Legal norms are integral parts of management system, the operation of the cold chain logistics management system of agricultural products takes influenced and restriction of the environment of laws and regulations. At present, it has established the legal framework with Law of Agricultural Produce Quality Safety, Farm Bill and other basic law as the main body, continuously strengthens the cohesion of department regulations and the operation of local laws and regulations, to speed up the legalization process of cold chain logistics management of agricultural products. The core construction of standard system is aim to develop and promote a batch of specific operation and technology standard of cold chain logistics of agricultural products, to establish the whole quality control system on the basis of HACCP, to develop specific operation and technology standard of agricultural products' cold chain logistics that in line with international standards.

Our cold chain logistics management system of agricultural products has the following basic features:

The first is the diversification of management. There are many links, participation, small scale, scattered and other characteristics in the agricultural production and circulation, it decides the main body is diversified in the cold chain logistics management of agricultural products. The main body of cold chain logistics management of agricultural products is made of the government at all levels that leading management, the industry association of auxiliary participation and industry management, business enterprise, etc. On one hand, the cold chain logistics management of agricultural products strength the government mandatory regulation; on the other hand, it gradually authorizes industry association industry management, gives full play to the role of the industry self-discipline, actively guides enterprise self-control and encourages social supervision.

The second is thorough regulatory. Along with the social attention to the quality and safety of agricultural products, consumers are demanding products whole process of the supply chain that from farm to table should be strictly regulated. Law of Agricultural Produce Quality Safety says: "Administrative departments of agriculture of the people's governments at or above the county level are responsible for the supervision and administration of the quality and safety of agricultural products, the relevant departments of the people's governments at or above the county level depend the responsibilities to devise the labor, and take responsible for the work of the quality and safety of agricultural products." It confirms the relevant departments' safety regulatory functions and responsibilities of the cold chain logistics of agricultural products, gradually establishes a management system in the whole process of the "from farm to table" [2] .

The third is the information is visualized all the way. The process and span of the cold chain logistics management of agricultural products is giant, many participative and administrative main bodies, transmission of information easies to interrupt. Law of Agricultural Produce Quality Safety says: "The competent agricultural administrative departments of the State Council and the people's governments of provinces, autonomous regions and municipalities directly under the central government shall publish information on the quality and safety of agricultural products in accordance with their respective functions and powers." In addition, the business and the departments of agriculture take responsibility for expanding the pilot range of full traceability information platform of the cold chain logistics of agricultural products, gradually perfecting the multistage information tracing system. Except the government departments extensively release cold chain logistics information of agricultural products, the main participative bodies of cold chain logistics gradually strengthen information system construction, continuously strengthen information exchange and communication, strengthen the information sharing of cold chain logistics of agricultural products. The fourth is management legalization and standardization. Legalization and standardization is the basic safeguard for the cold chain logistics management of agricultural products, and is the external driving force for specific management of the main body of cold chain logistics management of agricultural products. In recent years, our country continues to strengthen the construction of laws, regulations and standardization, makes the behavior of cold chain logistics management of agricultural products fully on the basis of legal and standard, and makes the legal business enterprises have full legal protection.

STRUCTURE ANALYSIS OF COLD CHAIN LOGISTICS MANAGEMENT SYSTEM OF AGRICULTURAL PRODUCTS

Cold chain logistics management system of agricultural products is a complex system. Its structure includes organizational system of cold chain logistics of agricultural products, supervision system of cold chain logistics of agricultural products, legal Standard system of cold chain logistics of agricultural products and industry public service system. In addition, the supervision system includes market access system, testing system and tracing system. The legal Standard system includes legal system, technical standards, specifications, application guidelines, etc. Aim to accurately grasp the internal logic relationship of each part and the operation mechanism through an analysis of internal structure and function of the cold chain logistics management system of agricultural products (Figure 1).

Analysis of the Supervision System of Cold Chain Logistics of Agricultural Products

The supervision system of agri-food cold chain logistics is generic terms of effective market activities. In the management activities, the government department or social organization of cold chain logistics of agricultural products through certain legal and institutional arrangements to maintain the normal order of market. In the process of cold chain logistics of agricultural products, agricultural products from production, processing, transportation, storage and other link to the consumer at last, based on each link of "from farm to table" to control market access, detection and tracing, to form the regulation and supervision for the operational main body of the market, to ensure the safety of service process of cold chain logistics of agricultural products. The supervision system includes tracing system, market access system, authentication system, inspection and testing system, etc. Looking from the content of the supervision system, supervision system of cold chain logistics of agricultural products includes macro supervision and concrete market supervision. The macro supervision is for the development of market of cold chain logistics and concrete market supervision. The concrete market supervision mainly includes: supervise the implementation of laws and regulations, maintain the order of fair competition in the market, to protect the benefit and interests of all kinds of market participator; strengthen the supervision of quality and safety of cold chain logistics of agricultural products, to maintain the interests of consumers; and strengthen the supervision of production safety of all kinds of cold chain logistics, to maintain the interests of labor [3] .

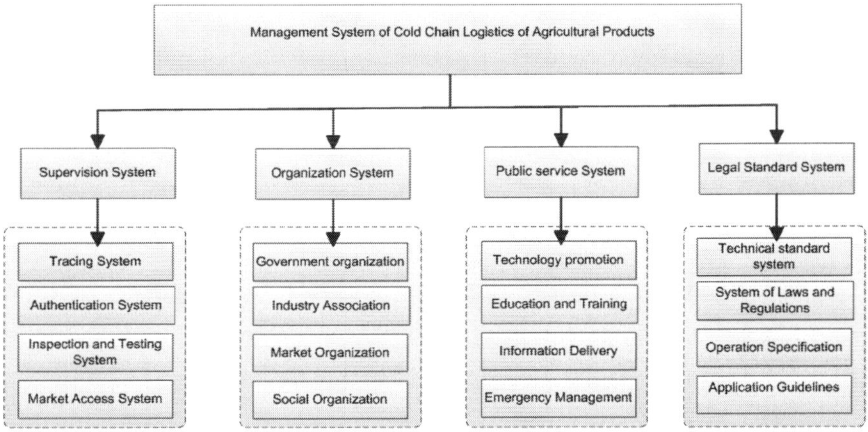

Figure 1. Framework of management system of cold chain logistics of agricultural products.

Analysis of the Organization System of Cold Chain Logistics of Agricultural Products

The organization system of cold chain logistics of agricultural products is the carrier of activities of cold chain logistics management. According to the structure of the cold chain logistics system, the related functional departments of agricultural products' cold chain logistics of the government construct the basic frame of organization system.

This structure includes both horizontal government institutions, function configuration and the longitudinal central and local institutions, function configuration, mainly including government departments at all levels, health department, quality checking department, the ministry of commerce and industry, business department, the ministry of environmental protection, etc. and includes these departments' corresponding extension organization that set up in the province, city and county respectively. In addition, the organizations system of cold chain logistics of agricultural products includes industry association organization, market organization and other auxiliary institutions. According to the organization's operation, in order to realize the management function of normal and efficient that by a certain legal and system design to form the management system and mechanism of formation, and the person that in the management system.

Analysis of the Cold Chain Logistics Industry Public Service System of Agricultural Products

Public service system of cold chain logistics is surrounding the problem of solving the industry developing, the government provides all kinds frameworks of public service, as well as provides the means and methods of public service, etc.

According to the content and structure of the public service system, the public service of cold chain logistics of agricultural products in China mainly includes: industry public education and training services, industry public infrastructure services, industry public information services, etc. In addition, the establishment of a sound emergency management mechanism of cold chain logistics of agricultural products has become an important part of the cold chain logistics management. The establishment of emergency mechanism to handling the safety incidents of agricultural product has become the international practice, China should establish the emergency public service system of cold chain logistics, such as emergency awareness, emergency laws and regulations, emergency plan, store of emergency cold chain logistics resources and so on.

Analysis of Laws, Regulations and Standards System of Cold Chain Logistics of Agricultural Products

Laws, regulations and standards system of cold chain logistics of agricultural products combines the characteristic of the industry development of cold chain logistics of agricultural products and established the policy legal system of cold chain logistics of agricultural products of adapted market economy system, to provide legal basis and support for the participator of cold chain logistic and management activities.

Policy and legal system of cold chain logistics of modern agricultural products uses the national basic laws and regulations, rules and regulations of government of all levels to concretely lead and regulate the constitution of specialized laws and regulations, industry technology and management standards of cold chain logistics of agricultural products. Looking from the point of function, Laws, regulations and standards system of cold chain logistics of agricultural products is the main method of regulating the market main body, it is the supporting system of building the market legal environment of cold chain logistics of agricultural products as well. Looking from the point of running, to set up the whole management system of cold chain logistics of agricultural products on the basis of the laws and regulations of cold chain logistics of agricultural products supporting the system, in order to realize the seamless connection between the supervision of whole process and information of cold chain logistics activities, and ensures the quality safety of agricultural products, give the government, enterprises and the consumer with effective legal protection.

THE OPERATIONAL MECHANISM OF COLD CHAIN LOGISTICS MANAGEMENT SYSTEM OF AGRICULTURAL PRODUCTS

In the constructional aspect of management system, drawing on the experience of cold chain logistics management of foreign agricultural products, the government and social forces coordinately take participate in governance has become the mainstream management system. In the construction of cold chain logistics management system of agricultural products in China, the government management department according to the characteristics of cold chain logistics activities, a clear division of government and society, and gradually form a sub full management system. At the same time, in the field of the government can not cover, or in the field of low efficiency in the role of government, the government gradually authorizes and plays the force of social organizations such as industry associations and other social organizations to participate in the industry management activities. The automatic control management of enterprises in the business activities gradually strengthens, market self-control awareness of supply and demand enterprises continually strengthens. Along with the social public and consumer awareness raising, the initiative of the government, social organizations, enterprises and other management entities and market players to participate in the supervision improves. Through the multi agent participation, the cold chain logistics management system of agricultural products is formed in the horizontal and vertical multi levels.

In the constructional aspect of laws, regulations and standards, China has formed the law system that such as Farm Bill, The Law of Agricultural Product Quality and Safety and other basic law system as main body, the basic legal construction has been strengthened, the making of regulations of departments focus on linking up each other, the establishment of local regulations focus on linking up the regulations that from the central to the local of all levels, increasing the integrity and operability of the legal system. In order to accelerate the construction of cold chain logistics standard system which is the main unit of the association, developing and promoting a batch of specific operation and technology standard of cold chain logistics of agricultural products, establishing the whole quality control system on the basis of HACCP, but laws, regulations, standards, guidelines and other mutual support and cohesion mechanism should be further strengthened.

In the regulatory system, the domestic government and the association gradually carry out HACCP, GMP, GAP (good agricultural practices), ISO (International Standard Organization) and other quality and safety certification system and market access system, enterprises begin to attach importance to the construction of standards and certification. Government and social forces increases investment year by year, improves the infrastructure building of the supervision and inspection of the cold chain logistics production, processing, storage, transportation, transit, import and export, and other major aspects. Overall, the cold chain logistics certification system just in the start, the market access system is still in the exploration stage, the construction of inspection and testing institutions to be strengthened, the information traceability system has a significant role in demonstration.

In the main body of market operation, because of the production and the circulation subject take small business as the main body, The operation of cold chain logistics of agricultural products still takes production, processing, distribution and other sub operations as main operational body, the whole process of cold chain logistics has developed in large chain enterprises and agricultural products leading enterprises. As a result, the main market operational body of cold chain logistics of agricultural products is still in the situation of the common participation that conjoining the operational enterprises of production and supply with the logistics enterprises of specific cold chain supply (Figure 2).

THE GOAL OF CONSTRUCTING THE COLD CHAIN LOGISTICS MANAGEMENT SYSTEM OF AGRICULTURAL PRODUCTS IN THE CITY

Drawing lessons from the city of agricultural products cold chain logistics management experience in the international, the main behavior of the logistics market, strengthen monitoring and early warning, strengthen the whole process, make market access strictly and other key links have to be closely surrounded. Through improving agricultural products cold chain logistics management laws and regulations system, standard system, management system, testing system, certification system, technology supporting system, information service system, and establishing emergency mechanism and other agricultural products' cold chain logistics security support system, government, industry, consumers, media, education and scientific research institutions and other relevant parties can take many aspects, many angles, many levels mutually supporting measures, in order to establish and improve the agricultural products cold chain logistics safety control system, and establish the whole process control system from farm to table, to ensure the safety of agricultural products cold chain logistics [4] .

1) To establish a clear and coordinated management system of agricultural products cold chain logistics between government agencies. The basic requirement of the system of agricultural products cold chain management is the government's position to be accurate, from the farmland to the dining table to implement the full responsibility of the parties to be clear, all levels of management agencies has to be lean and efficient.

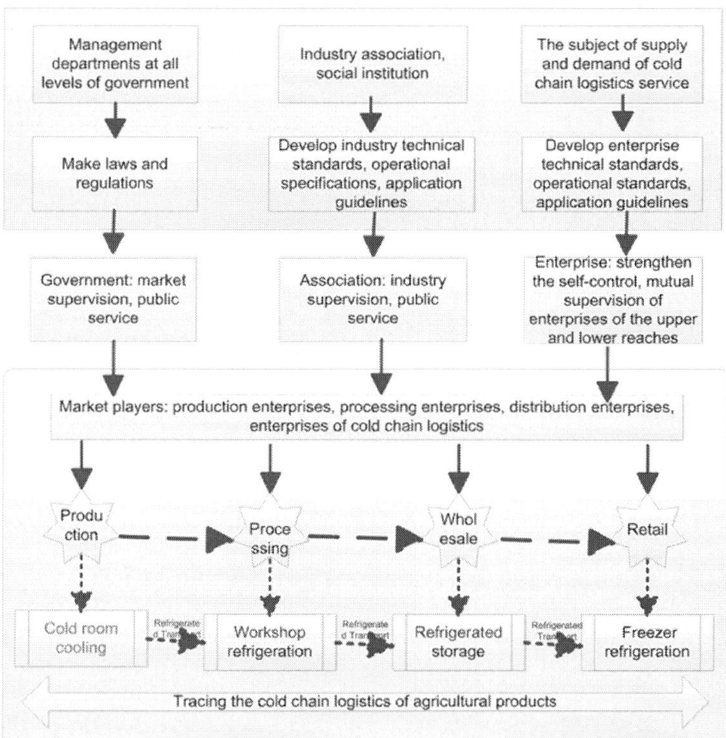

Figure 2. The operational mechanism of cold chain logistics management system of agricultural products.

2) Improving the technology level of agricultural products cold chain logistics management. The main factors affecting the safety of agricultural products cold chain logistics in China, and determined by the key technology. Research on the safety of cold chain logistics of agricultural products in the stage, the selection and the gradual development, priority develop energy saving, environmental protection, safety and cold chain logistics equipment technology, in order to further develop a more reliable, fast, portable, accurate detection technology, to speed up the development of agricultural production, processing, storage, packaging and transportation process of safety control technology. Preliminary establish cold chain logistics safety science and technology system to adapt to the development of modern agricultural products.
3) Improving the standard system of agricultural products cold chain logistics. In strengthening the unified management and give full play to the role of the relevant departments, to set up a set of agricultural products, which is in line with China's national conditions and international standards of cold chain logistics standard system. Activity using the international standards and foreign advanced standards actively and increase the intensity of international standards.
4) Establish a unified, authoritative and efficient inspection system for cold chain logistics inspection system of agricultural products. Drawing lessons from foreign experience and according to the principle of overall planning, reasonable layout, it has to establish a mutual coordination, division of labor is reasonable, function, technology, advanced, functional, personnel matching, running efficient cold chain logistics inspection system of agricultural products. In the detection range, it can meet the needs of production, processing and circulation of the whole process of the implementation of safety testing. In the detection ability, it also can meet the national standards, industry standards and relevant international standards for the agricultural products cold chain logistics parameters testing requirements. Relying on the existing monitoring and detection of resources, to further improve the main production base, processing base, distribution center, transit center, import and export port inspection and testing capabilities, improve the level of supervision, to ensure product's quality and safety.

5) To establish a unified and standardized system of agricultural products cold chain logistics certification and accreditation. In order to strengthen the whole process safety control, the material production, processing, transportation, sales enterprises in the food raw can promote the GPS system and HACCP system certification.
6) To establish and improve the emergency response mechanism of cold chain logistics of agricultural products. The establishment of emergency mechanism for handling the safety of agricultural products has become an international practice, China should establish a system of agricultural products cold chain logistics security emergency response mechanism from the establishment of laws and regulations system, improving the information collection, processing and dissemination mechanism.
7) To establish a unified and coordinated system of laws and regulations. We should learn from the experience of international agricultural products cold chain logistics laws and regulations, to establish the basic framework of China's agricultural products cold chain logistics safety regulations system, and to improve the existing laws and regulations system, giving law enforcement authorities more power, strengthen legislation and law enforcement supervision, etc.
8) Establish the cold chain logistics distribution system of agricultural products of urban and rural areas. Agricultural products cold chain logistics, from the farm to the table, as the main production base of the vast rural areas of the cold chain logistics system is an important part of the cold chain logistics. At the same time, the cold chain distribution system for the city's consumption driven type is also an essential part of cold chain logistics. Therefore, the construction of cold chain logistics system in urban and rural areas is the need to adapt to the development of modern agricultural products circulation, At the same time, it is an important task to construct the cold chain logistics distribution system for agricultural products, especially the construction of cold chain distribution system.

CONCLUSION

From what has been discussed above, firstly, according to the definition, activities and characters of agri-food cold chain logistics, the content of agri-food cold chain logistics management system is classified in three levels of management activities. There are macro management and part of the industry management of the government-led agricultural cold chain logistics activity, industry management of industry association as the main body, and micro enterprise operation management of enterprise as the main body. Secondly, cold chain logistics management system of agricultural products is a complex system. Its structure includes organizational system of cold chain logistics of agricultural products, supervision system of cold chain logistics of agricultural products, legal standard system of cold chain logistics of agricultural products and industry public service system. In addition, the supervision system includes market access system, testing system and tracing system. The legal standard system includes legal system, technical standards, specifications, application guidelines, etc. The aim is to accurately grasp the internal logic relationship of each part and the operation mechanism through an analysis of internal structure and function of the cold chain logistics management system of agricultural products. Finally, from the experience of international agricultural products cold chain logistics management, the regulatory authorities should strengthen the supervision of market behavior and early warning, strict market access, and the key to form a closed loop. Through improving agricultural products cold chain logistics management laws and regulations system, standard system, management system, testing system, certification system, technology supporting system, information service system, and establishing emergency mechanism and other agricultural products cold chain logistics security support system, government, industry, consumers, media, education and scientific research institutions and other relevant parties can take many aspects, many angles, many levels mutually supporting measures, in order to establish and improve the agricultural products cold chain logistics safety control system, and establish the whole process control system from farm to table, to ensure the safety of agricultural products cold chain logistics.

ACKNOWLEDGEMENTS

The fund project, the National Social Science Fund "The research of fresh agricultural products cold chain logistics management system for the needs of the city" (11BJY111).

REFERENCES

1. Reddy, C.M., Malliyala, S., Naresh, Y., Raghunandan, H.V. and Jinadatharaya, H. (2012) Good Cold Chain Management Practices. Journal of Pharmacy Research, 10, 5043-5047.
2. Li, C.J., Yang, C. and Liang, J. (2009) Analysis on the Dynamic Mechanism of Supervision of Agricultural Product Quality Safety. Journal of Guangxi University of Finance and Economics, 22.
3. Jin, F.Z. (2011) Thinking on the Construction of Supervision and Business Support System of Agricultural Product Quality Safety. Quality and Safety of Agricultural Products, 6, 8-10.
4. Zhang, B. and Weng, X.W. (2012) Function and Complexity of Supervision System of Agricultural Product Quality and Safety and Its Solution. Agricultural Modernization Analysis, 33, 59-63.

CHAPTER 13

Sea-Port Operational Efficiency: An Evaluation of Five Asian Ports Using Stochastic Frontier Production Function Model

Otieno Robert Kennedy[1], Khin Lin[1], Hualong Yang[1], Banomyong Ruth[2]

[1] Transportation Management College, Dalian Maritime University, Dalian, China

[2] Department of International Business, Logistics & Transport, Thammasat University, Bangkok, Thailand

ABSTRACT

Sea-port operational efficiency is critical factor for handling of goods in the international supply chains, and is viewed to impact transportation and logistics which play an important role in trade exchange with other countries. It is important to evaluate operational efficiency of sea-ports to reflect their

Citation: H. Yang, K. Lin, O. Kennedy and B. Ruth, Sea-Port Operational Efficiency: An Evaluation of Five Asian Ports Using Stochastic Frontier Production Function Model, Journal of Service Science and Management, Vol. 4 No. 3, 2011, pp. 391-399. doi:10.4236/jssm.2011.43045.

status and reveal their position in this competitive environment. Moreover, knowing impacts of efficiency of sea-ports on the supply chain is vital for business survival. This study uses stochastic frontier and inefficiency models to analyze sea-port operational efficiency and Delphi technique to seek expert respondents' opinion on its characteristics. The research also uses structural equation modeling to build a model of sea-port operational efficiency as a further step to examine the significance of the characteristics. The results of this study emphasize the need to improve sea-port operational efficiency, and indicate which characteristics should be given more attention.

Keywords:- Sea-port Operational Efficiency, Supply Chain, Stochastic Frontier Model, Delphi Technique, Asian Ports

INTRODUCTION

Sea-ports have been considered to be important parts of international supply chains [1]. They hold a very important role and are the most critical nodes in the supply chain [2]. It is widely believed that sea-ports form a vital link in the overall trading chain [3]. Sea-ports are a component of freight distribution as they offer a maritime to land interface for cross-border businesses. Therefore, efficiency of sea-port operation is vital for supply chains. A lot of research has been done in the area of sea-port operational efficiency. Many of such research dwell on tactical means of bolstering sea-port operational efficiency [3-6]. Some researchers regard sea-port as Third Party Logistics (3PL) provider that intervenes in a series of different companies and supply chains [7]. Three different channels: trade channel, logistics channel and supply chain channel were identified by [7] as a new framework of measuring performance of sea-ports. However, there still exists a gap in assessing the sea-port operational efficiency. The question: "What characteristics are key to improving sea-port operational and to what extent they can bolster efficiency?" has not been adequately addressed in literature. Some research dwell on one or two aspects at a time leaving out other aspects.

This research seeks to address this concern by examining sea-port operational efficiency, establishing determinants of such efficiency for its evaluation and building its model. Since various aspects of efficiency do not lend themselves to precise analytical techniques but can benefit from subjective judgments on collective basis [8], Delphi technique was chosen as a feasible method for identifying key factors that are significant to sea-port operational efficiency.

The paper will be comprised of five main sections. Following the introductory section, the paper will present reviews on related literature concerning sea-port operational efficiency and logistics flexibility. The section will be devoted to defining it and outlining its theoretical precepts. Next section will present the selected research methodology followed by results of the research as well as their analyses and related discussions. The paper will further present implications of research findings and discussions of limitations of the current study as well as recommendations for further research. Finally, there will be summary and conclusions.

LITERATURE REVIEW

Sea-port operation is defined as cargo handling (or moving) activity, performed by a designed company (gang or team), consisting of labor and machines. It is also defined as the operation of a wharf and other port facilities, operation of port passenger transport service, operation of cargo loading/unloading, haulage and warehousing services within a port area and so on [9].

Presently, there is difficulty in defining port efficiency due to non-universal definition of what indicates an efficient port or what port efficiency entails [10]. An efficient sea-port should be one that is competent in operations [10]. Based on this definition, efficiency of sea-port operations is determined by duration (time) of ship's stay in a port, quality of cargo handling and quality of service to inland transport vehicle during passage through the port [11]. Quality of cargo handling is in the form of berth throughput [10] and quality of service to inland vehicle is dependent on port infrastructure. Productivity has been identified as a measure of sea-port operational efficiency [3].

Many researchers have used various approaches to evaluate sea-port efficiency. Annual firm level surveys have been employed as indicators of sea-port operational efficiency, but "there was almost no information on how port efficiencies evolve over time from these studies" [11, p. 3]. A number of studies have used data on inputs, outputs and production function theory, by means of data envelopment analysis (DEA), to estimate the most efficient production frontier across a set of sea-ports [6,12, 13]. The approaches using this method have the advantage of economies of scale derived from econometric evidence but the drawback is that they typically assume constant return to scale [11]. To address the issue of error estimation and statistical confidence, another approach, econometric estimation of cost functions, was developed by [11]. The method, however, has "difficulties with data requirements, particularly measurement of labor, capital and other requirements" [11, p. 5] which limit its application to many sea-ports at a time.

Some research has been done on the contribution of port ownership to efficiency. Transformation from public to private ownership is believed to improve sea-port operational efficiency even without change in level of competition [14]. Some researchers contended this position [14] and have opinion that principal-agent problems may also arise in the private sector as a result of capital market imperfections [3]. Reference [15] applied stochastic production function to evaluate technical efficiency but did not show that port ownership has significant effects on sea-port operational efficiency. Moreover, [16] developed stochastic frontier model and carried out comparison of efficiency level of 40 container terminals, but also failed to establish the relationship between terminal ownership, operations and efficiency level. On the contrary, a number of studies have shown relationship between port ownership and sea-port operational efficiency [3,11,17]. Relative efficiency of a number of Asian ports was assessed by [17] using a combination of cross-sectional and panel data versions of stochastic frontier model and the finding was that there seems to be some support that privatization should have some relationship with improvement in efficiency [3]. These efforts by the researchers show that port ownership is a likely determinant of operational efficiency.

It has been found that size of sea-port has positive effects on its efficiency [18]. Also, it has been shown that ports with larger throughput seem to have certain performance advantage over those with smaller throughput [17]. In research on 15 sea-ports [19] showed that port efficiency has no clear relationship with its size and function (hub or feeder) [3].

METHODOLOGY AND DATA

Methodology

Researchers identified research tools and strategies that will be employed and related their application to specified research objectives. Questions to be addressed by this research set up the direction that the research will take and are tied to research objectives. The research questions were: 1) What are the operational efficiencies of a set of sea-ports? 2) What are the key characteristics of sea-port operational efficiency? 3) What is the model of sea-port operational efficiency? Based on the research questions, the objectives of this research were to: evaluate operational efficiencies of a number of selected sea-ports; examine the characteristics of sea-ports' operational efficiency and build its structural model.

Stochastic frontier production function model in [19] was used to evaluate the efficiency of selected sea-ports. This method was selected because of its ability to estimate technical inefficiency [19] and simultaneously estimates parameters of an inefficiency model with those of stochastic frontier production model [3]. Delphi technique was employed to seek expert respondents' opinion on the characteristics of sea-port operational efficiency.

Stochastic Frontier and Inefficiency Models

Stochastic frontier, also known as composed error, model $y_i = g(x_i, \beta) + \varepsilon_i \quad (i = 1, 2, 3, \cdots, N)$, [20] where y_i is the output for observation i, x_i is vector of inputs for observation *i*, β is the vector of parameters, e_i is error term for observation *i,* postulates that the error term e_i is made up of two independent components, $e_i = vi - u_i$ where, *vi* is a two- sided error term representing statistical noise in any relationship $u_i \geq 0$ is one-sided error term representing technical inefficiency.

The exponential form of the proposed model giving production function in Equation (1) as,

$$y_{it} = \exp\left(x_{it}\beta + v_{it} - u_{it}\right)$$

where, y_{it} is the production at the tth observation (t = 1, 2, …, T) for the ith firm (i = 1, 2, …, N); x_{it} is logarithm of input variables v_{it} is random error assumed to be truncated normal distribution with respect to mean and variance, $N(0, \sigma_v^2)$ and independently distributed of non-negative random variable, u_{it} . The truncated normal distribution using Wald or generalized likelihood-ratio test [20] is specified in this research to justify the selection of distribution form for technical inefficiency effects [3].

Regression of effects of inefficiency on the variables that explain inefficiency is given by Equation (2) as,

$$u_{it} = z_{it}\delta + W_{it}$$

where zit is a vector of explanatory variables; δ is a vector of unknown scalar parameters; W_{it} is truncation of normal distribution, $N(0, \sigma_v^2)$, such that the point of truncation is such that point of truncation is $-z_{it}\delta$ [3].

To avoid serial correlation among random errors, this part of the research will employ cross-sectional data and use cross-sectional analysis to address concerns about correlations of inefficiencies and input choices [15,17]. We propose maximum likelihood method for simultaneous estimation of parameters of stochastic frontier model and those of technical inefficiencies model [3]. The likelihood function is expressed in terms of variance parameters $\sigma_s^2 = \sigma_v^2 + \sigma^2$ and $\gamma = \sigma^2/\sigma_s^2$ inefficiency can therefore be defined in terms of the ratio between observed output and potential output given input x_{it} as,

$$TE_{it} = y_{it}/\exp(x_{it}\beta + v_{it}) = \exp(-z_{it}\delta - W_{it}) \quad (3)$$

Delphi Technique

Delphi, a systematic interactive forecasting technique which depends on a panel of independent, carefully selected expert respondents [21], was used to identify characteristics of efficient sea-port operations. Delphi was used because researchers felt that expert opinion was the best available evidence. The method has the ability to provide anonymity to respondents and controlled feedback process as well as allows application of variety of statistical analysis techniques to interpret data [8].

A group of selected 32 expert respondents in port management, shipping and logistics field drawn mainly from China, Hong Kong, Singapore and Korea participated in the study by answering questionnaires sent through email. Sample size was kept reasonably small, so as to do justice to the rich evidence given by qualitative studies [22]. Table 1 shows sample of expert respondents who participated in this part of the research. The process was carried out in three rounds as recommended by literature [21]. Caution was exercised to deal with concerns of Delphi such as time consumption [21], molding of opinion, subjectivity versus objectivity and the assumption that the participants have equivalent knowledge and experience [21].

Round 1 questionnaire was unstructured with questions and statements phrased to increase chances of accuracy of responses [21]. In Round 1 the respondents were asked to identify key characteristics of sea-port operational efficiency and provide their comments as to why they thought the identified items were important. After Round 1 deadline, two weeks as recommended by literature between rounds [21], results were analyzed and a summary of the same was included in the design of Round 2 questionnaire. A review of questionnaire statements was done to remove any possible influences by monitor team's views [9]. The questionnaire was sent to the experts to refine ideas, explore agreements and disagreements and to probe strengths and weaknesses of opinions.

After deadline of Round 2, responses were analyzed and Round 3 questionnaire was prepared. In this round respondents were asked to revise judgments or specify why they remain out of consensus [21]. Round 2 and Round 3 question was, "To what extent do you agree that the following factors contribute to sea-port operational efficiency? Please, provide your comments or additional factors that in your view are significant to sea-port efficiency." The experts were required to rate the characteristics using Likert-scale (1 = strongly disagree, 2 = disagree, 3 = moderately agree, 4 = strongly agree, 5 = very strongly agree).

Data

Since the main activity of container ports is handling containers only one output will be identified in this study. The total throughput is a good measurement for the output of a container terminal [3]. Table 2 shows container throughput of five container ports being investigated. Literature argues that only the input factors: quay length; terminal area; and the number of quay cranes are relevant variables affecting container terminal operational efficiency [17]. Table 3 shows these input factors for the five ports studied.

Table 1. Sample of Delphi respondents.

No. of port/firm employees	No. of employees represented	% employees the ports/firms in all	Respondents expected (f_e)	Respondents who participated (f_0)
100 - 249	1,505	43	22	14
250 - 499	770	22	11	7
500 - 999	525	15	8	5
1,000+	700	20	10	6
		χ^2 test (= =7.089, df = 3, p = 0.0691)		
Job title				
CEO/president	805	23	11	7
Vice president	525	15	8	5
Manager	1,890	54	27	17
Director	280	8	4	3
		χ^2 test (= =6.533, df = 3, p = 0.0884)		
Total	3,500	100	50	32

χ^2 is obtained using the formula: $\frac{\sum(f_o - f_e)^2}{f_e}$

Table 2. Port container throughput in TEUs.

Asian Port	2005	2006	2007	2008	2009	2010
Singapore	23,192,000	24,792,400	27,932,000	29,918,200	25,866,400	28,400,000
Hong Kong	22,602,000	23,538,580	23,998,000	24,248,400	21,040,000	23,699,000
Shanghai	18,084,000	21,710,000	26,168,000	28,006,000	25,002,000	29,100,000
Shenzhen	16,197,173	18,468,900	21099,000	21,416,000	18,250,100	22,510,000
Busan	11,840,000	12,030,000	13,261,000	13,425,000	11,954,000	14,180,000

Sources: China Port Industry Report, 2010; Container Throughput Hong Kong [available online] www.pdc.gov.hk/docs/Hkport.pdf. United Nations (2010) Review of Maritime Transport, Chapter 5.

Table 3. Input factors of five major Asian ports.

	Quay Length (km)	Terminal Area (hectares)	No. of Quay cranes
Singapore	16.945	600	190
Hongkong	19	285	93
Shanghai	20	401	240
Shenzhen	5.543	344	71
Busan	9.95	292.5	70

Source: Global Container Terminal Operators Annual Review, 2010.

We will also use the port ownership [3] to analyze port ownership structure. According to [3] the degree of private sector participation is given as 0/3 for purely public ownership; 1/3 for public regulator and landownership while private sector acts as operators; 2/3 is given for public being regulator while the private sector perform the role of landowner and operator and 3/3 is given for purely private ownership.

Cobb-Douglas and Translog functional forms for stochastic production function forms are tested based on maximum likelihood method by applying FRONTIER package version 4.1. The following, Equation (4), is the stochastic production function to be tested,

$$\begin{aligned} \ln Y_i = {} & \beta_0 + \beta_1 \ln(X_{1i}) + \beta_2 \ln(X_{2i}) + \beta_3 \ln(X_{3i}) \\ & + \beta_4 \ln(X_{1i})^2 + \beta_5 \ln(X_{2i})^2 + \beta_6 \ln(X_{3i})^2 \\ & + \beta_7 \ln(X_{1i})\ln(X_{2i}) + \beta_8 \ln(X_{1i}) + \beta_9 \ln(X_{3i}) \\ & + \beta_9 \ln(X_{2i})\ln(X_{3i}) + v_i - u_i \end{aligned} \tag{4}$$

Technical inefficiencies are defined by,

$$u_i = \delta_0 + \delta_1 z_{1i} + \delta_2 z_{2i} + \delta_3 z_{2i}^2 + W_i \qquad (5)$$

where,

ln: natural logarithm;

Y_i : total throughput in TEU on container port (terminal) i in a given year;

X_{1i} : terminal quay length in metres of port i;

X_{2i} : terminal area in hectares of port i;

X_{3i} : number of quay cranes used in port i;

z_{1i} : size of port i, dummy binary variable that distinguishes whether annual port throughput exceeds 15 mil- lion TEUs or not (i.e. 1 if throughput is ≥15 million TEUs, and 0 otherwise);

z_{2i} : the extent of private sector participation.

Generalized likelihood method was used to test functional forms. The method is as follows, Likelihood Ratio LR = –2{ln[L(H0)] – ln[L(H1)]}, where L(H0) and L(H1) are values of likelihood function under null hypothesis (H0: β4 = β5 = β6 = β7 = β8 = β9 = 0) and the alterna tive H_1, respectively. The presence of inefficiency effects of u_i was examined using one-sided generalized likelihood-ratio statistics [23,24].

RESULTS, DISCUSSIONS AND STRUCTURAL MODEL

Results of Stochastic Production and Inefficiency Models

Empirical results based on data from the five sea-ports are shown in Table 4. All beta were statistically significant at $p < 5\%$, showing that the three inputs: total quay length, terminal area and quay cranes, have significant effects on production, consistent with result of [3] and [18]. The estimate of č is 0.8283 implying that 82.83% of the total variability is associated with technical efficiency of production and it is very significant, $p < 1\%$.

The coefficient, d_2, is negative implying that there is positive relationship between technical efficiency and privatization in sea-ports. These results concur with those found by [3].The coefficient d_3 is positive implying an inverted U-shaped relationship of z_{2i}^2 with sea-port privatization. The best level of privatization for the seaports studied is given by: $z_{2i}^2 = \left| \frac{\delta_2}{2\delta_3} \right|$ obtained from

$$\frac{\partial U_i}{\partial z_{2i}} = \delta_2 + 2\delta_3 z_{2i} = 0 .$$

Table 4. Final estimates of time-invariant technical efficiencies for period 2005 to 2010.

	Stochastic Frontier Model				Inefficiency Model		
	Coefficient	Standard error	t-ratio		Coefficient	Standard error	t-ratio
β_0	13.758	2.214	6.213	δ_0	0.457	0.205	2.229
β_1	0.172	0.078	2.202	δ_1	0.609	0.195	3.123
β_2	0.278	0.124	2.242	δ_2	0.667	0.800	0.833
β_3	0.010	0.003	3.333	δ_3	0.444	0.871	0.510
Variance parameter σ^2	0.0732	0.0429	1.7068	Variance parameter γ	0.8283	0.1080	7.6696

Note: approximate critical value for t-ratio at p = 5% is 2.131 and at p = 1% is 2.947 log likelihood function = 17.332313; LR test of the one-sided error = 23.041894; [note that this statistic has a mixed chi-square distribution]; number of iterations = 13; (maximum number of iterations set at: 100).

Substituting the values of the coefficients gives the best level of privatization as 0.751 which falls between Private/Public (0.667) and Private (1.000). The estimate of coefficient of port size, δ_1 is negative meaning that large sea-ports are more efficient than smaller ones.

Technical efficiencies using production function of the ports were found and the yearly trends are as in Figures 1 and 2. In Figure 1 inefficiency effects were considered, for $H_0(\delta_1 = \delta_2 = \delta_3 = 0)$ and for ownership effects, for $H_0(\delta_2 = \delta_3 = 0)$ the results are as in Figure 2. Since null hypotheses: no inefficiency, $\delta_0 = \delta_1 = \delta_2 = \delta_3 = 0$,

Likelihood Ratio (LR) = 30.5544 greater than critical value x^2 (5%) = 10.371; no inefficiency effects, $\delta_1 = \delta_2 = \delta_3 = 0$, LR = 21.1298 greater than critical value $\chi^2(5\%) = 7.81$ and ownership effects (no privatization), $\delta_2 = \delta_3 = 0$, LR = 23.0419 greater than critical value $\chi^2(5\%) = 5.99$, the null hypotheses were rejected [24].

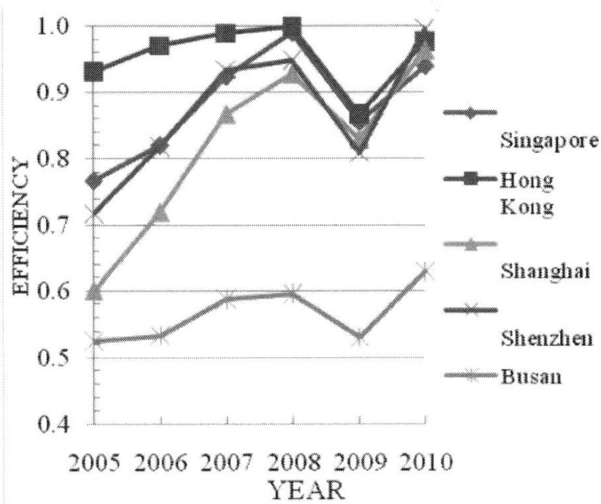

Figure 1. Technical efficiencies (inefficiency effects).

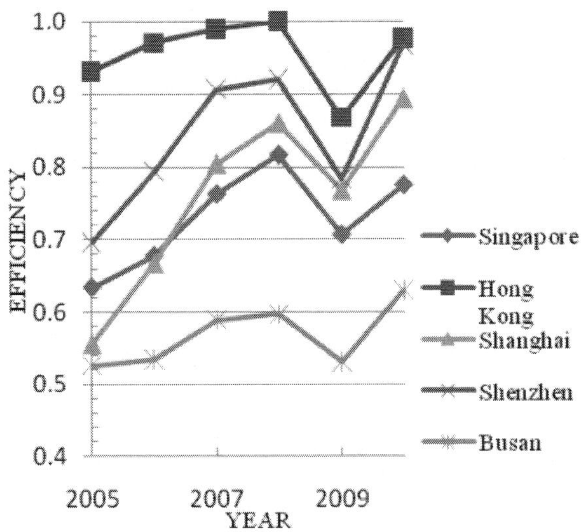

Figure 2. Technical efficiencies (ownership effects).

When inefficiency effects are considered, port of Singapore is found to be more efficient than that of Shanghai that catches up and overtakes it in the second half of the year 2009.

However, when ownership effects are considered, the port of Shanghai overtakes that of Singapore just before mid 2006, to be the third most efficient after Shenzhen and Hong Kong. Between 2008 and 2009 all the trends were on downward trend. This is possibly due to global recession. After 2009, the trends were upwards with Shenzhen having highest rise in efficiency followed by Shanghai port. Time invariant efficiencies were found as follows: Singapore, 0.9116; Hong Kong, 0.9443; Shanghai, 0.9029; Shenzhen, 0.9412 and Busan, 0.5963. Mean technical efficiency for all the ports was 0.8593.

Results of Delphi Survey

Round 1:- Round 1 responses were analyzed after all the 32 experts had submitted their responses (after a period of two weeks). Items generated by respondents were checked for their similarity and were refined with great care to avoid losing their initial meanings. Finally, a summary of eight items identified by respondents was drawn. Items generated in Round 1 were: "Quality of port infrastructure i.e. information systems, storage area" (PE1); "Size of sea-port, terminal area, quay length, quay cranes, berths" (PE2); "Quality of cargo/container handling" (PE3); "Port throughput" (PE4); "Measures to reduce ship turn-round time" (PE5); "Level of private sector participation" (PE6); "Nature of value added services" (PE7) and "Port charges and other costs incurred by port users" (PE8).

Round 2:- The results of Round 2 are presented in Table 5. Literature recommends using median with inter-quartile range when applying Likert scale [25]. The results show that both PE1 and PE8 were rated highest followed by PE3. The range of respondent rating was same for all the characteristics except PE2 whose ratings had inter-quartile range of 1.75 showing that there was comparatively less consensus on its level of contribution.

Round 3:- The Round 3 results in Table 5 show that PE1, PE3 and PE8 were rated highest, the rating of PE3 improved from median of 4.5 in Round 2 to 5 in Round 3. Consensus also improved for both PE2 and PE4, evidenced by lower inter-quartile range in Round 3 than in Round 2. These results show that respondents either strongly or very strongly agree that the factors contribute to efficiency of sea-port operations.

Structural Model

Model of sea-port operational efficiency was built using the characteristics. Testing of model fit was done using Normed Fit Index (NFI), Incremental Fit Index (IFI), Tucker-Lewis Index (TLI), Root-Mean-Square Error Approximation (RMSEA) and Standardized Root Mean Residual (SRMR). NFI, IFI and TLI ≥ 0.9 imply acceptable model.

Table 5. Median and inter-quartile range of Rosunds 2 and 3 responses.

Item	Round 2				Round 3			
	Median	Quartiles		Inter-quartile range	Median	Quartiles		Inter-quartile range
		First	Third			First	Third	
PE1	5	4	5	1	5	4	5	1
PE2	4	3	4.75	1.75	4	3.25	4.75	1.5
PE3	4.5	4	5	1	5	4	5	1
PE4	4	3	4	1	4	4	4.75	0.75
PE5	4	4	5	1	4	4	5	1
PE6	4	4	5	1	4	4	5	1
PE7	4	4	5	1	4	4	5	1
PE8	5	4	5	1	5	4	5	1

RMSEA and SRMR ≤ 0.08 show an acceptable model. For RMSEA and SRMR = 0.00 the model is perfect [26]. Factor of loading for characteristics in model should be ≥ 0.7 for acceptable models [27]. Reliability of characteristics and that of model estimates was tested by determining Cronbach's alpha [28]. Reliability occurs for Cronbach's alpha ≥ 0.7 [26]. Cronbach's alpha for characteristics in the models was 0.735. Convergent validity of model estimates was measured by Average Variance Extracted (AVE) with acceptable values of ≥ 0.5 [27]. The factors of loading, λ, necessary for calculation of AVE were obtained using principal component analysis (PCA) extraction capability of SPSS® software version 19.

Structural Model of Port Operational Efficiency

Figure 3 shows model of sea-port operational efficiency. The characteristic "Quality of cargo-handling" (PE3) and "Port throughput" (PE5) have the highest regression weights of 3.00 and 2.38, respectively. The mean rating by respondents appear as 4.41 and 4.34 respectively for the two characteristics with their residual error terms e3 and e5 being the lowest at 0.10 and 0.19 respectively.

Model fit indices were as follows: NFI = 0.901; IFI = 1.101; TLI = 1.187; RMSEA = 0.000; SRMR = 0.0307; χ^2 {N = 32, df = 8, p = 0.0342} = 16.637. The fit indices show that the model is acceptable.

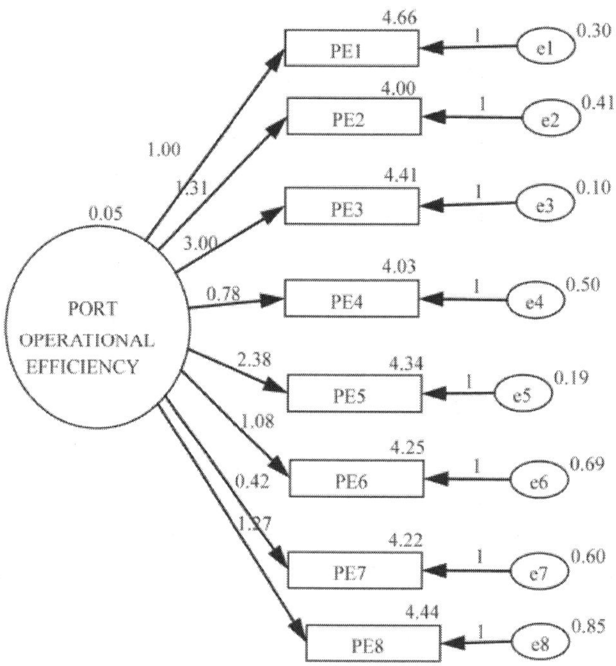

Figure 3. Structural model of port efficiency.

IMPLICATIONS OF RESEARCH FINDINGS

The empirical results provide some valuable implications for port authorities, operators, shipping companies and logistics providers. It is found that, generally, large seaports are more efficient than smaller ones possibly because of the quality of port infrastructure, storage and cargo-handling. Apart from port size, the level of privatization is also an important factor for efficiency. It is, however, noted that full privatization is not effective in increasing efficiency of port operations; meaning that the relationship between private sector participation and seaport operational efficiency is not linear. It is shown that the best extent of privatization is between public/private (0.67) and Private (1.00) mode, according to literature. This means that private sector participation should be limited to "landowner and operator" functions while port authorities should take the role of regulator.

Survey results for Delphi technique show that the respondents had consensus that all the eight factors identified (see Table 5) were important determinants of operational efficiency of sea-ports. They had opinion that port infrastructure, quality of cargo-handling and port charges including other costs are the top ranking determinants of port efficiency. These results imply that port authorities should charge reasonable amounts since shippers use costs in selecting port to use, therefore it is a measure of competitiveness as supported by [3]. Model results in show that sea-port management need tò focus mainly on boosting quality of container-handling, putting measures aimed at reducing ship turn-round time, improving quality of port infrastructure and equipment. This is evidenced by the regression weights of the characteristics in the model.

LIMITATIONS AND FURTHER RESEARCH

It is worth noting that this research did not examine the effects of cost due to unavailability of data. Another limitation is that analyses were limited to containers and left out cargo which could have provided interesting form of results. The third limitation was that this research relied on Delphi interviews to examine the determinants of efficiency perhaps a different approach could yield another scenario. Further research is therefore recommended to address these issues and shed more light in this area, and possibly with a wide range of ports.

SUMMARY AND CONCLUSIONS

The objective of this paper is to evaluate operational efficiencies of a number of selected sea-ports; examine the characteristics significant to sea-ports' operational efficiency and build a structural model of sea-port operational efficiency. The obtained results provide valuable implications to port authorities, operators and business practitioners depending on port. The results show that port size and infrastructure, private sector participation and quality of both cargo-handling and logistics services are important determinants of efficiency.

ACKNOWLEDGEMENTS

We thank the editor in chief and the reviewers very much for their constructive pieces of advice. This research is supported by the National Natural Science Foundation of China (Grant Nos. 70972008 and 70971014).

REFERENCES

1. B. Slack and A. Fremont, "Transformation of Port Terminal Operations: From the Local to the Global," Transport Reviews, Vol. 25, No. 1, 2005, pp. 117-130.
2. L. Wang, "Study of Port Logistics Marketing under the Environment of Supply Chain," International Journal of Business and Management, Vol. 6, No. 3, 2011, p. 267.
3. J. Tongzon and W. Heng, "Port Privatization, Efficiency and Competitiveness: Some Empirical Evidence from Container Ports (Terminal)," Transportation Research, Part A, Vol. 39, No. 5, 2005, pp. 405-424.
4. J. Zhu, "Data Envelopment Analysis with Preference Structure," Journal of the Operational Research Society, Vol. 47, No. 1, 1996, pp. 136-150.
5. H. S. Lee, M. T. Chou and S. G. Kuo, "Evaluating Port Efficiency in Asia Pacific Region with Recursive Data Envelopment Analysis," Journal of the Eastern Asia Society for Transportation Studies, Vol. 6, 2005, pp. 544-599.
6. K. Lin and H.-L. Yang, "Port Throughput Analysis of China and Five Member Countries of ASEAN," International Conference on Intelligent Computation Technology and Automation, Vol. 2, 2010, pp. 914-917.
7. H. A. Linstone and M. Turoff, "The Delphi Method: Techniques and Applications," Addison-Wesley, Reading, 1975.
8. Business Alert—China, "Shenzhen Requires Operation Licence for Port Enterprises," Business Alert, No. 07, July 2004.
9. P. A. Osaretin, "Efficient Port Operations and Benefits of Actors—A Case Study of Tin Can Island Port, Nigeria," Masters Dissertation, Göteborg University, Gothenburg, 2006.
10. G. De Monie, "Measuring and Evaluating Port Performance and Productivity," CEPAL Review, No. 99, 2009, p. 173.
11. B. A. Blonigen and W. W. Wilson, "New Measures of Port Efficiency Using International Trade Data," NBER Working Paper No. 12052, 2006.
12. J. Tongzon, "Efficiency Measurement of Selected Australian and Other International Ports using Data Envelopment Analysis," Transportation Research Part A: Policy and Practice, Vol. 35, No. 2, 2001, pp. 113-128. doi:10.1016/S0965-8564(99)00049-X

13. E. Martinez-Budria, R. Diaz-Armas, M. Navarro-Ibanez, and T. Ravelo-Mesa, "A Study of the Efficiency of Spanish Port Authorities Using Data Envelopment Analysis," International Journal of Transport Economics, Vol. 26, No. 2, 1999, pp. 237-253.
14. S. Estrin and V. Perontin, "Does Ownership Always Matter?" International Journal of Industrial Organization, Vol. 9, No. 1, 1991, pp. 55-72. doi:10.1016/0167-7187(91)90005-6
15. Z. Liu, "The Comparative Performance of Public and Private Enterprises," Journal of Transportation Economics and Policy, Vol. 29, No. 3, 1995, pp. 263-274.
16. J. Van Den Broeck, G. Koop, J. Osiewalski and J. Steel, "Stochastic Frontier Models: A Bayesian Perspective," Journal of Econometrics, Vol. 61, No. 2, 1994, pp. 273- 303. doi:10.1016/0304-4076(94)90087-6
17. K. Cullinane, D. W. Song and R. Gray, "A Stochastic Frontier Model of the Efficiency of Major Container Terminals in Asia: Assessing the Influence of Administrative and Ownership Structures," Transportation Research, Part A, Vol. 36, No. 8, 2002, pp. 743-62.
18. T. Notteboom, C. Coeck, and J. Van Den Broeck, "Measuring and Explaining the Relative Efficiency of Container Terminals by Means of Bayesian Stochastic Frontier Models," International Journal of Maritime Economics, Vol. 2, No. 2, 2000, pp. 83-106.
19. G. E. Battese and T. J. Coelli, "A Model for Technique Inefficiency Effects in a Stochastic Frontier Production Function for Panel Data," Empirical Economics, Vol. 20, No. 2, 1995, pp. 325-332. doi:10.1007/BF01205442
20. T. J. Coelli, D. S. P. Rao, C. J. O'Donnell and G. Battese, "An Introduction to Efficiency and Productivity Analysis," 2nd Edition, Springer, Berlin, 2005.
21. C.-C. Hsu and B. A. Sandford, "The Delphi Technique: Making Sense of Consensus," Practical Assessment Research & Evaluation, Vol. 12, No. 10, 2007, pp. 1-8. http://pareonline.net/ getvn.asp?v=12&n=10
22. J. Ritchie and J. Lewis, "Qualitative Research Practice," A Guide for Social Science Students.
23. T. J. Coelli, "Estimators and Hypothesis Tests for a Stochastic Frontier Function: A Monte Carlo Analysis," Journal of Productivity Analysis, Vol. 6, No. 3, 1995, pp. 247-268. doi:10.1007/BF01076978

24. D. A. Kodde and F. C. Palm, "Wald Criteria for Jointly Testing Equality and Inequality Restrictions," Econometrica, Vol. 54, No. 5, 1986, pp. 1243-1248.

25. N. Mogey, "So You Want to Use Likert Scale?" 1999. http://www.icbl.hw.ac.uk/ltdi/cookbook/info_likert_scale/ index.html

26. S. A. Mulaik, L. R. James, J. Van Alstine, N. Bennett, S. Lind and C. D. Stilwell, "Evaluation of Goodness-of-Fit Indices for Structural Equation Models," Psychological Bulletin, Vol. 105, No. 3, 1989 pp. 430-445. doi:10.1037/0033-2909.105.3.430

27. R. Zinbarg, W. Revelle, I. Yovel and W. Li, "Cronbach's, Revelle's and McDonald's: Their Relations with Each other and Two Alternative Conceptualizations of Reliability," Psychometrika, Vol. 70, No. 1, 2005, pp. 123-133.

CHAPTER 14

The Design of Urban Traffic in Ferizaj through Operational Research

Bashkim Çerkini, Fakije Zejnullahu, Dhuratë Hyseni

Faculty of Engineering and Informatics, University of Applied Sciences in Ferizaj, Ferizaj, Kosovo.

ABSTRACT

Cities are undergoing rapid changes continuously due to the high demands of this era, and simultaneously affect several life fields, namely urban transport. High demands are triggering people to travel more frequently than ever; thus, they utilize public transportation more and private transportation less as the latter ceased to fulfill all the transportation needs. Thus, public transportation demand has been increasing greatly alongside citizens' needs. Nonetheless, many cities lack proper urban traffic planning and organization, while some lack an urban transport service. Ferizaj, a city in Kosovo, is one of the cities that lacks an urban traffic designation; hence, this paper

Citation: Çerkini, B. ,Zejnullahu, F. and Hyseni, D. (2020), The Design of Urban Traffic in Ferizaj through Operational Research. Journal of Computer and Communications, 8, 40-48. doi: 10.4236/jcc.2020.812004.

presents a designed urban traffic model, precisely suitable to fulfill the urban transportation need for Ferizaj city. This model is designed under the utilization of applied mathematics' techniques and operational research. Several factors have been considered, following the geographical distribution of the population, existing roads, and residents' needs. Consequently, the Solver program has been used as an optimization tool to find the shortest path and most economical paths, added in the discussion part. Besides, the likelihood of the designed urban traffic model's application in Ferizaj is discussed, considering its viability and application conditions. This study presents mathematical constraints to design a model of the bus network in Ferizaj through Solver. We have used mathematical optimization methods, graph theory, the simulation model through the Solver computer program for network minimal distances and presenting the first model of the Urban traffic network in Ferizaj.

keywords:- Transport, Solver, Algorithm, Optimization and Nodes

INTRODUCTION

The development of public urban transport is an important topic in modern society. As cities grow, so do people's need for transportation, and not all of these trips can be made by private transportation. Therefore, there is a need to provide a well-organized urban public transport to efficiently meet the demands of citizens. Planning and optimization of public urban transport is needed to make the best possible use of economic resources, better functionality for citizens, and preserve the environment. Thus, the requirement of this paper constitutes basic information for proper transport planning using operational research techniques.

Sustainable and integrated transport is associated with several challenges, especially in urban public transport within cities and peripheral regions. Many regions worldwide face the problems of increasing individual motorized transport, causing many negative phenomena such as traffic jams, parking problems in city centers and increased pollution from the number of emissions. Governments and local authorities are trying to solve these problems by making various investments in public transport.

The research in urban public transport aims to optimize urban traffic lines, find shorter routes, and reduce travel time that overall brings economic and environmental benefits. A combination of mathematical models and operational research techniques are used to optimize the urban public transport network.

Ferizaj has not had an urban bus transportation network within the city yet. Traveling in an urban network represents an interaction between the demand for transport and means of transport, road networks, population distribution as well as the existing road infrastructure. In this research paper, Solver is used to finding a solution that satisfies the constraints and minimizes the objective cell value. The mathematical relationships between the objective and constraints and the decision variables are used to get the solution for optimized Urban Traffic in Ferizaj. With the Simplex LP Solving method, we could find a globally optimal solution given enough time [1].

Many researchers have analyzed the urban public transport service problems using various mathematical and operational research methods.

In [2] an optimization model is developed for the bus transit network based on the road network and the origin-destination area (OD). The model aims to achieve minimum transfers and maximum passenger flow for line length and non-linear rates as constraints. The Coarse-Grain Parallel Ant Colony Algorithm (CPACA) was used to solve this problem. To effectively search for optimal global solutions, we use a pheromone distribution orientation rule to regulate ant path search activities according to objective value. Parallel Ant Colony Algorithm (ACA) is used to shorten the computation time. The model was tested with Dalian city survey data. The results show that in an optimized bus network with fewer transfers and travel time, the application of CPACA effectively increases the calculation speed and quality.

[3] Ant Colony Optimization algorithms are used for more efficient road traffic. In traffic management, these algorithms have growing popularity, with their ability to find optimal solutions in situations where traditional methods fail to find any good solution. A transport system where ACO has been used to help with the transport network problems is on the London Underground network. The network consists of 270 connected stations along with a combined total of 250,000 miles of roads and transports around 1.107 million people to London each year. With this size and volume of users, disturbances in specific areas can greatly impact the network. Therefore, finding an effective diversion around any particular problem in a short period of time is vital to prevent blockages that can spread throughout the system.

In [4] the problem of finding the shortest path is solved through the Ant Colony Optimization algorithm (ACO), which is based on the Ant Colony metaheuristic method. One of the most important elements of any algorithm based on the Ant Colony paradigm system's use is the rules for creating paths leading from a starting point to the endpoint.

[5] optimizes the distribution of bus routes in Wuhan duke to alleviate or solve the problems exposed by the extension of existing roads in the city, e.g., reducing overlapping roads, increasing network coverage, and reducing the main road's burden. This research has used a method for designing various transit routes based on stops, which treats some railway routes as constraints. Genetic algorithms have been applied to search for the optimal combination of candidate pathways. Evaluation based on optimization results is generated to analyze whether the expected improvements have been achieved, especially in the short term.

In [6] some of the basic line planning models, identification of their characteristics, mathematical approaches, and line planning algorithms are presented. Furthermore, similar topics are highlighted, such as current and future research directions, and the line planning process is examined. It is assumed that the infrastructure has already been provided and introduced as a public transport network. In particular, it is assumed that the stations are fixed, and the set of possible node pairs will be given. These can be roads (by bus) or the rail system (rail, tram, or underground). Line planning includes the defined number of lines and line routes. They also include determining the frequency of lines, i.e., how often these services should be provided. The problem of line planning is finding the line system, ensuring that urban public transport is convenient for passengers and small costs.

[7] proposes a methodology for the optimal development of transit networks, which minimizes transit transfers and total user costs while maximizing service coverage, the information provided on transit demand and the transit network road network area. Research provides an effective mathematical and computer tool with minimal support of the heuristic method. The methodology includes the representation of road transit networks and the solution of search spaces, the objective functions that represent the user's total costs, and the users' lack of willingness to make transfers. The methodology has been tested with solutions to problems published at this stage and has been implemented on a large scale in network optimization in Miami-Dade County, Florida.

METHODOLOGY

The methodology used in this paper is the collection and processing of data related to urban transport and the definition of optimization criteria [8], the use of mathematical methods, and the Solver program for the design of urban transport, which has not yet been implemented in the city of Ferizaj.

Mathematical optimization methods, graph theory, and solver programs are used. Sustainable and integrated transport is associated with several challenges, especially in urban public transport within cities and peripheral regions.

The paper is based on the following objectives:

- Minimize the length of roads between origin and destination for certain paths
- Proposal for the urban public transport network in Ferizaj.

In Figure 1, the Map of Ferizaj is presented from Google Map. We have placed the numbers at each intersection and measurements between the points obtained in the geoportal. The geoportal is a web portal used to find and access geographic information via the Internet. Each intersection has been assigning a random number that we will consider from now on as nodes.

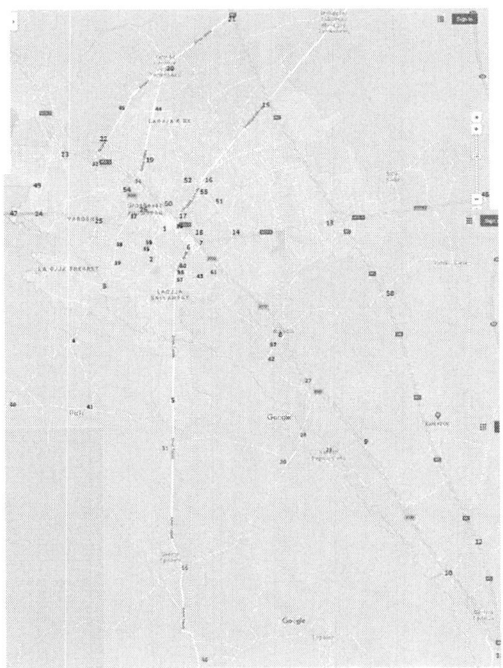

Figure 1. Ferizaj roads with intersection with numbers.

In Figure 2 all nodes are connected based on existing paths, and we have created the graph where all the work from now on is based on this graph through graph theory.

All paths between nodes are measured, and we have at our disposal the values of the lengths of all the paths for use in this paper. Using mathematical constraints, we have placed all this data in the Solver program to find the shortest possible routes and, consequently, the urban traffic network model in general in the city of Ferizaj.

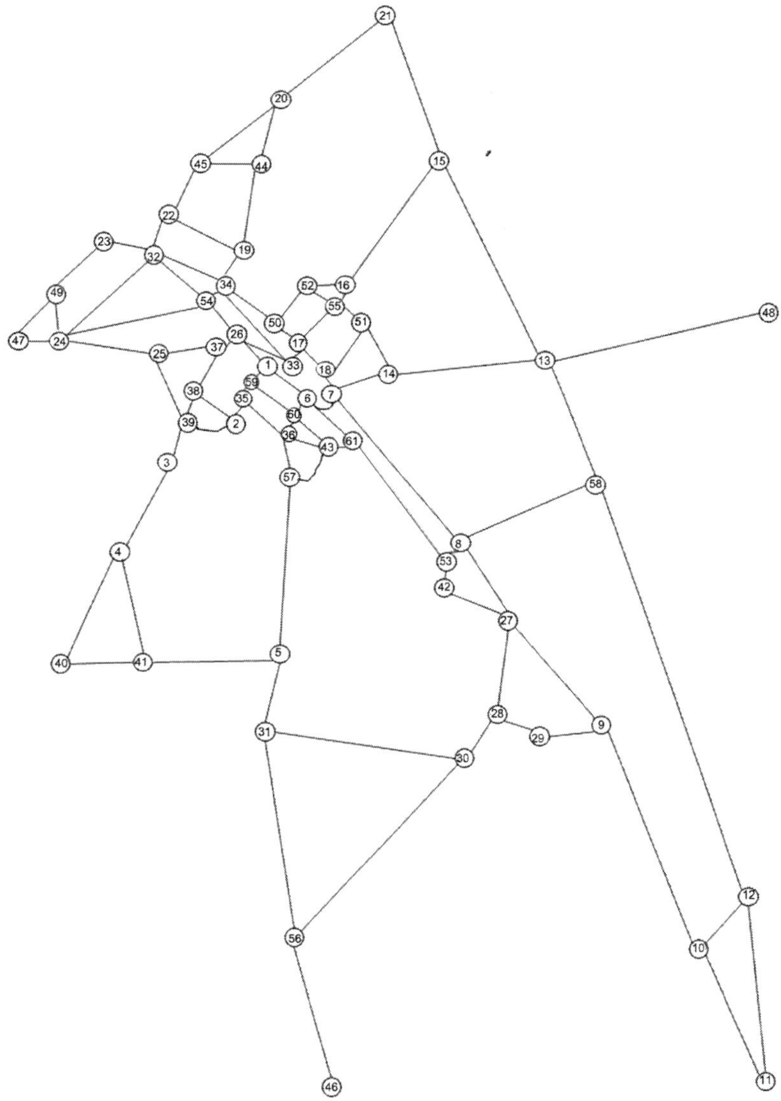

Figure 2. Directed graph with nodes with numbers.

THE DESIGN OF BUS TRANSPORTATION NETWORK IN FERIZAJ WITH SOLVER

Solver is an optimization tool used through Microsoft (MS) Excel that uses genetic algorithms (GA) and linear programming technology that quickly solves various problems in finance, distribution, resource allocation, manufacturing, engineering, etc [9]. Virtually any type of problem that can be modeled in MS excel can be solved with Solver, including the impossible ones and complex nonlinear problems [10].

Solver represents optimization through the fastest and most advanced genetic algorithms ever used. Through the most powerful application of optimization techniques based on algorithms, the solver can find optimal solutions to problems that are considered unsolvable for linear and nonlinear optimizers. [11]. Optimization is a process of finding the best solution to a problem that can have many possible solutions.

The transport network is presented with a graph $G(V, E)$ with a set of nodes and of streets [12]. Notice the number of joints with $|V| = n$ and the number of roads $|E| = m$ [13]. We will define the variables 0 and 1 for u_{ij}:

$$u_{ij} = \begin{cases} 1, & \text{If there is a path between nodes } i \text{ and } j \\ 0, & \text{others} \end{cases}$$

If there is a path between nodes v_i and v_j then this is applied:

$$u_{ij}+u_{ji}=1$$

Since there is a starting and a final node, the eventual solution of the shortest path is a direct path between them [14]. Nodes x_i and x_j may be limited as below:

$$u_{ij}+u_{ji}\leq 1$$
$$i,j \in V$$

Every bus line has only one starting node described as "line in" and one end node described as "line out". Joints in between have both "line in", and "line out" and off-road nodes have neither the "line in" nor "line out". So, we define "net flow" as "line out" minus "line in". Because of the nodes in the applicable path there is an "out line" and the nodes in the applicable path have no "line in", so we have:

$$\sum_j u_{ij} \leq 1, \forall i \in V$$

In order to calculate the distance of each applicable road we need to find the shortest path:

$$\min S = \sum_i^n \sum_j^n l_{ij} u_{ij}$$

We can present the problem of the shortest road model as follows:

$$\text{s.t.} \sum_j u_{ij} - \sum_j u_{ji} = \begin{cases} 1, & \text{is starting node} \\ -1, & i \text{ is last node} \\ 0, & \text{others} \end{cases}$$

$$u_{ij} \in \{0,1\}, \forall (i,j) \in E,$$

The new model of the Urban Traffic Network in Ferizaj will contain eight bus lines. All these nodes with the corresponding numbers can be found in the maps of Figure 1 and Figure 2 to find all the bus lines.

In Table 1 are shown eight bus lines with origin and destination nodes.

After we enter all needed data's we will start to find shortest path between origin and destination nodes for all eight bus lines.

The objective in this case is to minimize total costs and is placed in the Target Cell which we call "Optimization Goal". In Table 2 is shown the net flow of the starting node is 1 and the last node is −1.

The column "Road Selected by Solver" have all results presented with "0" or "1". "0" is when the Solver does not select the road, whereas "1" is selected by the Solver [15]. Out of eight bus lines, we have shown results only for bus line 1 between nodes 1 and 40 in Table 3. This table shows results with "1", which is the road selected by the Solver, and none of the results with "0". The total minimum length for bus line 1 is 4456 meters.

Similar action was conducted for all eight bus lines, the obtained results of which are presented in Table 4 through the Solver.

Table 1. Bus lines with origine and destination nodes.

Bus Line	Origin Node	Destination Node
1	1	40
2	6	46
3	7	11
4	7	48
5	26	15

6	34	21
7	26	47
8	26	49

Table 2. Starting and end point.

Node	Net flow	1 Starting Point, −1 End Point	All Paths around the Node
1	1.00	1	1-59, 1-26, 1-6
40	−1.00	−1	40-4, 40-41

Table 3. Path between node 1 and 40 selected by Solver.

From Node	To Node	Length of the Road in meter	Road Selected by Solver	
1	59	310	1	
59	35	53	1	
35	2	120	1	
2	39	520	1	
39	3	573	1	
3	4	1660	1	
4	40	1220	1	
Target Cell			4456 m	

Table 4. Selected urban transport bus network in Ferizaj through Solver.

Selected urban transport bus network in Ferizaj through Solver		
Bus Line	Paths between Nodes	Length in Meter
1	1, 59, 35, 2, 39, 3, 4, 40	4456
2	6, 60, 36, 57, 5, 31, 56, 46	5568
3	7, 8, 27, 9, 10, 11	6354
4	7, 14, 13, 48	1640

5	26, 33, 17, 55, 16, 15	2296
6	34, 19, 44, 20, 21	2858
7	26, 54, 24, 47	2410
8	26, 54, 24, 49	2213

CONCLUSIONS

As stated already, public urban transport is a critical modern society topic nowadays. It facilitates the daily transport of citizens providing services that private transport cannot. In this regard, since no public transport is available in Ferizaj yet, only some independent bus lines from city to villages around trips, this called for a designation of public transport. Hence, this paper presented a designation of a public transport model for Ferizaj by considering the existing roads and citizens' needs. This public transport model was designed under mathematical optimization methods, graph theory, and solver program as an optimization tool to find the shortest and most economical paths. The designation process has been presented through several graphics above, in detail, results of which are promising to fulfill and facilitate the citizens' transportation needs. Essentially, public transportation in Ferizaj can play an important role in improving and providing different benefits, including direct benefits to the passengers and indirect benefits to society.

Nonetheless, this research had its own limitations. The actual bus station location has not been studied to collect data on whether it meets the requirement to have the new public urban transport design applied. Nevertheless, this paper, despite bringing an innovative solution to Ferizaj's public transport, paves the way for future studies regarding the analysis of bus station location.

CONFLICTS OF INTEREST

The authors declare no conflicts of interest regarding the publication of this paper.

REFERENCES

1. Optimizimi i Rrjetit të Transportit Urban (Rasti i Prishtinës) (2016) Bashkim Çerkini. https://docs.google.com/a/fshn.edu.al/viewer?a=v&pid=sites&srcid=ZnNobi5lZHUuYWx8ZnNobnxneDo1N2ZlNDNlYzg1ZDdmZWY3
2. Yu, B., Yang, Z.Z., et al. (2005) Optimizing Bus Transit Network with Parallel Ant Colony Algorithm. Proceedings of the Eastern Asia Society for Transportation Studies, 5, 374-389.
3. Burrows, P., Reed K., Templer, K. and Walker, J. (2012) Efficient Traffic Routing Using ACO. https://www.semanticscholar.org/paper/Efficient-Traffic-Routing-using-ACO-Burr ows-Reed/ace5876581abc06e1fc48cfa9ead675b4ad695fa
4. Głabowski, M., Musznicki, B., Nowak, P. and Zwierzykowski, P. (2012) Shortest Path Problem Solving Based on Ant Colony Optimization Metaheuristic. Image Processing & Communication, 17, 7-18. https://doi.org/10.2478/v10248-012-0011-5
5. Ning, Z. (2011) Bus Routes Optimization in Wuhan, China. http://essay.utwente.nl/84980/1/ning.pdf
6. Schöbel, A. (2012) Line Planning in Public Transportation: Models and Methods. OR Spectrum, 34, 491-510.
7. Zhao, F. (2006) Large-Scale Transit Network Optimization by Minimization User Cost and Transfers. Journal of Public Transportation, 9, 107-129. https://doi.org/10.5038/2375-0901.9.2.6
8. Klier, M.J. and Haase, K. (2008) Line Optimization in Public Transport Systems. In: Kalcsics, J. and Nickel, S., Eds., Operations Research Proceedings 2007, Springer, Berlin, 473-478. https://doi.org/10.1007/978-3-540-77903-2_73
9. Salini, R., Xu, B. and Lenngren, C.A. (2015) Application of Artificial Intelligence for Optimization in Pavement Management. International Journal of Engineering and Technology Innovation, 5, 189.
10. Çerkini, B., Kosova, R. and Prifti, V. (2014) Transportation Cost Optimization using Evolver, IP-Solver and MS Foundation. Durrës, Albania, 7 November 2014, 23-31. https://knowledgecenter.ubt-uni.net/conference/2014/all-events/55/
11. Fan, W. and Machemeh, R.B. (2006) Optimal Transit Route Network Design Problem with Variable Transit Demand: Genetic Algorithm

Approach. Journal of Transportation Engineering, 132, 40-51. https://doi.org/10.1061/(ASCE)0733-947X(2006)132:1(40)

12. Çerkini, B. and Mitre, T. (2016) Design of Public Urban Transport Network through Software Evolver in Pristina. Journal of Multidisciplinary Engineering Science and Technology, 3, 4103-4106. http://www.jmest.org/wp-content/uploads/JMESTN42351409.pdf
13. Çerkini, B., Bajrami, R., Kosova, R. and Shehu, V. (2015) Transportation Cost Optimization. Academic Journal of Interdisciplinary Studies, 4, 42. https://doi.org/10.5901/ajis.2015.v4n2s1p42
14. Mason, A.J. (2012) Opensolver-An Open Source Add-in to Solve Linear and Integer Progammes in Excel. Operations Research Proceedings 2011, Springer, Berlin, 401-406. https://doi.org/10.1007/978-3-642-29210-1_64
15. Fylstra, D., Lasdon, L., Watson, J. and Waren, A. (1998) Design and Use of the Microsoft Excel Solver. INFORMS Journal on Applied Analytics, 28, 29-55. https://doi.org/10.1287/inte.28.5.29

Section 4: Operational Research Applications in Logistics

CHAPTER 15

Application of the Two Nonzero Component Lemma in Resource Allocation

Morteza Seddighin

Mathematics Department, Indiana University East, Richmond, IN, USA

ABSTRACT

In this paper we will generalize the author's two nonzero component lemma to general self-reducing functions and utilize it to find closed from answers for some resource allocation problems.

Keywords:- Two Nonzero Component Lemma, Resource Allocation, The Distribution of the Search Effort

INTRODUCTION

The technique we will use in this paper was first applied by this author to problems in matrix inequalities and matrix optimization. Historically, many

Citation: Seddighin, M. (2014), Application of the Two Nonzero Component Lemma in Resource Allocation. Journal of Applied Mathematics and Physics,2, 653-661. doi:10.4236/jamp.2014.27072.

researchers have established matrix inequalities by variational methods. In a variational approach one differentiates the functional involved to arrive at an "Euler equation" and then solves the Euler Equation to obtain the minimizing or maximizing vectors of the functional. The same technique is also often used in matrix optimization. Solving the Euler equations obtained are tedious and generally provide little information. Others have established inequalities for matrices and operators by going through a two-step process which consists of first computing upper bounds for suitable functions on intervals containing the spectrum of the matrix or operator and then applying the standard operational calculus to that matrix. This method, which we refer to as "the operational calculus method", has the following two limitations: First, it does not provide any information about vectors for which the established inequalities become equalities (a matrix optimization problem). Second, the operational calculus method is futile in extending Kantorovich-type inequalities to operators on infinite dimensional Hilbert spaces. See [1] for examples using each of the two methods mentioned above. In his study of matrix inequalities and matrix optimization, the author has discovered and proved a lemma called the Two Nonzero Component Lemma or TNCL for short. In this paper we will state an extension of the author's Two Nonzero Component Lemma and utilize it to solve some resource allocatoin problems. While resource allocatoin problems are not generally formulated in terms of matrices, as we will see, there are some similarities between matrix optimization problems and resource allocation problems Let us first state the TNCL as it was used in author's previous papers on matrix inequalities and matrix optimization.

THE TWO NONZERO COMPONENT LEMMA

It was in his investigation on problems of antieigenvalue theory that the author discovered and proved the Two Nonzero Component Lemma (see [2] -[4]). Although this lemma is utilized effectively by the author in matrix theory, it is by nature a dimension reducing optimization lemma which has potential applications in all areas of mathematics and physics. While TNCL was implicitly used in all of the papers just cited, it was not until 2009 that the author stated a formal description of the lemma in his paper titled, "Antieigenvalue Techniques in Statistics" (see [4]). Below is the statement of the lemma. For the proof of the lemma please see the author's work cited above.

Lemma 1 (The Two Nonzero Component Lemma) Let l_1^+ be the set of all sequences with nonnegative terms in the Banach Space l_1. That is, let

$$l_1^+ = \{\mathbf{t} = (t_i) \varepsilon l_1, t_i \geq 0\}. \quad (1)$$

Let

$$F(x_1, x_2, \cdots x_m) \quad (2)$$

be a function from R^m to R. Assume $g_k(\mathbf{t}) = \sum c_i^k t_i$ *for* $(c_i^k) \in l_1^+$, $\mathbf{t} \in l_1^+$, , and $1 \leq k \leq m$. Then the minimizing vectors for the function

$$F(g_1(\mathbf{t}), g_2(\mathbf{t}), \cdots g_m(\mathbf{t})) \quad (3)$$

on the convex set $C = \{(t_i) \in l_1 : \sum t_i = 1\}$ have at most two nonzero components.

What make the lemma possible are the following two facts: First, the fact that the set

$$C = \{(t_i) \in l_1^+ : \sum t_i = 1\} \quad (4)$$

is convex. Second, a special property that the functions

$$F(g_1(\mathbf{t}), g_2(\mathbf{t}), \cdots g_m(\mathbf{t})) \quad (5)$$

involved possess. If we set

$$D(t_1, t_2, t_3, \cdots) = F(g_1(\mathbf{t}), g_2(\mathbf{t}), g_3(\mathbf{t}), \cdots) \quad (6)$$

then all restrictions

$$D(t_1, t_2, \cdots, t_{i-1}, 0, t_{i+1}, \cdots) \quad (7)$$

of

$$D(t_1, t_2, t_3, \cdots) \quad (8)$$

obtained by setting one component t_i equal to zero, have the same algebraic form as $D(t_1, t_2, t_3, \cdots)$ itself. We call functions with this property self-reducing functions. Please note that TNCL is valid for both finite and infinite variable cases. Let us look at an example of a self-reducing function where there are only a finite number of variables $t_1, t_2, t_3 \cdots t_n$ involved. Considered the function

$$\frac{\beta_1 t_1 + \beta_2 t_2 + \cdots + \beta_n t_n}{|\lambda_1|^2 t_1 + |\lambda_2|^2 t_2 + \cdots + |\lambda_3|^2 t_3} \quad (9)$$

where $\beta_1, \beta_2, \beta_3, \cdots \beta_n$ are real numbers and $\lambda_1, \lambda_2, \cdots \lambda_n$ are complex numbers (this function appeared in [2]).

Let

$$D(t_1, t_2, \cdots, t_n) = \frac{\beta_1 t_1 + \beta_2 t_2 + \cdots + \beta_n t_n}{|\lambda_1|^2 t_1 + |\lambda_2|^2 t_2 + \cdots + |\lambda_3|^2 t_3} \quad (10)$$

then we have

$$D(0, t_2, \cdots, t_n) = \frac{\beta_2 t_2 + \cdots + \beta_n t_n}{|\lambda_2|^2 t_2 + \cdots + |\lambda_3|^2 t_3} \quad (11)$$

which has the same algebraic form as

$$D(t_1, t_2, \cdots, t_n) = \frac{\beta_1 t_1 + \beta_2 t_2 + \cdots + \beta_n t_n}{|\lambda_1|^2 t_1 + |\lambda_2|^2 t_2 + \cdots + |\lambda_3|^2 t_3}. \quad (12)$$

Indeed, for any j, $1 \le j < n$;; all restrictions of the function

$$D(t_1, t_2, \cdots, t_n) = \frac{\beta_1 t_1 + \beta_2 t_2 + \cdots + \beta_n t_n}{|\lambda_1|^2 t_1 + |\lambda_2|^2 t_2 + \cdots + |\lambda_3|^2 t_3} \quad (13)$$

obtained by setting an arbitrary set of j components of $D(t_1, t_2, \cdots, t_n)$ equal to zeros have the same algebraic form as $D(t_1, t_2, \cdots, t_n)$.

Obviously, not all functions have this property. For instance, for the function $G(t_1, t_2) = 2t_1 + t_1 t_2$, we have

$G(t_1, 0) = 2t_1$, which does not have the same algebraic form as $G(t_1, t_2)$. Note that functions $g_i(t)$ appearing in the statement of TNCL above are finite or infinite linear combinations of $t_1, t_2, t_3, \cdots$ The lemma was originally stated this way because when we deal with a matrix or operator optimization problem each $g_i(t)$ is either a finite or infinite linear combination of variables $t_1, t_2, t_3, \cdots$.

Example 2 In Theorem 1 of [4] we used TNCL to find the minimum of a Rayleigh quotient. A Rayleigh quotient of a positive operator C over positive operators A and B is a quotient of the form

$$\frac{(Cf, f)}{(Af, f)^{\frac{1}{2}} (Bf, f)^{\frac{1}{2}}} \quad (14)$$

The unit vectors f for which the minimum in

$$\inf_{Af\neq 0, Bf\neq 0} \frac{(Cf,f)}{(Af,f)^{\frac{1}{2}}(Bf,f)^{\frac{1}{2}}} \tag{15}$$

is attained are called stationary values for (14). In Theorem 1 of [4] the minimum of (14) was found by converting the problem to the problem of finding the minimum of

$$\frac{\left(\sum_{i=1}^{n}\lambda_i t_i\right)^2}{\left(\sum_{i=1}^{n}\alpha_i t_i\right)\left(\sum_{i=1}^{n}\beta_i t_i\right)} \tag{16}$$

Subject to $\sum_{i=1}^{n} t_i = 1$. (17)

In this case $g_1(t)=\sum_{i=1}^{n}\lambda_i t_i$, $g_2(t)=\sum_{i=1}^{n}\alpha_i t_i$, and $g_3(t)=\sum_{i=1}^{n}\beta_i t_i$ The sets $\{\lambda_i\}$, $\{\alpha_i\}$, $\{\beta_i\}$ are the set of eigenvalues of C, A, and B respectively.

A GENERALIZATION OF TNCL (GTNCL)

In this section we will show how a generalization of TNCL can be formulated. In the proof of the TNCL in [2] and [3] we took advantage of the facts that the set

$$C=\left\{(t_i)\in l_1^+ : \sum t_i = 1\right\}$$

is a convex set and the function

$$F(g_1(\mathbf{t}), g_2(\mathbf{t}), \cdots g_m(\mathbf{t}))$$

is a self-reducing function. A function

$$F(t_1, t_2, \cdots t_n)$$

can be a self-reducing function without being composed of linear combinations of the form $g_k(\mathbf{t})=\sum c_i^k t_i$.

Example 3 Consider the function

$$F(t_1,t_2,\cdots t_n)=\left(1-e^{-t_1}\right)+\left(1-e^{-t_2}\right)+\left(1-e^{-t_3}\right)+\cdots+\left(1-e^{-t_n}\right).$$

This function is self-reducing but is not a composition of linear combinations. A close look at our arguments in [3] shows that the only property used was the fact that the function to be minimized was self-reducing and the set $\sum_{i=1}^{n} t_i = 1$ was convex. Therefore, we can state the following lemma which is a generalization of TNCL.

We state the lemma for the case that the number of variables in finite (a case which occurs in many applied problems) but the arguments used in [3] show that the lemma is also valid in the case where the number of variables is infinite.

Lemma 4 If the function

$$F(t_1, t_2, \cdots t_n) \quad (18)$$

is a positive self-reducing function on the convex set

$$C = \{(t_i) \in l_1^+ : \sum t_i = 1\} \quad (19)$$

then the minimizing vectors

$$t = (t_1, t_2, \cdots t_n)$$

have at most two nonzero components.

We call the lemma stated above the General Two Nonzero Component Lemma or GTNCL for short.

Remark 5 We can also use TNCL and GTNCL to find the maximum of a positive self-reducing function on (19). To see this please note that if

$$F(t_1, t_2, \cdots t_n) \quad (20)$$

is a positive self-reducing function so is

$$\frac{1}{F(t_1, t_2, \cdots t_n)} \quad (21)$$

and maximum of (20) on (19) is the reciprocal of the minimum of (21) on (19).

A general resource allocation problem is stated as

$$\text{minimize } f(x_1, x_2, \cdots, x_n)$$

$$\text{subject to } \sum_{i=1}^{n} x_i = N, x_i \geq 0, \ i = 1, 2, \cdots, n.$$

which can be converted to

$$\text{minimize } g(t_1, t_2, \cdots, t_n)$$

$$\text{subject to } \sum_{i=1}^{n} t_i = 1, t_i \geq 0, \ i = 1, 2, \cdots, n.$$

In the following sections we will use GTNCL to compute a closed form answer for the distribution of the search effort problem.

THE DISTRIBUTION OF THE SEARCH EFFORT

This problem is formulated as

$$\text{minimize} \sum_{i=1}^{n}\left(1-e^{-\alpha x_i}\right)p_i$$

$$\text{subject to} \sum_{i=1}^{n} x_i = N,\ x_i \geq 0,\ i = 1,2,\cdots,n.$$

where α is a positive number, p_i is the probability of an object being at position i and $\left(1-e^{-\alpha x_i}\right)p_i$ is the conditional probability of detecting the object at position i. If we define

$$u_i = \frac{x_i}{N}$$

then the distribution of the search effort problem will be transformed into

$$\text{minimize} \sum_{i=1}^{n}\left(1-e^{-\beta u_i}\right)p_i$$

$$\text{subject to} \sum_{i=1}^{n} u_i = 1,\ x_i \geq 0,\quad i = 1,2,\cdots,n.$$

Theorem 6 The minimum of

$$\sum_{i=1}^{n}\left(1-e^{-\beta x_i}\right)p_i \tag{22}$$

subject to

$$\sum_{i=1}^{n} u_i = 1,\ u_i \geq 0,\quad i = 1,2,\cdots,n \tag{23}$$

is either

$$\left(1-e^{-\beta}\right)p_i \tag{24}$$

for some $1 \leq i \leq n$ or

$$p_i + p_j + e^{-\frac{1}{2}\beta}\left(p_i\sqrt{\frac{p_j}{p_i}} + p_j\sqrt{\frac{p_i}{p_j}}\right) \tag{25}$$

for a a pair of i and j, $1 \le i \le n \;\; and \;\; 1 \le j \le n$.

Proof. Since

$$\sum_{i=1}^{n}\left(1-e^{-\beta u_i}\right)p_i \tag{26}$$

is a self -reducing function, the GTNCL can be used to find the minimum of this function subject to

$$\sum_{i=1}^{n} u_i = 1, u_i \ge 0, \quad i = 1, 2, \cdots, n \tag{27}$$

Since α is positive so isβ. Suppose $u_i \ge 0, \quad i = 1, 2, \cdots, n$ are components of a minimizing vector on the feasible set (27). By GTNCL either there is an i, $1 \le i \le n$ so that $u_i = 1$ and $u_j = 0$ for $j \ne i$ or there is a pair of i and j, $1 \le i \le n, \; 1 \le j \le n$ such that $u_i \ne o, \; u_j \ne o$ and $u_k = o$ for $k \ne i, \; k \ne j, \; 1 \le k \le n$. In the first case the minimum of (26) on (27) is obviously (24). In the second case the minimum of (26) on (27) is the same as the minimum of

$$\left(1-e^{-\beta u_i}\right)p_i + \left(1-e^{-\beta u_j}\right)p_j \tag{28}$$

subject to

$$u_i + u_j = 1 \tag{29}$$

Expression (28) can be simplified to

$$p_i + p_j - p_i e^{-\beta u_i} - p_j e^{-\beta u_j} \tag{30}$$

Substituting $u_j = 1 - u_i$ in (30) we have

$$p_i + p_j - p_i e^{-\beta u_i} - p_j e^{-\beta(1-u_i)} \tag{31}$$

If we differentiate (31) with respect to u_i and put the derivative equal to zero we have

$$p_j \beta e^{\beta(u_i - 1)} - p_i \beta e^{-\beta u_i} = 0 \tag{32}$$

If we solve (32) for u_i, we obtain

$$u_i = \frac{1}{2\beta}\left(\beta + \ln p_i - \ln p_j\right) = \frac{1}{2\beta}\left(\beta + \ln \frac{p_i}{p_j}\right) \tag{33}$$

Substituting u_i from (33) in (29) gives us

$$u_j = \frac{1}{2\beta}\left(\beta + \ln \frac{p_j}{p_i}\right) \tag{34}$$

If we substitute (33) and (34) in (30) we have

$$p_i + p_j + p_i e^{-\frac{1}{2}\left(\beta + \ln \frac{p_i}{p_j}\right)} + p_j e^{-\frac{1}{2}\left(\beta + \ln \frac{p_j}{p_i}\right)}. \tag{35}$$

The last expression is equivalent to

$$p_i + p_j + e^{-\frac{1}{2}\beta}\left(p_i\sqrt{\frac{p_i}{p_j}} + p_j\sqrt{\frac{p_j}{p_i}}\right). \tag{36}$$

Please note that the derivative of the function

$$p_j \beta e^{\beta(u_i - 1)} - p_i \beta e^{-\beta u_i} \tag{37}$$

with respect to u_i is

$$\beta^2 p_j e^{\beta(u_i - 1)} + \beta^2 p_i e^{-\beta u_i} \tag{38}$$

which is positive for u_i, $0 \le u_i \le 1$. Therefore, by the second derivative test (36) is a minimum value not a maximum value for the objective function in the resource allocation problem.

Although the GTNCL states two components u_i and u_j are nonzero but in general we do not know exactly which pair of p_i and p_j expresses (36). When applying TNCL to problems of matrix optimization, the author was able to determine exactly which component or which pair of components of the optimizing vectors are nonzero (see [5]) under certain conditions. The same can be done here.

Theorem 7 Suppose the probabilities p_i, $1 \le i \le n$ are ordered such that

$$p_1 \le p_2 \le p_3 \cdots \le p_n$$

Then the minimum of

$$\sum_{i=1}^{n}\left(1 - e^{-\beta u_i}\right) p_i$$

subject to

$$\sum_{i=1}^{n} u_i = 1, u_i \geq 0, \quad i = 1, 2, \cdots, n$$

is

$$\left(1-e^{-\beta}\right) p_i$$

Proof. Assume

$$p_1 \leq p_2 \leq p_3 \cdots \leq p_n. \tag{39}$$

Since $\beta > 0$ (39) implies that

$$\min\left\{\left(1-e^{-\beta}\right) p_i\right\}_{i=1}^{n} = \left(1-e^{-\beta}\right) p_1$$

Furthermore, in (25) assume $i < j$. Let $x = \frac{p_j}{p_i}$ and $k = e^{-\frac{1}{2}\beta}$. Obviously x ≥ 1. Now (25) can be written as

$$p_i\left[1 + x + k\left(\frac{\sqrt{x}}{x} + x\sqrt{x}\right)\right] \tag{40}$$

Note that k > 0. If we define

$$f(x) = 1 + x + k\left(\frac{\sqrt{x}}{x} + x\sqrt{x}\right) \tag{41}$$

then

$$\frac{df(x)}{dx} = \frac{1}{2x^{\frac{3}{2}}}\left(3kx^2 - k + 2x^{\frac{3}{2}}\right) \tag{42}$$

Since x ≥ 1 then

$$3kx^2 - k + 2x^{\frac{3}{2}} > 3k - k + 2 = 2k + 2 > 0 \tag{43}$$

This shows that f (x) is an increasing function on [1,∞) .. Hence on the finite set

$$\left\{\frac{p_{i+1}}{p_i}, \frac{p_{i+2}}{p_i}, \cdots \frac{p_{n-1}}{p_i}\right\}$$

the function

$$f(x) = 1 + x + k\left(\frac{\sqrt{x}}{x} + x\sqrt{x}\right)$$

has its minimum at $\frac{p_{i+1}}{p_i}$. Therefore, if two components u_i and u_j are nonzero, we must have $j = i + 1$ and in this case the minimum of the objective function is

$$p_i + p_{i+1} + e^{-\frac{1}{2}\beta}\left(p_i\sqrt{\frac{p_{i+1}}{p_i}} + p_j\sqrt{\frac{p_i}{p_{j+1}}}\right)$$

for some i, $1 \le i \le n$. Next notice that

$$\left(1 - e^{-\beta}\right)p_1 < p_1 < p_i + p_{i+1} + e^{-\frac{1}{2}\beta}\left(p_i\sqrt{\frac{p_{i+1}}{p_i}} + p_j\sqrt{\frac{p_i}{p_{i+1}}}\right)$$

For $1 \le i \le n$. Hence the minimum of the objective function is $\left(1 - e^{-\beta}\right)p_1$.

Since both TNCL and GTNCL are valid for infinite number of variables, these techniques can be used to solve resource allocatoin problems involving an infinite number of variables as well. For example, in the distribution of the search effort problem we can assume the search is for an object on the plane that can be potentially detected at an infinite set of locations (such as points with integer x and y coordinates).

OPTIMAL PORTFOLIO SELECTION

There are other resource allocation problems that we are able to tackle with GTNCL, One of these problems is the problem of optimal portfolio selection. One model for this general problem is formulated as finding the maximum of

$$\frac{\left(\sum_{i=1}^{n} R_i x_i\right)}{\left(\sum_{i,j} \sigma_{ij} x_i x_j\right)^{\frac{1}{2}}}$$

subject to $\sum_{i=1}^{n} x_i = 1,\ x_i \ge 0,\ i = 1, 2, \cdots, n.$

(see [6]). Here R_i is expected return on security i and σ_{ij} is the covariance between securities i and j.

If the correlation coefficients between i and j are constant, with ρ, with $0 < \rho < 1$, then $\sigma_{ii} = \sigma_i^2$,

$\sigma_{ij} = \rho\sigma_i\sigma_j$ and from Karush-Kuhn-Tucker conditions the problem is reduced to

$$\text{maximize} \sum_{i=1}^{n} \left\{ \left(R_i - \frac{1}{2}\sigma \right) x_i - \frac{(1-\rho)}{2} \sigma_i^2 x_i^2 \right\} \tag{44}$$

$$\text{subject to} \sum_{i=1}^{n} x_i = 1,\ x_i \geq 0,\ i = 1, 2, \cdots, n.$$

Notice that (44) is a self-reducing function and one can again apply the GTNCL to find a maximum value for it.

The problems of distribution of search effort and optimal portfolio selection are both examples of separable resource allocation problems. A separable resource allocation problem is a problem where we want to minimize or maximize

$$\sum_{i=1}^{n} f_i(x_i)$$

$$\text{subject to} \sum_{i=1}^{n} x_i = N,\ x_i \geq 0,\ \ i = 1, 2, \cdots, n.$$

where each f_i is continuously differentiable over an interval including [0, N]. The GTNCL can be applied to such a problem if $f_i(0) = 0$, for each i, $1 \leq i \leq n$. This condition is not satisfied for a number of resource allocation problems including optimal sample allocation in stratified sampling, and production planning (see [5]).

Remark 8 In a broader sense, each Kantorovich-type matrix optimization problem such as the one in Example 2 can be regarded as a resource allocation problem where our resource is just the set of pure numbers on the interval [0,1].. For instance in Example 2 the problem is reduced to finding minimum of

$$\frac{\left(\sum_{i=1}^{n} \lambda_i t_i \right)^2}{\left(\sum_{i=1}^{n} \alpha_i t_i \right) \left(\sum_{i=1}^{n} \beta_i t_i \right)}$$

$$\text{Subject to} \sum_{i=1}^{n} t_i = 1$$

Each $t_i = \|z_i\|^2$, where each z_i is a component of a minimizing vector f of norm 1 for the operator optimization problem (15). Indeed nonzero components of a minimizing vector f are important in applications. Historically, the optimization problem (15) was first discussed by R. Cameron and B. Kouvartakis in an effort to minimize the norm of output feedback controllers used in pole placement (see [7]).

Morteza Seddighin Remark 9. The author is not aware of any other theorem that provides closed form answers for resource allocation problems. The results we obtain might be of interest for instance in signal analysis where one needs to minimize the resource spent finding a signal that is probable to detected at a certain location. Computer models are used for solving such problems and it is interesting to investigate how consistent the results of computer models are with our results here. Also, many theories in portfolio selection suggest that diversification maximizes the profit. At the first glance this might sound inconsistent with the results one might obtain using the two nonzero theorem. However, we have to remember that the over theory also ensures diversification increases profit. If the profit is maximized for one or two securities, then the more the number of securities, the more pairs of securities we have.

REFERENCES

1. Gustafson, K. and Rao, D. (1997) Numerical Range. Springer, New York.http://dx.doi.org/10.1007/978-1-4613-8498-4
2. Gustafson, K. and Seddighin, M. (1989) Antieigenvalue Bounds. Journal of Mathematical Analysis and Applications, 143, 327-340. http://dx.doi.org/10.1016/0022-247X(89)90044-9
3. Seddighin, M. (2002) Antieigenvalues and Total Antieigenvalues of Normal Operators. Journal of Mathematical Analysis and Applications, 274, 239-254. http://dx.doi.org/10.1016/S0022-247X(02)00295-0
4. Seddighin, M. (2009) Antieigenvalue Techniques in Statistics. Linear Algebra and Its Applications, 430, 2566-2580.http://dx.doi.org/10.1016/j.laa.2008.05.007
5. Seddighin, M. and Gustafson, K. (2005) On the Eigenvalues which Express Antieigenvalues. International Journal of Mathematics and Mathematical Sciences, 2005, 1543-1554. http://dx.doi.org/10.1155/IJMMS.2005.1543
6. Ibaraki, T. and Katoh, N. (1988) Resource Allocation Problems. The MIT Press, Cambridge.
7. Cameron, R. and Kouvartakis, B. (1980) Minimizing the Norm of Output Feedback Controllers Used in Pole Placement: A Dyadic Approach. International Journal of Control, 32, 759-770.http://dx.doi.org/10.1080/00207178008922889

CHAPTER 16

Energy Efficient Non-Cooperative Methods for Resource Allocation in Cognitive Radio Networks

Enrico Del Re[1,2], Renato Pucci[1,2], Luca Simone Ronga[2]

[1]University of Florence, Florence, Italy

[2]CNIT, Florence Research Unit, Florence, Italy

ABSTRACT

In a cognitive radio network wherein primary and secondary users coexist, an efficient power allocation method represents one of the most important key aspects. This paper provides a novel approach based on a game theory framework to solve this problem in a distributed and fair way. Formulated as an optimization problem, the resource allocation problem between secondary users and primary users can be modeled and investigated with the Game Theory, and in particular S-Modular Games, since they provide useful tools for the definition of multi objective distributed algorithms in the

Citation: E. Del Re, R. Pucci and L. Ronga, Energy Efficient Non-Cooperative Methods for Resource Allocation in Cognitive Radio Networks, Communications and Network, Vol. 4 No. 1, 2012, pp. 1-7. doi: 10.4236/cn.2012.41001.

context of radio communications. This paper provides also a performance comparison among the proposed game and two other algorithms, frequently used in this context: Simulated Annealing and Water Filling.

Keywords:- Cognitive Radio Networks; Resource Allocation

INTRODUCTION

Cognitive Radio represents a promising paradigm aimed to optimize the radio spectrum efficiency. In a cognitive radio network, two kind of users can exist: primary (noncognitive) users and secondary (cognitive) users. Even though primary and secondary users coexist within the same network and sharing the same frequency bands, primary users may be unaware of the presence of secondary users. Contrary to primary users, secondary users are smart, since they are intelligent and interact with selfish network users, choosing best operating parameters on the base of the sensed spectrum. Due to the natural radio environment changes, spectrum sharing schemes change frequently, accordingly with the users allocated resource. In this scenario, a game theoretic framework allows to study, model and analyze cognitive radio networks in a distributed way. Such attractive feature allows to achieve the flexibility and the efficient adaptation to the operative environment that were previously mentioned. Due to the players behavior, noncooperative game theory is closely connected to mini/max optimization and typically results in the study of various equilibria, most notably the Nash equilibrium [1]. Developed cognitive radio strategy has been formulated according the mathematical discipline of Game Theory, with particular reference to S-Modular Games [2].

Non-cooperative games have been proposed for spectrum sharing in [3], which reports a detailed survey on game theoretic approaches for dynamic spectrum sharing in cognitive radio networks, by in-depth theoretic analysis and an overview of the most recent practical implementations. In [4], the authors investigate the issue about the spectrum sharing between a decentralized cognitive network and a primary system, comparing a suboptimal distributed non-cooperative game theoretic power control algorithm with the optimal solution power control algorithm and the power control algorithm proposed in [5]. Besides the above referred papers, in [6] and [7] it is also discussed the power control problem in spectrum sharing model. In [8,9] authors proposed different game-theoretic approaches to maximize energy efficiency of the users within wireless networks, making the utility functions being inversely proportional to the transmit power.

This paper extends the above described results providing a distributed game-theoretic approach to obtain a quasi-optimal power allocation method that maximize the energy efficiency of each user, within the coexistence of primary and secondary users. The proposed method take into account throughput fairness among secondary users.

The paper is organized as follows: in Section 2 the proposed system model and applicative scenario are presented. The game description and the NE existence and uniqueness is discussed in Section 3, while in Section 4 the Water-Filling algorithm and a his energy efficient modified version is reported. In 5 the results from computer simulation are commented. Finally some conclusions are expressed in Section 6.

SYSTEM MODEL

In this work we consider a Cognitive Radio context inspired by a tactical/military scenario where a primary system (owner of the spectrum rights) coexisting with one or more secondary systems and sharing the same frequency band. This kind of situation is very interesting and, at the same time, very frequent, i.e. during coalition deployment of forces or in case of coexistence of humanitarian and military convoys, especially when mobility is taken into account. It is to be noted that such kind of context is a suggested scenario by EDA and NATO [10] to provide a better reuse of the frequency resource among several nations (coexistence of networks) and give a great help to accommodate dynamism of the operational deployments. Note that, considering a primary system in the network, the proposed scheme includes the possibility to existence of more than one primary user.

In the proposed system, each user is characterized by a dedicated sender and receiver, thus each communicating couple consists of a transmitter site TX_i and a receiver site RX_i, as shown in Figure 1. In the most general context, in this work we consider the transmitters and the receivers positions completely independent the ones from the others. Moreover, we use a discrete-time model, based on iterations (which we ahead refer as t). Indeed, for every single iteration, all users act only once and until the next iteration they can't do anything else.

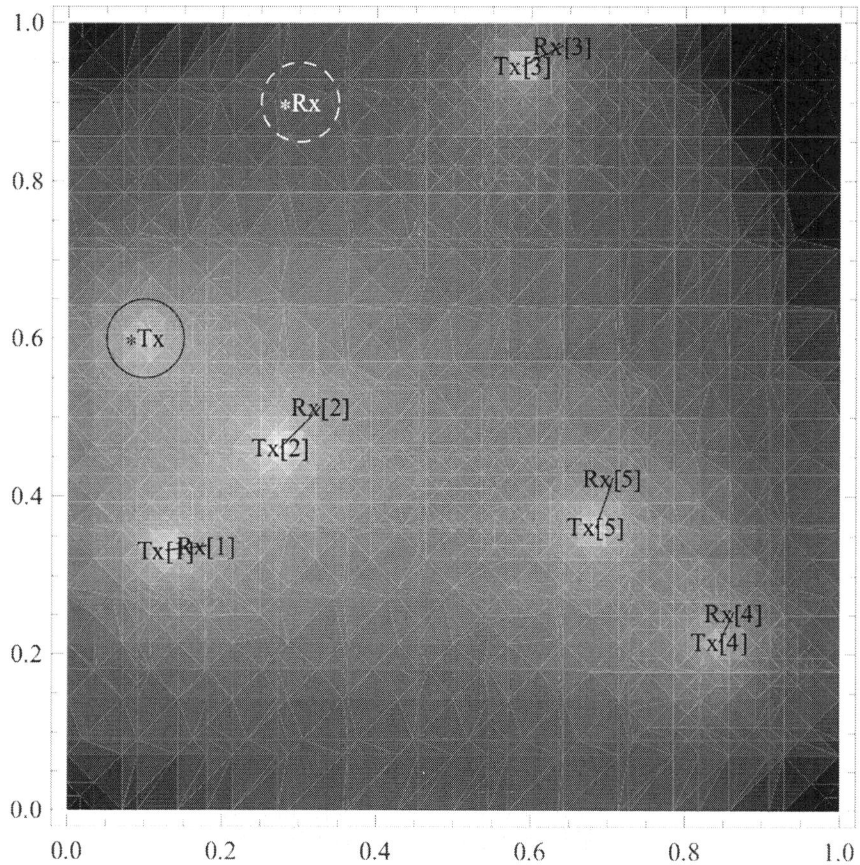

Figure 1. Scenario with one primary users pair (represented by circles) and 5 secondary users pairs; shaded colors represent the path loss component of the channel.

Each user broadcast a pilot signals at the first iteration of the algorithm in order estimate the channel gain coefficients, that are assumed not changing during the execution of the algorithm.

Since in the above described scenario primary users may be unaware of the presence of secondary users, in the proposed system there can't be a "direct" cooperation among primary and secondary users. However, by definition, primary users should not undergo a degradation of the required QoS due to the presence of secondary users. For this reason, we propose the following the solution: the primary AP selects and broadcasts periodically on the shared channel a reasonable interference cap on the total interference it willing to tolerate. Together with the interference cap, the value of the total interference received by the primary receiver is transmitted. Thanks to this solution, we introduce a sort of "indirect" unaware cooperation among the two kind of users. The direct result of the introduction of a chosen interference cap is a limitation of the total transmit power of the secondary users on the shared channel. Thanks to the introduction of the interference cap, for the simplicity of exposition, hereinafter we will consider only one primary transmitter-receiver pair, since the proposed scheme can be easily extended to include more than one primary user.

Viewed from the perspective of secondary users, each of them will choose the more suitable transmission power in order to achieve the best transmission quality, respecting the interference cap (broadcasted by primary AP) and ensuring low interference to other secondary users. Due to the consistent decisions made by primary and secondary users, game theory represents an inbred framework to study, analyze and predict the behavior of this system. For simplicity of exposition, we will consider a fixed primary interference cap and therefore a fixed maximum transmission power for the secondary users; this assumption can be made without altering the validity of the system, since variations of this value are relatively slow compared with the time of convergence of the algorithm. In case of wireless networks with high primary mobility and/or more strict delay needs, a delay efficient approach should be followed, see [11].

THE PROPOSED GAME

Game Description

The non-cooperative game proposed in this paper can be modeled as game with N secondary users, namely the players of the game, operating on one radio resource. This game can be easily extended considering M radio resources (i.e. subcarriers of the same multi-carrier channel or different channels) following the approach proposed in [12], where subcarrier allocation is based on the normalized channel gain. Formally, the proposed noncooperative game can be modeled as follows:

- Set of Players: $N = \{1, 2, 3, \cdots, N\}$ where $i \in N$ is the i-th secondary user.
- Set of Strategies: $S = \{P_{min}, \cdots, P_{max}\}$.
- Utility function: $u_i(p)$ where $i \in N$ is the i-th secondary user.

Following the approach proposed by [8,13], we take into account the energy efficiency problem at the physical layer, considering an utility function expressed in bit/Joule as performance measure of the model. Each player tries to maximize the following utility function:

$$u_i\left(p(t),p(t-1)\right)=W\frac{R_i f(\gamma_i)}{p_i(t)}-\Omega_i\left(p(t-1)\right)p_i(t) \quad (1)$$

where p is the complete set of strategies of all secondary users, W is the ratio between the number of information bits per packet and the number of bits per packet, R_i is the transmission rate of the i th user in bits/sec, $f(\gamma_i)$ is the is the efficiency function (depending on the considered modulation), that represents a stochastic modeling of the number of bits that are successfully received for each unit of energy drained from the battery for the transmission.

Thanks to the efficiency function, the utility function each user tries to maximize is related to its instantaneous signal to noise plus interference ratio (SINR), defined as:

$$\gamma_{i,c}(t)=10\log\left(\frac{g_{i,i}p_i(t)}{I_i^r\left(p_{-i}(t-1),P\right)}\right) \quad (2)$$

where with the notation p_{-i} we refer to all components of p not belonging to user i, p_i is the power allocated from the secondary transmitter $i\,(TX_i)$, $g_{j,i}$ is the path gain from TX_j to RX_i, $g_{12,i}$ is the interference channel from the primary to the secondary receiver RX_i, p is the primary transmitted power, while σ^2 is the AWGN component at RX_i.

$I_i^r\left(p_{-i}(t-1),P\right)$ represents the total interference received by the i th user and it can be wrote as:

$$I_i^r\left(p_{-i}(t-1),P\right)=\sum_{k\neq i}g_{k,i}p_k(t-1)+\sigma^2+g_{12,i}P \quad (3)$$

The path gain can be written as [14]:

$$g_{i,j}=\frac{K}{\left[\left(x_i-x_j\right)^2+\left(y_i-y_j\right)^2\right]^d} \quad (4)$$

where K = 0.097 and d = 4.

The adopted channel model is composed by a small scale fading and a path-loss component. In particular, the path-loss model is the Okomura-Hata model, while the small scale fading is modeled as a Rayleigh process.

Since the above defined utility function depends on the path gains, each secondary user need to know it. In order to solve this problem that could have a strong impact on the signalling process, we assume that each receiver periodically send out a beacon, thanks to which transmitters can measure path gains.

In order to make the NE of the game as efficient as possible (moving it to the Pareto Optimum), we consider the adaptive pricing function $\Omega_i(p_{i,-i})$ that generates pricing values basing on the interference generated by network users. Thus, users that cause high interference transmitting at high power will obtain high value of pricing, due to the fact that $\Omega(p)$ is strictly increasing with p. The pricing function is written as follows:

$$\Omega_i(p(t-1)) = \beta - \delta \exp\left(-\mu \frac{p_i(t-1) \sum_{i=1, k \neq i}^{N} g_{k,i}}{I_i^r(p_{-i}(t-1), P)} \right) \quad (5)$$

where:

- $\beta > 1$ is the maximum pricing value
- $\delta > 1$ is the price weight of the generated interference
- $\mu > 1$ is the sensitivity of the users to interference.

These three parameters are very useful to adapt the pricing function to the considered wireless network requirements; i.e. we can make the algorithm converge faster decreasing the value of δ or force all secondary users to transmit at lower power levels increasing their sensitivity to the interference [13].

Existence and Uniqueness of NE

A Nash Equilibrium [15] offers a stable outcome and it can be guaranteed to exist, under certain conditions, but does not necessarily mean the best payoff for all the players involved, especially in presence of pricing techniques. In the literature there are lots of mathematical methods to demonstrate the existence and uniqueness of NE, like graphical [13,16], quasi-concavity curve [17] and super-modularity [8,18].

Supermodular Games are an interesting class of games that exhibits strategic complementarity. There are several compelling reasons like existence of pure strategy Nash equilibrium, dominance resolvability, identical bounds on joint strategy space etc. that make them a strong candidate for resource allocation modeling. Supermodular games are based on the concept of "supermodularity", which is used in the social sciences to analyze how one agent's decision affects the incentives of others.

S-Games are normal form games $\Gamma = \langle N, S, \{f_i\} \rangle$ where N is the set of users, S the strategy space, f_i the set of utility functions and $\forall i \in N$ these conditions are satified:

1) The strategy space S_i of user i is a complete lattice.
2) f_i is supermodular in s_i.
3) f_i presents increasing differences in s.

The proposed utility function in Equation (1) can be easily demonstrated to be supermodular, since:

1) The strategy space P is a complete lattice;
2) $$\frac{\partial^2 u_i(p)}{\partial p_i \partial p_j} \geq 0 \quad (6)$$

for all $p \in P$ and $i \neq j$.

3) The utility function has the increasing difference property.

For the details of proofs we refer to [8], under the proposed conditions. Uniqueness of the NE can be also demonstrated following the same approach, since we use a Best Response rule. Even if our proposed pricing function is more complicated, in comparison with the above-cited work, the demonstration procedure does not change. Indeed, the pricing function $\Omega(p(t), p(t-1))$ can be considered linear in p(t), since the coefficient of p(t) at time t is a constant.

ENERGY EFFICIENT ITERATIVE WATER-FILLING ALGORITHM

Water filling is a frequently used algorithm in power allocation methods. This algorithm starts from the idea that a vase can be filled by a quantity of water equal to the empty volume of the vase. It is well-known in the literature that power allocation in parallel uncoupled channels can follow the water filling principle in order to maximize data-rate. A channel can be filled by an amount of power depending on the existing noise level. A multiuser scenario cannot be modeled as the parallel uncoupled channels case, but it has to be modeled following the approach of an interference channels.

On the base of these considerations, an iterative water filling procedure can be obtained; each user updates its transmission power level as follows:

$$P_i^{(t)} = \left(P_{max}^{(t)} - \frac{\gamma_i(t)}{p_i(t)} \right)^+, i = 1 \cdots N \quad (7)$$

where $P_i^{(t)}$ is the power level assigned at the user i in the iteration t and P_{max} is maximum power that can be transmitted in the channel (the water level). Because of, if $a^+ = \max\{0, a\}$, if $\frac{\gamma_i(t)}{p_i(t)} > P_{max}$, then $P_i^{(t)} = 0$ is assigned to the user i.

Iterative Water-Filling gets excellent performances in presence of low interference environments and/or limited number of users. However for increasing values of interference, the algorithm get worst; indeed, users experimenting the best channel conditions will transmit at high power levels, while users experimenting bad channel conditions (i.e. being the receiver close to another transmitter) will receive high interference values and then they will be inactivated. For this reason, EEIWF turns out to be unfair.

For a fixed target data-rate, we can identify a minimum target value for the SINR. In this case, Iterative Water-Filling is energy inefficient, due to the fact that the algorithm tries to maximize the total transmission power, achieving SINR values that are greater than the target value. For this reason, we propose the following energy efficient modified version of the algorithm, called Energy Efficient Iterative Water-Filling (EEIWF). For each iteration, P_{max} is updated as follows:

- If $SINR^{(t)} - SINR^{(t-1)} \geq 0$, $P_{max}^{(t)} = \frac{P_{max}^{(t-1)}}{k}$
- If $SINR^{(t)} - SINR^{(t-1)} < 0$, $P_{max}^{(t)} = P_{max}^{(t-1)}$

where $k > 1$ represents the reduction factor and it controls the convergence speed of the algorithm. Note that for $k = 2$ the algorithm becomes the bisection method.

Such approach allows us to maintain the fixed datarate, using the lowest total transmission power level, taking into account its trend in Figure 2.

In the case of $N < 8$ number of user, the SINR trend for decreasing values of P_{max} is a monotonous decreasing function. Otherwise, when $N \geq 8$, a reduction of P_{max} should also improve SINR final value.

SIMULATION RESULTS

In this paragraph we show the results of the simulations that we run in order to verify the behavior of a cognitive network based on the our proposed game-theoretical framework. In the subsection 5.1 the convergence of the algorithm is shown, while in subsection 5.2 a comparison between proposed game and heuristic power allocations will be presented.

Convergence of the Algorithm

The operating context is a terrain square area of 1 km edge, with a suburban path-loss profile. Primary transmitter and receiver positions are fixed; secondary transmitters are independently located in the area, while the secondary receivers positions are placed randomly in a 200 m diameter circle around the respective transmitters. Each secondary user transmits isotropically with $p_i \leq p_{max}$, where $p_{max} = 1$ dBm on the base of a fixed interference cap. Moreover, we consider a noise power $\sigma^2 = -10$ dBm, frequency $f = 1$ Ghz, $W = 4/5$, a common rate $R_i = 10^4$ bit/s, $\beta = 10^4$, $\delta = 10^4$ and $\mu = 10^{-2}$.

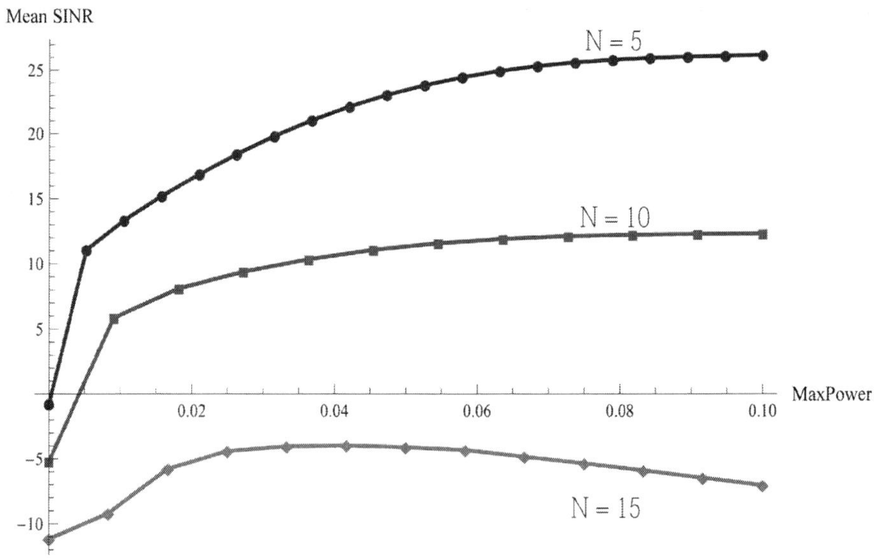

Figure 2. SINR trends for increasing values of maximum transmission power for different number of users.

Results of a simulation with a primary user and N = 5 secondary users show a fast convergence of the transmitted power levels and SINR experimented by secondary users. For increasing numbers of secondary users in the networks, the algorithm still maintains a very short time of convergence, see Figure 3; in the particular case of very bad location of secondary users (i.e. users concentrated in a small area), a possible growth of the time convergence may be avoided decreasing the value of δ parameter.

Performance Comparison

In order to obtain a qualitative evaluation of the proposed game, we decide to compare its performance with both EEIWF and an optimal centralized heuristic power allocation system, like Simulated Annealing (SA) [19]. The mean value of the SINR (experimented by secondary users) has been chosen as the performance index for the three optimization methods. We run the simulations for increasing number of secondary users, while all the other parameters of the system remain the same of the previous shown configuration.

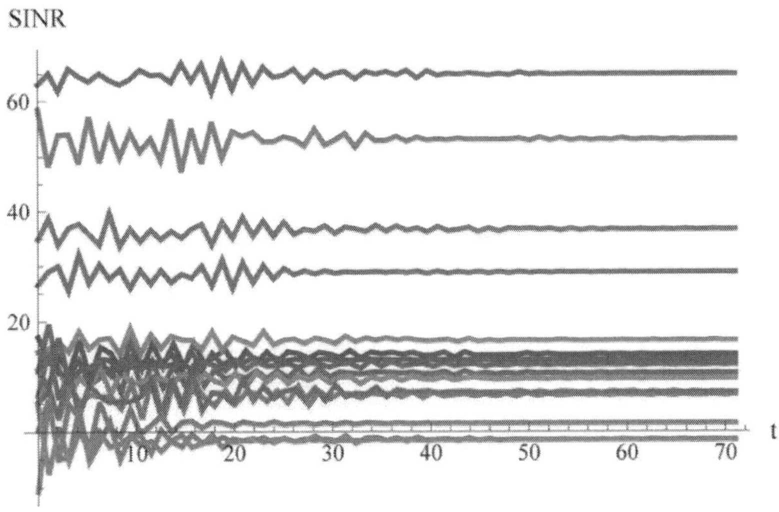

Figure 3. SINR convergence in a 15-user simulation with $\delta = 10^4$.

The simulation results illustrated in Figure 4 show clearly that the SA and the proposed game have quite the same performance, since the maximum difference between the mean value of the SINR obtained by the SA and the game is −3 dB. On the other hand, Water-Filling obtains lower mean SINR levels and performance worsens for increasing number of users in the network.

In addition to the SINR, the energy efficiency of the three considered methods is an another important key feature that we need to investigate. If the SINR performance are quite the same for the proposed game and the SA, on the contrary we can observe a great difference in terms of power allocations. Indeed, Figure 5 shows that, for a 15-user simulation, SA allocation uses approximately 80% of power more than game allocation. For what concern the EEIWF, while some users are switched off, the others transmit at highest levels, compared with the other two proposed allocations. In Figure 5 the power allocation of the proposed game is shown in purple, in yellow is reported the additional power allocated by SA (with respect to game) and in blue the excess additional power allocated by EEIWF (with respect to SA).

CONCLUDING REMARKS

In this paper we provide an energy efficient game theoretic framework to solve the resource allocation problem in a cognitive network, wherein primary and secondary users coexist. The power allocation problem is solved thanks to the application of S-Modular Games. Transmission power of secondary users is upper bounded by the interference cap, defined as the total interference that primary users willing to tolerate, without loosing their required QoS.

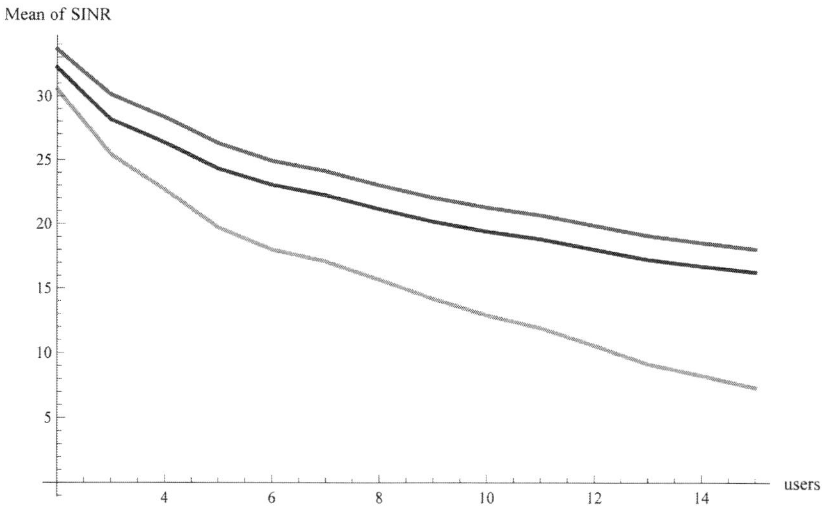

Figure 4. Trends of SINR mean values for increasing number of secondary users in the network; SA in red, Game in blue, EEIWF in green.

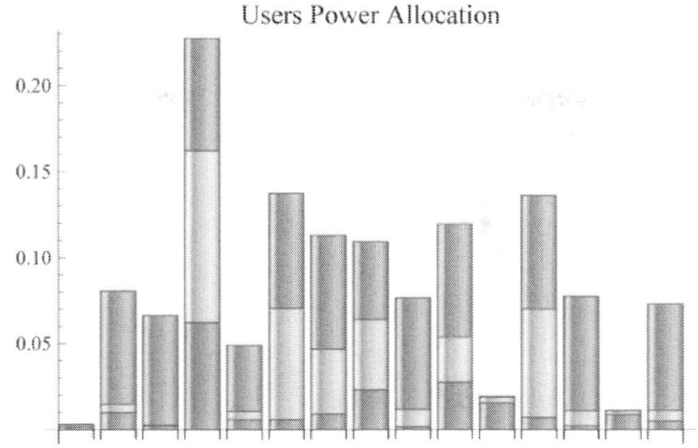

Figure 5. Example of power allocation for a 15 users network; SA in yellow, Game in purple, Water-Filling in blue.

Moreover, secondary users are discouraged to transmit at high power levels, since they are charged on the base of the interference they generate, thanks to the introduction of a pricing function inside of the utility function. Tuning utility function parameters, the proposed game is able to adapt his performance (in terms of time of convergence) to every kind of network configuration. Indeed, simulation results show a fast convergence of the algorithm for any number of considered users in the cognitive network.

In this work, a performance comparison among the proposed game, an optimal centralized resource allocation method (Simulated Annealing) and an Energy Efficient version of the Water Filling is also included. Simulation results show clearly that game theory obtains better performance than water filling and the proposed game converges to the same SINR values obtained from the heuristic optimization method. However, unlike these, the proposed game results to be the most energy efficient, also for a large number of considered users. Further investigations will be made in order to quantify and analyze the signaling process among secondary users.

ACKNOWLEDGEMENTS

The work presented in this paper is part of the CORASMA project (COgnitive RAdio for dynamic Spectrum MAnagement) promoted the European Defense Agency. Enrico Del Re, Renato Pucci,, Luca Simone Ronga We would like to thank Pierpaolo Piunti for the contribution in Water-Filling algorithm.

REFERENCES

1. L. S. Ronga and E. Del Re, "S-Modular Games for Distributed Power Allocation in Cognitive Radio Systems," Proceedings of the PIMRC 2009, Tokyo, 13-16 September 2009.
2. D. M. Topkis, "Equilibrium Points in Nonzero-Sum N- Person Submodular Games," SIAM Journal on Control and Optimization, vol. 17, No. 5, 1979, pp. 773-778. doi:10.1137/0317054
3. B. Wang, Y. Wu and K. J. R. Liu, "Game Theory for Cognitive Radio Networks: An Overview," Computer Networks, vol. 54, No. 14, 2010, pp. 2537-2561. doi:10.1016/j.comnet.2010.04.004
4. R. Luo and Z. Yan, "Power Allocation Using Non-Cooperative Game Theoretic Analysis in Cognitive Radio Networks," Proceedings of WiCOM, Chengdu, 23-25 September, 2010.
5. D. Li, X. Dai and H. Zhang, "Game Theoretic Analysis of Joint Rate and Power Allocation in Cognitive Radio Networks," International Journal of Communications, Network and System Sciences, Vol. 1, 2009, pp. 1-89. doi:10.4236/ijcns.2009.21001
6. G. Bansal, Md. J. Hossian and V. K. Bhargava, "Adaptive Power Loading for OFDM-Based Cognitive Radio Systems," IEEE Conference on Communications, Glasgow, 24-28 June 2007, pp. 5137-5142. doi:10.1109/ICC.2007.849
7. J. Huang, R. Berry and M. L. Honig, "Auction-Based Spectrum Sharing," ACM/Springer Mobile Networks, 2010.
8. C. U. Saraydar, N. B. Mandayam and D. Goodman, "Efficient Power Control via Pricing in Wireless Data Networks," IEEE Transaction on Communications, Vol. 50, No. 2, 2002, pp. 291-303. doi:10.1109/26.983324
9. R. D. Yates, "A Framework for Uplink Power Control in Cellular Radio Systems," IEEE Journal on Selected Area of Communications, Vol. 13, No. 7, 1995, pp. 1341-1347. doi:10.1109/49.414651
10. NATO IST-077 (RTG-035), "Final Technical Activity Report of IST077-RTG035," Cognitive Radio in NATO, NATO, 2011.
11. F. Meshkati, H. V. Poor and S. C. Schwartz, "Energy Efficiency-Delay Tradeoffs in CDMA Networks: A Game Theoretic Approach," IEEE Transactions on Information Theory, Vol. 55, No. 7, 2009, pp. 3220-3228. doi:10.1109/TIT.2009.2021374

12. D. Wu, D. Yu and Y. Cai, 'Subcarrier and Power Allocation in Uplink OFDMA Systems Based on Game Theory," Proceedings of IEEE International Conference Neural Networks & Signal Processing, Zhenjiang, 7-11 June 2008.
13. C. K. Tan, M. L. Sim and T. C. Chuah, "Fair Power Control for Wireless Ad Hoc Networks Using Game Theory with Pricing Scheme," IET Communications, Vol. 4, No. 3, 2008, pp. 322-333.
14. H. Arslan, "Cognitive Radio, Software Defined Radio, and Adaptive Wireless Systems," Springer, Berlin, 2007. doi:10.1007/978-1-4020-5542-3
15. D. Fudenberg and J. Tirole, "Game Theory," MIT press, Cambridge, 1991.
16. F. Nan, M. Siun-Chuon and N. B. Mandayam, "Pricing and Power Control for Joint Network-Centric and UserCentric Radio Resource Management," IEEE Transactions on Communications, Vol. 52, No. 9, 2004, pp. 1547- 1557. doi:10.1109/TCOMM.2004.833191
17. H. Zhu, Z. Ji and K. J. R. Liu, "Fair Multiuser Channel Allocation for OFDMA Networks Using Nash Bargaing Solutions and Coalitions," IEEE Transactions on Communications, Vol. 53, No. 8, 2005, pp. 1366-1376. doi:10.1109/TCOMM.2005.852826
18. H. Jianwei, R. A. Berry and M. L. Honig, "Game Theoretic Model for Radio Resource Management in HIPERLAN Type 2 Networks," IEEE Journal in Selected Areas in Communications, Vol. 24, No. 6, 2006, pp. 1074-1084.
19. S. Kirkpatrick, et al., "Optimization by Simulated Annealing," Science, New Series, Vol. 220, No. 4598, 1983, pp. 671-680.

CHAPTER 17

An Alternative Interpretation of Mixed Strategies in N-Person Normal Form Games Via Resource Allocation

Ahmad Nahhas and H. W. Corley

Center on Stochastic Modeling, Optimization, and Statistics, The University of Texas at Arlington, Arlington, TX, USA

ABSTRACT

In this paper, we give an interpretation of mixed strategies in normal form games via resource allocation games, where all players utilize the same resource. We define a game in normal form such that each player allocates to each of his pure strategies a fraction of the maximum resource he has available.

However, he does not necessarily allocate all of the resource at his disposal. The payoff functions in the resource allocation games vary with how each player allocates his resource. We prove that a Nash equilibrium

Citation: Nahhas, A. and Corley, H. (2018), An Alternative Interpretation of Mixed Strategies in n-Person Normal Form Games via Resource Allocation. Theoretical Economics Letters, 8, 1854-1868. doi: 10.4236/tel.2018.810122.

always exists in mixed strategies for n-person resource allocation games. On the other hand, we show that a mixed Berge equilibrium may not exist in such games.

Keywords:- Game Theory, Mixed Strategy, Resource Allocation

INTRODUCTION

Game theory is the study of mathematical decision making among multiple players. Each player makes an individual choice according to his notion of rationality and to his expectations of the other players' choices. Game theory can be non-cooperative as described in [1] or cooperative as described in [2]. In non-cooperative game theroy, the concept of the Nash equilibrium (NE) was introduced in [3] and [4]. The proof of the existence was based on the Kakutani and the Brouwer fixed point theorems [5]. Another solution concept, the Berge equilibrium, was introduced in [6] and formalized by [7]. A strategy is considered to be a Berge equilibrium if some or all players other than player i cannot increase the expected payoff for player i by changing strategies. The Berge equilibrium was extended to mixed strategies in [8], where it was also shown that a mixed Berge equilibrium may not exist.

The computation of equilibria points is an essential component of game theory research and is well studied in the literature. For example, a nonlinear programming approach to find an NE for three player games was developed in [9] and for n-person games in [10]. The nonlinear programming approach for finding an NE was extended in [11] to find a generalized equilibrium that includes the case of an MBE.

Such equilibria have been widely in economics. However, game theory faces some challenges such as making it more useful in applications as noted in [12]. Thus there is a need for a behavior interpretation of a strategy. For example, [13] gives a behavioral interpretation of a dominant strategy, while belief hierarchies play a prominent role in game theory as described in [14]. Moreover, there has been a growing use of epistemic game theory. [15] gives a historical overview of the transition from classical game theory in the sense of Nash to epistemic game theory.

The purpose of this paper is to deal with the difficulties associated with mixed strategies. See [16] for an extensive literature review on the concept of mixed strategies, which require a randomizing process as described in [17] and [18]. According to [19], randomization lacks behavioral support. [20] gives two interpretations for mixed strategies. The first is based on the purification theorem of [21]. Purification refers to how mixed strategies reflect the player's lack of knowledge of other players' information and decision-making process. The second interpretation is that a mixed strategy represents the fraction of a large population that adapts each of the pure strategies. In [22] mixed strategies are interpreted as the belief player i's opponents have on what strategy player i will choose. Therefore, as argued in [16], mixed strategies are used in games where a pure NE does not exist, but their use is principally to provide an elegant mathematical theory. In practice, they are problematic with no generally accepted interpretation. We offer here a simple, intuitive interpretation for a certain class of games.

In this paper, we construct resource allocation games (RAGs) such that the equilibria strategies represent the fraction of a resource each player allocates to each of his pure strategies. In particular, we consider the NE and the MBE. The purpose of RAGs is to give a physical interpretation of the concept of mixed strategies, as distinguished from other interpretations. Our interpretation is as follows. The probability that a player chooses a pure strategy equals the fraction of the resource the player allocates to that pure strategy over the total amount of the resource the player allocates to all his pure strategies. The underlying idea is as follows. Suppose there is a single resource used by all players. Suppose player 1 had 100 units of the resource to be used up in whole or in part. If he considered using 40 units on implementing strategy 1 and 60 units on implementing strategy 2, whatever they are, then his associated mixed game theoretic strategywould be (0.4, 0.6), Thus in a RAG, he would choose both strategies as opposed to one chosen by, say, randomization. In other words, in a resource allocation game, one pure strategy does not preclude another. This simple idea is generalized here. A related notion was studied in [23] for infinitely repeated non-cooperative games played at discrete instants called stages. The payoffs in [23] were linear in the frequency that they had been played previously. Our approach differs significantly. For example, here a mixed strategy may or may not maximize the payoff functions for each player.

The organization of this paper is as following. In Section 2, we present the notation used. In Section 3, we prove the existence of an NE for a RAG. In Section 4, we present a nonlinear program to find an NE analytically. In Section 5, we consider the case of the MBE and present a nonlinear program to find one if one exists. In Section 6, we give some numerical examples and show that an MBE may not exist. In Section 7, we state our conclusions.

PRELIMINARIES

In this section we define the notation used. Let the RAG $\Gamma = \langle I, (S_i)_{i\in I}, (f_i)_{i\in I} \rangle$ be an n-person resource allocation game in normal form with exactly one resource. The set $I = 1, \cdots, n$ is the set of the n-players. Let R_i be the resource available for player i, so the n-tuple $R = (R_1, \cdots, R_n)$ represents the amount of the resource each player has available. Define min $R_i^{\min} > 0$ to be the minimum amount of the resource R_i player i needs to allocate and $\alpha_i^{\min} = \frac{R_i^{\min}}{R_i}$. The setof the m_i pure strategies for player i is $S_i = \left(s_i^1, \cdots, s_i^{m_i}\right)$.

Each player playeri allocates from his resource R_i the fraction α_i^j to his pure strategy $s_i^j, j = 1, \cdots, m_i$. The set of all possible allocations for each player i is

$$\Delta_i = \left\{ \boldsymbol{\alpha}_i = \left(\alpha_i^1, \cdots, \alpha_i^{m_i}\right) : \alpha_i^j \geq 0, j = 1, \cdots, m_i, \alpha_i^{\min} \leq \sum_{j=1}^{m_i} \alpha_i^j \leq 1 \right\}. \tag{1}$$

Note that Δ_i is compact and convex for each player $i \in I$. Let $\Delta_{-i} = \Delta_1 \times \cdots \times \Delta_{i-1} \times \Delta_{i+1} \times \cdots \times \Delta_n$ and $\Delta = \Delta_1 \times \cdots \times \Delta_n$. The probability the player i chooses strategy s_i^j is $\frac{\alpha_i^j}{\sum_{j=1}^{m_i} \alpha_i^j}$. Hence a mixed strategy for player i is the m_i-tuple

$$\left(\frac{\alpha_i^1}{\sum_{j=1}^{m_i} \alpha_i^j}, \cdots, \frac{\alpha_i^{m_i}}{\sum_{j=1}^{m_i} \alpha_i^j} \right),$$

where $\alpha_i^{\min} \leq \sum_{j=1}^{m_i} \alpha_i^j \leq 1$ and $\alpha_i^j \geq 0, j = 1, \cdots, m_i$. A pure strategy j is an allocation $\alpha_i^{\min} \leq \alpha_i^j \leq 1$ where the player i allocates α_i^j to his pure strategy j and allocates 0 to the rest of his pure strategies. The payoff function for each player is $f_i^{j,k}(\boldsymbol{\alpha})$, where $\boldsymbol{\alpha} = (\boldsymbol{\alpha}_1, \cdots, \boldsymbol{\alpha}_n)$ and $\boldsymbol{\alpha}_{-i} = (\boldsymbol{\alpha}_1, \cdots, \boldsymbol{\alpha}_{i-1}, \boldsymbol{\alpha}_{i+1}, \cdots, \boldsymbol{\alpha}_n)$.

The payoff functions $f_i^{j,k}(\boldsymbol{\alpha}), j = 1, \cdots, m_i, k = 1, \cdots, m_{-i}$ are assumed to becontinuous in $\alpha_i^j \in [0,1], j = 1, \cdots, m_i, \forall i \in I$.

The set of joint pure strategies of all players other than player i, is the Cartesian product of the sets of pure strategies of all players other than player $S_{-i} = \times_{j\in I-\{i\}} (S_j)$ and is denoted by $S_{-i} = \left\{ s_{-i}^1, \cdots, s_{-i}^{m_{-i}} \right\}$, where

$m_{-i} = \prod_{j \in I - \{i\}} m_j$.

The joint probability

$\alpha_{-i}^k = \prod_{p \in I - \{i\}} \frac{\alpha_p^k}{\sum_{j=1}^{m_p} \alpha_p^j}, k = 1, \cdots, m_{-i}$ is the probability that all the players other than player i choose the joint pure strategy s_{-i}^k . It is the product of the fraction that each player in $I - \{i\} = \{1, \cdots, i-1, i+1, \cdots, n\}$ allocates to his corresponding strategy. We apply the identities proved in [8] to Γ. The following identities represent the expected payoff for player i. If player i allocates $\alpha_i^{\min} \le \alpha_i^j \le 1$ to his strategy j and he allocates 0 to his other pure strategies while the rest of players choose the allocation α_{-i} is

$$F_i^j(\alpha) = \sum_{k=1}^{m_{-i}} \alpha_{-i}^k f_i^{j,k}(\alpha) \tag{2}$$

If player i chooses the mixed allocation α_i and the rest of players choose the allocation α_{-i}, then the expected payoff for player i is

$$F_i(\alpha) = \sum_{j=1}^{m_i} \sum_{k=1}^{m_{-i}} \frac{\alpha_i^j}{\sum_{j=1}^{m_i} \alpha_i^j} \alpha_{-i}^k f_i^{j,k}(\alpha). \tag{3}$$

Table 1 shows an example of a 2-person RAG.

In this paper, it is useful to consider the following two cases.

1) Case 1. Each player i allocates all of his resource R_i. In other words $\sum_{j=1}^{m_i} \alpha_i^j = 1, \forall i \in I$. In this case, each player i chooses strategy j with the probability α_i^j .
2) Case 2. Each player i does not necessarily allocate all of his resource R_i. Hence $\alpha_i^{\min} \le \sum_{j=1}^{m_i} \alpha_i^j \le 1, \forall i \in I$. In this case, each player i chooses strategy j with the probability

$$\frac{\alpha_i^j}{\sum_{j=1}^{m_i} \alpha_i^j}, j = 1, \cdots, m_i .$$

Table 1. Example of a RAG.

	s_2^1	s_2^2
s_1^1	$f_1^{1,1}\left(\alpha_1^1 R_1, \alpha_2^1 R_2\right)$, $f_2^{1,1}\left(\alpha_1^1 R_1, \alpha_2^1 R_2\right)$	$f_1^{1,2}\left(\alpha_1^1 R_1, \alpha_2^2 R_2\right)$, $f_2^{1,2}\left(\alpha_1^1 R_1, \alpha_2^2 R_2\right)$
s_1^2	$f_1^{2,1}\left(\alpha_1^2 R_1, \alpha_2^1 R_2\right)$, $f_2^{2,1}\left(\alpha_1^2 R_1, \alpha_2^1 R_2\right)$	$f_1^{2,2}\left(\alpha_1^2 R_1, \alpha_2^2 R_2\right)$, $f_2^{2,2}\left(\alpha_1^2 R_1, \alpha_2^2 R_2\right)$

$R_i, \forall i \in I$ is considered fixed in these two cases. However, in the second case each player i may not use all of his resource. Note that the first case is a special case of the second case. In particular if $R_i^{\min} = R_i$, then the second case becomes the first case. We formalize this previous statement as follows.

Lemma 1. $\forall i \in I$ let $R_i^{\min} = R_i$. Then Case 1 and Case 2 are equivalent. Proof. Let $R_i^{\min} = R_i, \forall i \in I$. Hence $\alpha_i^{\min} = 1$ and $\sum_{j=1}^{m_i} \alpha_i^j = 1$. It follows immediately that Case 2 reduces to Case 1.

EXISTENCE OF AN NE FOR A RAG

In this section we prove the existence of an NE in Case 1 and Case 2 above. Hence we seek to find a mixed strategy such that a player i chooses strategy j with a probability

$$\frac{\alpha_i^j}{\sum_{j=1}^{m_i} \alpha_i^j}, j = 1, \cdots, m_i, \forall i \in I.$$

We next restate the definition of an NE in terms of allocation.

Definition 1. A strategy * α is an NE if and only if

$$F_i\left(\boldsymbol{\alpha}^*\right) = \max_{j=1,\cdots,m_i} F_i^j\left(\boldsymbol{\alpha}^*\right), \forall \boldsymbol{\alpha}_i \in \Delta_i, \forall i \in I. \tag{4}$$

Intuitively, in an NE for the game Γ, no player can improve his expected payoff with a unilateral change in strategy, i.e., a unilateral reallocation of his previously allocated resource level. We now prove the existence of an NE in a finite n-person Γ. It suffices to prove the existence for case 2 since it subsumes case 1 by Lemma 1 when $R_i^{\min} = R_i$. The proof of the next theorem is similar to the proof of the existence of an equilibrium in [4]. Let

$$\Delta_i = \left\{ \alpha_i : \alpha_i^j \geq 0, j = 1, \cdots, m_i, \alpha_i^{\min} \leq \sum_{j=1}^{m_i} \alpha_i^j \leq 1 \right\} \quad (5)$$

And $\Delta = \Delta_1 \times \cdots \times \Delta_n$. The set Δ is compact and convex since the number of player is finite and each player has a finite number of strategies. Define the function $\phi = (\phi_1, \cdots, \phi_n) : \Delta \to \Delta$ where $\phi_i = (\phi_i^1, \cdots, \phi_i^{m_i})$ and

$$\phi_i^j = \frac{\alpha_i^j + \max\{0, F_i^j(\boldsymbol{\alpha}) - F_i(\boldsymbol{\alpha})\}}{1 + \sum_{j=1}^{m_i} \max\{0, F_i^j(\boldsymbol{\alpha}) - F_i(\boldsymbol{\alpha})\}}, i = 1, \cdots, n, j = 1, \cdots, m_i. \quad (6)$$

The functions ϕ_i^j are continuous since we assume that the $f_i^{j,k}(\boldsymbol{\alpha})$ are continuous in $\alpha_i^j \in [0,1], j = 1, \cdots, m_i, \forall i \in I$. Therefore by the Brouwer fixed point theorem there exists fixed points

$$\alpha_i^j = \frac{\alpha_i^j + \max\{0, F_i^j(\boldsymbol{\alpha}) - F_i(\boldsymbol{\alpha})\}}{1 + \sum_{j=1}^{m_i} \max\{0, F_i^j(\boldsymbol{\alpha}) - F_i(\boldsymbol{\alpha})\}}, i = 1, \cdots, n, j = 1, \cdots, m_i. \quad (7)$$

We now prove the following result.

Theorem 2. Every finite RAG Γ has an NE in mixed strategies. Proof. Let α be an NE. Then no player has an incentive to change his strategy based on the allocation α. Note that the function $\max\{0, F_i^j(\boldsymbol{\alpha}) - F_i(\boldsymbol{\alpha})\}$ represent player's i gain by choosing his pure strategy j given the previous allocation α. Hence $\max\{0, F_i^j(\boldsymbol{\alpha}) - F_i(\boldsymbol{\alpha})\} = 0, j = 1, \cdots, m_i, \forall i \in I$. Thus α is a fixed point. Conversely, let α be a fixed point. Then for each i let 1 be a pure strategy such that $\alpha_i^l > 0$, and $F_i^l(\boldsymbol{\alpha}) = \min_{j=1,\cdots,m_i} F_i^j(\boldsymbol{\alpha})$. Therefore, $\max\{0, F_i^l(\boldsymbol{\alpha}) - F_i(\boldsymbol{\alpha})\} = 0$, since $F_i^l(\boldsymbol{\alpha}) \leq F_i(\boldsymbol{\alpha})$ Note that from Equation (7),the right hand side is α_i^j only when the denominator equals 1. Hence $\sum_{j=1}^{m_i} \max\{0, F_i^j(\boldsymbol{\alpha}) - F_i(\boldsymbol{\alpha})\} = 0$. Hence no player has an incentive to change his strategy, and so α is an NE allocation to complete the proof. We next show how a standard n-person game in normal form with constant von Neumann-Morgenstern (VNM) utility functions is a special case of an allocation game as defined in this paper.

Theorem 3. The payoff matrix for a standard normal form game is a special case of the payoff matrix for a RAG.

Proof. Let $u_i(s_i^j, s_{-i}^k) = c_i^{j,k}, j = 1, \cdots, m_i, k = 1, \cdots, m_{-i}, \forall i \in I$ be constant VNM utilities for a normal form game. It suffices to show that for any player i the VNM utilities can be written as the payoffs for player i in an allocation game Γ.To do so simply let $f_i^{j,k}(\alpha) = c_i^{j,k} \times R_i, R_i = 1, \forall i \in I$. It follows that a standard normal form game with constant VNM utilities is a special case of the game Γto complete the proof.

In other words, for 1, R iIi = ∀∈ the payoff functions for each player i need not vary with the fraction each player allocates to each of his pure strategies. It follows that for any equilibrium, say an NE or an MBE, a normal form game with VNM utilities is a special case of an associated RAG. In the next section we consider the computation of an NE. The computation of an MBE will be considered in Section 5.

THE COMPUTATION OF AN NE FOR A RAG

In this section we provide a nonlinear programming approach to compute an NE for an n-person game by extending the nonlinear program in [10] to find an NE for the game Γ. Therefore an allocation α is an NE if and only the maximum of the following nonlinear program is zero.

Theorem 4. α* is an NE for Γ if and only if the maximum of the following nonlinear program is 0.

$$\text{Maximize } g(\alpha, \beta) = \sum_{i=1}^{n} \left[F_i(\alpha) - \beta_i \right]$$

subject to

$$\sum_{k=1}^{m_{-i}} \alpha_{-i}^k f_i^{j,k}(\alpha) \le \beta_i, j = 1, \cdots, m_i, \forall i \in I,$$

$$\alpha_i^j \ge 0, j = 1, \cdots, m_i, \forall i \in I,$$

$$\alpha_i^{\min} \le \sum_{j=1}^{m_i} \alpha_i^j \le 1, \forall i \in I.$$

(8)

Proof. Let α* be an NE where each player i allocates $\sum_{j=1}^{m_i} \alpha_i^{j^*}$ of his total resource R_i.Then $F_i(\alpha^*) = \max_j F_i^j(\alpha^*) = \beta_i^*$. Therefore $g(\alpha^*, \beta^*) = 0$. Furthermore, all constraints (8) are satisfied since from Definition 1 $\beta_i^* = \max_{j=1,\cdots,m_i} F_i^j(\alpha^*)$.

Conversely, let α^*, β^*, be a feasible point such that $g(\alpha^*, \beta^*) = 0$. It can easily be checked by the constraints (8) and Equation (3) that $F_i(\alpha^*) \le \beta_i^*$. Hence it must be the case that $F_i(\alpha^*) = \beta_i^*$. Otherwise $g(\alpha^*, \beta^*) \ne 0$which yields a contradiction. Moreover, from the constraints 8 $\beta_i^* = \max_{j=1,\cdots,m_i} F_i^j(\alpha^*)$.

Therefore α* is a NE by Definition 1. It is worth noting that the payoff functions at an NE may not be maximized. To maximize the payoff functions one needs to find an allocation such that the fraction each player i allocates to each strategy maximizes each of the player i's payoff functions. However, we analyze only the case where the payoff functions are monotonically nondecreasing functions in the fraction of the resource α_i^j. In this case, each of the payoff functions is maximized when the total resource R_i is allocated to that strategy. In other words,

$$F_i^j\left(\alpha_i^j = 1, \boldsymbol{\alpha}_{-i}\right) \geq F_i^j\left(\boldsymbol{\alpha}_i, \boldsymbol{\alpha}_{-i}\right), j = 1, \cdots, m_i, \forall \alpha_i \in \Delta_i, \forall i \in I$$

Lemma 5. Let $f_i^{j,k}(\boldsymbol{\alpha})$ be monotonically increasing functions in $\alpha_i^j, j = 1, \cdots, m_i, \forall i \in I$. Then the optimal strategy for each player is an NE if and only if the maximum of the following nonlinear program is 0.

$$\text{Maximize } g(\boldsymbol{\alpha}, \boldsymbol{\beta}) = \sum_{i=1}^{n}\left[F_i(\boldsymbol{\alpha}) - \beta_i\right]$$

subject to

$$F_i^j\left(\alpha_i^j = 1, \boldsymbol{\alpha}_{-i}\right) \leq \beta_i, j = 1, \cdots, m_i, \forall i \in I,$$

$$\alpha_i^j \geq 0, \forall i \in I, j = 1, \cdots, m_i,$$

$$\sum_{j=1}^{m_i} \alpha_i^j = 1, \forall i \in I. \tag{9}$$

Proof. Let α^* be an NE where the payoff is maximized. Then

$$\beta_i^* = \max_{j=1,\cdots,m_i} F_i^j\left(\alpha_i^{j^*} = 1, \boldsymbol{\alpha}_{-i}\right).$$

However, α^* is an NE. Hence from Theorem 4 the maximum of the nonlinear program is 0.

Conversely, let α^*, β^*, be a feasible point such that the maximum of (9) is 0. The functions f_i^j are monotonically nondecreasingin $\alpha_i^j, \forall i \in I, j = 1, \cdots, m_i$. Therefore, there exists a solution that zero-maximizes the objective function and satisfies all the conditions of the nonlinear program in Theorem 4. Hence the solution is an NE. Furthermore, the solution maximizes the expected payoff for each player of over all payoff functions and the proof is complete.

THE COMPUTATION OF AN MBE FOR A RAG

In this section, we consider the MBE. We present an approach similar to (8) for computing an MBE if one exists. However, the case of computing an

MBE isdifferent from the case of computing an NE since an MBE for n ≥ 3 may not exist as shown by the example of section 6. An MBE for a RAG is defined as follows.

Definition 2. A strategy α* is an MBE for Γ if and only if

$$F_i\left(\alpha^*\right) = \max_{k=1,\cdots,m_{-i}} \sum_{j=1}^{m_i} \frac{\alpha_i^{j^*}}{\sum_{j=1}^{m_i} \alpha_i^{j^*}} f_i^{j,k}\left(\alpha^*\right), \forall \alpha_{-i} \in \Delta_{-i}, \forall i \in I. \tag{10}$$

In other words, a strategy is an MBE if for all $i = 1, \cdots, n$, when player i does not change his strategy, no one or more other players can change strategies and increase player i's expected payoff. Thus in an MBE for the game Γ, no playerno player has an incentive for a unilateral change of his resource allocation if increasing the other players' expected payoffs is his completely unselfish objective. We now extend the nonlinear program presented in [11] to the RAG Γ.

Theorem 6. α* is an MBE for Γ if and only if the maximum of the following nonlinear program is 0.

$$\text{Maximize } h(\alpha,\beta) = \sum_{i=1}^{n}\left[F_i(\alpha) - \beta_i\right]$$

subject to

$$\sum_{j=1}^{m_i} \frac{\alpha_i^j}{\sum_{j=1}^{m_i} \alpha_i^j} f_i^{j,k}(\alpha) \le \beta_i, k = 1, \cdots, m_{-i}, \forall i \in I,$$

$$\alpha_i^j \ge 0, \forall i \in I, j = 1, \cdots, m_i,$$

$$\alpha_i^{\min} \le \sum_{j=1}^{m_i} \alpha_i^j \le 1, \forall i \in I. \tag{11}$$

Proof. Let α* be an MBE allocation. Then each player allocates to each strategy a fraction $\alpha_i^{j^*}$ of his resource that equals to the probability that the player uses that strategy. From Definition 2 one can check that $F_i\left(\alpha^*\right) = \beta_i^* = \max_{k=1,\cdots,m_{-i}} F_i^k\left(\alpha^*\right), \forall i \in I$. Hence all constraints are satisfied. Moreover, $h\left(\beta^*, \alpha^*\right) = 0$.

Conversely, let (α^*, β^*) be a feasible solution such that $h\left(\alpha^*, \beta^*\right) = 0$. From (11), it is easy to see that $F_i\left(\alpha^*\right) \le \beta_i^*, \forall i \in I$. But $h\left(\alpha^*, \beta^*\right) = 0$, so it must be that $F_i\left(\alpha^*\right) = \beta_i^*, \forall i \in I$ and $\beta_i^* = \max_{k=1,\cdots,m_{-i}} F_i^k(\alpha)$. Therefore $F_i\left(\alpha^*\right) = \max_{k=1,\cdots,m_{-i}} F_i^k\left(\alpha^*\right)$, and hence α* is an MBE by Definition 2.

EXAMPLES

In this section we present three examples. The first example is a 2-person RAG, while the second and third are 3-person RAGs.

Example 1

In this 2-person RAG each player has 2 strategies. For some single resource, player 1 has $R_1 = 30$ units and player 2 has $R_2 = 50$ units. The payoff matrix for each player is shown in Table 2. For this game, we consider Case 1 and Case 2 from Section 2. In the first case, each player uses his maximum resource. The following NLP finds an NE for Γ for Case 1.

Table 2. Example 1.

	s_2^1	s_2^2
s_1^1	$(3+\alpha_1^1 \times 30, 5+\alpha_2^1 \times 50)$	$(2+\alpha_1^1 \times 30, 8+\alpha_2^2 \times 50)$
s_1^2	$(2+\alpha_1^2 \times 30, 6+\alpha_2^1 \times 50)$	$(5+\alpha_1^2 \times 30, 4+\alpha_2^2 \times 50)$

$$(P1)\ \text{Maximize } g(\boldsymbol{\alpha},\boldsymbol{\beta}) = \frac{\alpha_1^1}{\alpha_1^1+\alpha_1^2}\frac{\alpha_2^1}{\alpha_2^1+\alpha_2^2}\left(3+\alpha_1^1 \times 30+5+\alpha_2^1 \times 50\right)$$

$$+\frac{\alpha_1^1}{\alpha_1^1+\alpha_1^2}\frac{\alpha_2^2}{\alpha_2^1+\alpha_2^2}\left(2+\alpha_1^1 \times 30+8+\alpha_2^2 \times 50\right)$$

$$+\frac{\alpha_1^2}{\alpha_1^1+\alpha_1^2}\frac{\alpha_2^1}{\alpha_2^1+\alpha_2^2}\left(2+\alpha_1^2 \times 30, 6+\alpha_2^1 \times 50\right)$$

$$+\frac{\alpha_1^2}{\alpha_1^1+\alpha_1^2}\frac{\alpha_2^2}{\alpha_2^1+\alpha_2^2}\left(5+\alpha_1^2 \times 30, 4+\alpha_2^2 \times 50\right)-\beta_1-\beta_2$$

subject to

$$\frac{\alpha_2^1}{\alpha_2^1+\alpha_2^2}\left(3+\alpha_1^1 \times 30\right)+\frac{\alpha_2^2}{\alpha_2^1+\alpha_2^2}\left(2+\alpha_1^1 \times 30\right) \le \beta_1$$

$$\frac{\alpha_2^1}{\alpha_2^1+\alpha_2^2}\left(2+\alpha_1^2 \times 30\right)+\frac{\alpha_2^2}{\alpha_2^1+\alpha_2^2}\left(5+\alpha_1^2 \times 30\right) \le \beta_1$$

$$\frac{\alpha_1^1}{\alpha_1^1+\alpha_1^2}\left(5+\alpha_2^1 \times 50\right)+\frac{\alpha_1^2}{\alpha_1^1+\alpha_1^2}\left(6+\alpha_2^1 \times 50\right) \le \beta_2$$

$$\frac{\alpha_1^1}{\alpha_1^1+\alpha_1^2}\left(8+\alpha_2^2 \times 50\right)+\frac{\alpha_1^2}{\alpha_1^1+\alpha_1^2}\left(4+\alpha_2^2 \times 50\right) \le \beta_2$$

$$\alpha_1^1+\alpha_1^2=1$$

$$\alpha_2^1+\alpha_2^2=1.$$

One solution to (P1) with $g(\alpha^*,\beta^*)=0$ and hence an NE is $\alpha_1^{1^*}=0.52$ $\alpha_1^{2^*}=0.48$, $\alpha_2^{1^*}=0.51$, $\alpha_2^{2^*}=0.49$, $\beta_1^*=17.99$, $\beta_2^*=30.77$. In Case 2 when each player allocates at least 0.4 of his resource, the following NLP finds an NE strategy for this problem.

$$(P2)\ \text{Maximize } g(\boldsymbol{\alpha},\boldsymbol{\beta})=\frac{\alpha_1^1}{\alpha_1^1+\alpha_1^2}\frac{\alpha_2^1}{\alpha_2^1+\alpha_2^2}\left(3+\alpha_1^1\times 30+5+\alpha_2^1\times 50\right)$$

$$+\frac{\alpha_1^1}{\alpha_1^1+\alpha_1^2}\frac{\alpha_2^2}{\alpha_2^1+\alpha_2^2}\left(2+\alpha_1^1\times 30+8+\alpha_2^2\times 50\right)$$

$$+\frac{\alpha_1^2}{\alpha_1^1+\alpha_1^2}\frac{\alpha_2^1}{\alpha_2^1+\alpha_2^2}\left(2+\alpha_1^2\times 30,6+\alpha_2^1\times 50\right)$$

$$+\frac{\alpha_1^2}{\alpha_1^1+\alpha_1^2}\frac{\alpha_2^2}{\alpha_2^1+\alpha_2^2}\left(5+\alpha_1^2\times 30,4+\alpha_2^2\times 50\right)-\beta_1-\beta_2$$

subject to

$$\frac{\alpha_2^1}{\alpha_2^1+\alpha_2^2}\left(3+\alpha_1^1\times 30\right)+\frac{\alpha_2^2}{\alpha_2^1+\alpha_2^2}\left(2+\alpha_1^1\times 30\right)\le\beta_1$$

$$\frac{\alpha_2^1}{\alpha_2^1+\alpha_2^2}\left(2+\alpha_1^2\times 30\right)+\frac{\alpha_2^2}{\alpha_2^1+\alpha_2^2}\left(5+\alpha_1^2\times 30\right)\le\beta_1$$

$$\frac{\alpha_1^1}{\alpha_1^1+\alpha_1^2}\left(5+\alpha_2^1\times 50\right)+\frac{\alpha_1^2}{\alpha_1^1+\alpha_1^2}\left(6+\alpha_2^1\times 50\right)\le\beta_2$$

$$\frac{\alpha_1^1}{\alpha_1^1+\alpha_1^2}\left(8+\alpha_2^2\times 50\right)+\frac{\alpha_1^2}{\alpha_1^1+\alpha_1^2}\left(4+\alpha_2^2\times 50\right)\le\beta_2$$

$$0.4\le\alpha_1^1+\alpha_1^2\le 1$$

$$0.4\le\alpha_2^1+\alpha_2^2\le 1.$$

One solution to (P2) with $g(\alpha^*,\beta^*)=0$ and hence an NE is $\alpha_1^{1^*}=0.45$, $\alpha_1^{2^*}=0$, $\alpha_2^{1^*}=0.23$, $\alpha_2^{2^*}=0.17$, $\beta_1^*=15.94$, $\beta_2^*=16.5$. Thus there is an MBE for this example. However, in general an MBE need not exist for $n\ge 3$ as shown in [8]. In such cased, the interpretation is that there may not exist an allocation such that every player other than player i allocates to each strategy a fraction equals to the probability of using that strategy that maximizes player i's payoff. In the next example, an MBE does not exist, However, an NE exists by Theorem 2.

Example 2

In this 3-person RAG each player has 2 strategies with $R_1=R_2=R_3$ and needs to allocate at least 0.2 of his maximum resource. The payoff matrix for each player is shown in Table 3. We now write the following NLP to find an MBE.

$$\text{Maximize } h(\boldsymbol{\alpha},\boldsymbol{\beta}) = \frac{\alpha_1^1}{\alpha_1^1+\alpha_1^2}\frac{\alpha_2^1}{\alpha_2^1+\alpha_2^2}\frac{\alpha_3^1}{\alpha_3^1+\alpha_3^2}\left(1+\alpha_2^1+\alpha_3^1+1+\alpha_1^1+\alpha_3^1+0\right)$$

$$+\frac{\alpha_1^1}{\alpha_1^1+\alpha_1^2}\frac{\alpha_2^1}{\alpha_2^1+\alpha_2^2}\frac{\alpha_3^2}{\alpha_3^1+\alpha_3^2}\left(1+\alpha_1^1+\alpha_2^1\right)$$

$$+\frac{\alpha_1^2}{\alpha_1^1+\alpha_1^2}\frac{\alpha_2^2}{\alpha_2^1+\alpha_2^2}\frac{\alpha_3^1}{\alpha_3^1+\alpha_3^2}\left(1+\alpha_1^2+\alpha_2^2\right)$$

$$+\frac{\alpha_1^2}{\alpha_1^1+\alpha_1^2}\frac{\alpha_2^2}{\alpha_2^1+\alpha_2^2}\frac{\alpha_3^2}{\alpha_3^1+\alpha_3^2}\left(1+\alpha_2^2+\alpha_3^2+1+\alpha_1^2+\alpha_3^2\right)$$

$$-\beta_1-\beta_2-\beta_3$$

subject to

$$\frac{\alpha_1^1}{\alpha_1^1+\alpha_1^2}\left(1+\alpha_2^1+\alpha_3^1\right)+\frac{\alpha_1^2}{\alpha_1^1+\alpha_1^2}(0)\le\beta_1$$

$$\frac{\alpha_1^1}{\alpha_1^1+\alpha_1^2}(0)+\frac{\alpha_1^2}{\alpha_1^1+\alpha_1^2}(0)\le\beta_1$$

Table 3. Example 2.

s_3^1	s_2^1	s_2^2
s_1^1	$(1+\alpha_2^1+\alpha_3^1, 1+\alpha_1^1+\alpha_3^1, 0)$	$(0,0,0)$
s_1^2	$(0,0,0)$	$(0,0,1+\alpha_1^2+\alpha_2^2)$
s_3^2	s_2^1	s_2^2
s_1^1	$(0,0,1+\alpha_1^1+\alpha_2^1)$	$(0,0,0)$
s_1^2	$(0,0,0)$	$(1+\alpha_2^2+\alpha_3^2, 1+\alpha_1^2+\alpha_3^2, 0)$

$$\frac{\alpha_1^1}{\alpha_1^1+\alpha_1^2}(0)+\frac{\alpha_1^2}{\alpha_1^1+\alpha_1^2}\left(1+\alpha_2^2+\alpha_3^2\right)\le\beta_1$$

$$\frac{\alpha_2^1}{\alpha_2^1+\alpha_2^2}\left(1+\alpha_1^1+\alpha_3^1\right)+\frac{\alpha_2^2}{\alpha_2^1+\alpha_2^2}(0)\le\beta_2$$

$$\frac{\alpha_2^1}{\alpha_2^1+\alpha_2^2}(0)+\frac{\alpha_2^2}{\alpha_2^1+\alpha_2^2}(0)\le\beta_2$$

$$\frac{\alpha_2^1}{\alpha_2^1+\alpha_2^2}(0)+\frac{\alpha_2^2}{\alpha_2^1+\alpha_2^2}\left(1+\alpha_1^2+\alpha_3^2\right)\le\beta_2$$

$$\frac{\alpha_3^1}{\alpha_3^1+\alpha_3^2}\left(1+\alpha_1^2+\alpha_2^2\right)+\frac{\alpha_3^2}{\alpha_3^1+\alpha_3^2}(0)\le\beta_3$$

$$\frac{\alpha_3^1}{\alpha_3^1+\alpha_3^2}(0)+\frac{\alpha_3^2}{\alpha_3^1+\alpha_3^2}(0)\le\beta_3$$

$$\frac{\alpha_3^1}{\alpha_3^1+\alpha_3^2}(0)+\frac{\alpha_3^2}{\alpha_3^1+\alpha_3^2}\left(1+\alpha_1^1+\alpha_2^1\right)\le\beta_3$$

$$\alpha_i^j\ge 0,\forall i\in I, j=1,\cdots,m_i$$

$$0.2\le\sum_{j=1}^{m_i}\alpha_i^j\le 1,\forall i\in I.$$

In this problem, an MBE does not exist. This fact follows from (P3) having the maximum objective function value not equal zero. Moreover, note that there is not any pure Berge equilibrium because whenever players 1 and 2 gets a positive payoff, player 3 gets a payoff 0 and vice versa. Furthermore, if any mixed strategy is used then for at least one player i, the players −i will choose with a positive probability a strategy where at least one player i gets a payoff 0. Hence the maximum of (P3) cannot be 0, and there is no MBE by Theorem 6. In contrast to the MBE, an NE always exists by Theorem 2. The following is the nonlinear program to find an NE for this game.

$$(P4)\ \text{Maximize } g(\boldsymbol{\alpha},\boldsymbol{\beta})=\frac{\alpha_1^1}{\alpha_1^1+\alpha_1^2}\frac{\alpha_2^1}{\alpha_2^1+\alpha_2^2}\frac{\alpha_3^1}{\alpha_3^1+\alpha_3^2}\left(1+\alpha_2^1+\alpha_3^1+1+\alpha_1^1+\alpha_3^1+0\right)$$

$$+\frac{\alpha_1^1}{\alpha_1^1+\alpha_1^2}\frac{\alpha_2^1}{\alpha_2^1+\alpha_2^2}\frac{\alpha_3^2}{\alpha_3^1+\alpha_3^2}\left(1+\alpha_1^1+\alpha_2^1\right)$$

$$+\frac{\alpha_1^2}{\alpha_1^1+\alpha_1^2}\frac{\alpha_2^2}{\alpha_2^1+\alpha_2^2}\frac{\alpha_3^1}{\alpha_3^1+\alpha_3^2}\left(1+\alpha_1^2+\alpha_2^2\right)$$

$$+\frac{\alpha_1^2}{\alpha_1^1+\alpha_1^2}\frac{\alpha_2^2}{\alpha_2^1+\alpha_2^2}\frac{\alpha_3^2}{\alpha_3^1+\alpha_3^2}\left(1+\alpha_2^2+\alpha_3^2+1+\alpha_1^2+\alpha_3^2\right)$$

$$-\beta_1-\beta_2-\beta_3$$

subject to

$$\frac{\alpha_2^1}{\alpha_2^1+\alpha_2^2}\frac{\alpha_3^1}{\alpha_3^1+\alpha_3^2}\left(1+\alpha_2^1+\alpha_3^1\right)\le\beta_1$$

$$\frac{\alpha_2^2}{\alpha_2^1+\alpha_2^2}\frac{\alpha_3^2}{\alpha_3^1+\alpha_3^2}\left(1+\alpha_2^2+\alpha_3^2\right)\le\beta_1$$

$$\frac{\alpha_1^1}{\alpha_1^1+\alpha_1^2}\frac{\alpha_3^1}{\alpha_3^1+\alpha_3^2}\left(1+\alpha_1^1+\alpha_3^1\right)\le\beta_2$$

$$\frac{\alpha_1^2}{\alpha_1^1+\alpha_1^2}\frac{\alpha_3^2}{\alpha_3^1+\alpha_3^2}\left(1+\alpha_1^2+\alpha_3^2\right)\le \beta_2$$

$$\frac{\alpha_1^2}{\alpha_1^1+\alpha_1^2}\frac{\alpha_2^2}{\alpha_2^1+\alpha_2^2}\left(1+\alpha_1^2+\alpha_2^2\right)\le \beta_3$$

$$\frac{\alpha_1^1}{\alpha_1^1+\alpha_1^2}\frac{\alpha_2^1}{\alpha_2^1+\alpha_2^2}\left(1+\alpha_1^1+\alpha_2^1\right)\le \beta_3$$

$$\alpha_1^1,\alpha_1^2,\alpha_2^1,\alpha_2^2,\alpha_3^1,\alpha_3^2\ge 0, \forall i\in I, j=1,\cdots,m_i$$

$$\alpha_1^1,\alpha_1^2,\alpha_2^1,\alpha_2^2,\alpha_3^1,\alpha_3^2\ge 0, \forall i\in I, j=1,\cdots,m_i$$

$$0.2\le \alpha_1^1+\alpha_1^2\le 1$$

$$0.2\le \alpha_2^1+\alpha_2^2\le 1$$

$$0.2\le \alpha_3^1+\alpha_3^2\le 1.$$

One solution to (P4) with $g(\alpha^*,\beta^*)=0$ and hence an NE is $\alpha_1^{1^*}=\alpha_1^{2^*}=\alpha_2^{1^*}=\alpha_2^{2^*}=\alpha_3^{1^*}=\alpha_3^{2^*}=0.1$, $\beta_1^*=\beta_2^*,\beta_3^*=0.3$, $\beta_1^*=\beta_2^*,\beta_3^*=0.3$.

Example 3

In this 3-person RAG each player has 2 pure strategies with $R_1=R_2=R_3=1$, and each player needs to allocate at least 0.2 of his maximum resource. The payoff matrices are shown in Table 4. This example has an MBE obtained from the following NLP.

$$(P5)\ \text{Maximize } h(\boldsymbol{\alpha},\boldsymbol{\beta})=\frac{\alpha_1^1}{\alpha_1^1+\alpha_1^2}\frac{\alpha_2^1}{\alpha_2^1+\alpha_2^2}\frac{\alpha_3^1}{\alpha_3^1+\alpha_3^2}$$

$$\times\left(2+\alpha_2^1+\alpha_3^1+1+\alpha_1^1+\alpha_3^1+2+\alpha_1^1+\alpha_2^1\right)$$

$$+\frac{\alpha_1^1}{\alpha_1^1+\alpha_1^2}\frac{\alpha_2^2}{\alpha_2^1+\alpha_2^2}\frac{\alpha_3^1}{\alpha_3^1+\alpha_3^2}$$

$$\times\left(1+\alpha_2^2+\alpha_3^1+2+\alpha_1^1+\alpha_3^1+1+\alpha_1^1+\alpha_2^2\right)$$

$$+\frac{\alpha_1^2}{\alpha_1^1+\alpha_1^2}\frac{\alpha_2^1}{\alpha_2^1+\alpha_2^2}\frac{\alpha_3^1}{\alpha_3^1+\alpha_3^2}$$

$$\times\left(1+\alpha_2^1+\alpha_3^1+2+\alpha_1^2+\alpha_3^1+1+\alpha_1^2+\alpha_2^1\right)$$

$$+\frac{\alpha_1^2}{\alpha_1^1+\alpha_1^2}\frac{\alpha_2^2}{\alpha_2^1+\alpha_2^2}\frac{\alpha_3^1}{\alpha_3^1+\alpha_3^2}$$
$$\times\left(2+\alpha_2^2+\alpha_3^1+1+\alpha_1^2+\alpha_3^1+2+\alpha_1^2+\alpha_2^2\right)$$
$$+\frac{\alpha_1^1}{\alpha_1^1+\alpha_1^2}\frac{\alpha_2^1}{\alpha_2^1+\alpha_2^2}\frac{\alpha_3^2}{\alpha_3^1+\alpha_3^2}$$
$$\times\left(1+\alpha_2^1+\alpha_3^2+2+\alpha_1^1+\alpha_3^2+1+\alpha_1^1+\alpha_2^1\right)$$

Table 4. Example 3.

s_3^1	s_2^1	s_2^2
s_1^1	$\left(2+\alpha_2^1+\alpha_3^1, 1+\alpha_1^1+\alpha_3^1, 2+\alpha_1^1+\alpha_2^1\right)$	$\left(1+\alpha_2^2+\alpha_3^1, 2+\alpha_1^1+\alpha_3^1, 1+\alpha_1^1+\alpha_2^2\right)$
s_1^2	$\left(1+\alpha_2^1+\alpha_3^1, 2+\alpha_1^2+\alpha_3^1, 1+\alpha_1^2+\alpha_2^1\right)$	$\left(2+\alpha_2^2+\alpha_3^1, 1+\alpha_1^2+\alpha_3^1, 2+\alpha_1^2+\alpha_2^2\right)$
s_1^2	s_2^1	s_2^2
s_1^1	$\left(1+\alpha_2^1+\alpha_3^2, 2+\alpha_1^1+\alpha_3^2, 1+\alpha_1^1+\alpha_2^1\right)$	$\left(2+\alpha_2^2+\alpha_3^2, 1+\alpha_1^1+\alpha_3^2, 2+\alpha_1^1+\alpha_2^2\right)$
s_1^2	$\left(2+\alpha_2^1+\alpha_3^2, 1+\alpha_1^2+\alpha_3^2, 2+\alpha_1^2+\alpha_2^1\right)$	$\left(1+\alpha_2^2+\alpha_3^2, 2+\alpha_1^2+\alpha_3^2, 1+\alpha_1^2+\alpha_2^2\right)$

$$+\frac{\alpha_1^1}{\alpha_1^1+\alpha_1^2}\frac{\alpha_2^2}{\alpha_2^1+\alpha_2^2}\frac{\alpha_3^2}{\alpha_3^1+\alpha_3^2}$$
$$\times\left(2+\alpha_2^2+\alpha_3^2+1+\alpha_1^1+\alpha_3^2+2+\alpha_1^1+\alpha_2^2\right)$$
$$+\frac{\alpha_1^2}{\alpha_1^1+\alpha_1^2}\frac{\alpha_2^1}{\alpha_2^1+\alpha_2^2}\frac{\alpha_3^2}{\alpha_3^1+\alpha_3^2}$$
$$\times\left(2+\alpha_2^1+\alpha_3^2+1+\alpha_1^2+\alpha_3^2+2+\alpha_1^2+\alpha_2^1\right)$$
$$+\frac{\alpha_1^2}{\alpha_1^1+\alpha_1^2}\frac{\alpha_2^2}{\alpha_2^1+\alpha_2^2}\frac{\alpha_3^2}{\alpha_3^1+\alpha_3^2}$$
$$\times\left(1+\alpha_2^2+\alpha_3^2+2+\alpha_1^2+\alpha_3^2+1+\alpha_1^2+\alpha_2^2\right)$$
$$-\beta_1-\beta_2-\beta_3$$

subject to

$$\frac{\alpha_1^1}{\alpha_1^1+\alpha_1^2}\left(2+\alpha_2^1+\alpha_3^1\right)+\frac{\alpha_1^2}{\alpha_1^1+\alpha_1^2}\left(1+\alpha_2^1+\alpha_3^1\right)\le\beta_1$$
$$\frac{\alpha_1^1}{\alpha_1^1+\alpha_1^2}\left(1+\alpha_2^2+\alpha_3^1\right)+\frac{\alpha_1^2}{\alpha_1^1+\alpha_1^2}\left(2+\alpha_2^2+\alpha_3^1\right)\le\beta_1$$
$$\frac{\alpha_1^1}{\alpha_1^1+\alpha_1^2}\left(1+\alpha_2^1+\alpha_3^2\right)+\frac{\alpha_1^2}{\alpha_1^1+\alpha_1^2}\left(2+\alpha_2^1+\alpha_3^2\right)\le\beta_1$$
$$\frac{\alpha_1^1}{\alpha_1^1+\alpha_1^2}\left(2+\alpha_2^2+\alpha_3^2\right)+\frac{\alpha_1^2}{\alpha_1^1+\alpha_1^2}\left(1+\alpha_2^2+\alpha_3^2\right)\le\beta_1$$

$$\frac{\alpha_2^1}{\alpha_2^1+\alpha_2^2}\left(1+\alpha_1^1+\alpha_3^1\right)+\frac{\alpha_2^2}{\alpha_2^1+\alpha_2^2}\left(2+\alpha_1^1+\alpha_3^1\right)\le\beta_2$$
$$\frac{\alpha_2^1}{\alpha_2^1+\alpha_2^2}\left(2+\alpha_1^2+\alpha_3^1\right)+\frac{\alpha_2^2}{\alpha_2^1+\alpha_2^2}\left(1+\alpha_1^2+\alpha_3^1\right)\le\beta_2$$
$$\frac{\alpha_2^1}{\alpha_2^1+\alpha_2^2}\left(2+\alpha_1^1+\alpha_3^2\right)+\frac{\alpha_2^2}{\alpha_2^1+\alpha_2^2}\left(1+\alpha_1^1+\alpha_3^2\right)\le\beta_2$$
$$\frac{\alpha_2^1}{\alpha_2^1+\alpha_2^2}\left(1+\alpha_1^2+\alpha_3^2\right)+\frac{\alpha_2^2}{\alpha_2^1+\alpha_2^2}\left(2+\alpha_1^2+\alpha_3^2\right)\le\beta_2$$
$$\frac{\alpha_3^1}{\alpha_3^1+\alpha_3^2}\left(2+\alpha_1^1+\alpha_2^1\right)+\frac{\alpha_3^2}{\alpha_3^1+\alpha_3^2}\left(1+\alpha_1^1+\alpha_2^1\right)\le\beta_3$$

$$\frac{\alpha_3^1}{\alpha_3^1+\alpha_3^2}\left(1+\alpha_1^1+\alpha_2^2\right)+\frac{\alpha_3^2}{\alpha_3^1+\alpha_3^2}\left(2+\alpha_1^1+\alpha_2^2\right)\le\beta_3$$
$$\frac{\alpha_3^1}{\alpha_3^1+\alpha_3^2}\left(1+\alpha_1^2+\alpha_2^1\right)+\frac{\alpha_3^2}{\alpha_3^1+\alpha_3^2}\left(2+\alpha_1^2+\alpha_2^1\right)\le\beta_3$$
$$\frac{\alpha_3^1}{\alpha_3^1+\alpha_3^2}\left(2+\alpha_1^2+\alpha_2^2\right)+\frac{\alpha_3^2}{\alpha_3^1+\alpha_3^2}\left(1+\alpha_1^2+\alpha_2^2\right)\le\beta_3$$
$$\alpha_i^j\ge 0,\forall i\in I, j=1,\cdots,m_i$$
$$0.2\le\sum_{j=1}^{m_i}\alpha_i^j\le 1,\forall i\in I.$$

One solution to (P5) with $h(\alpha^*,\beta^*)=0$ and hence an MBE is $\alpha_1^{1^*}=\alpha_1^{2^*}=\alpha_2^{1^*}=\alpha_2^{2^*}=\alpha_3^{1^*}=\alpha_3^{2^*}=0.125$, $\beta_1^*=\beta_2^*$, $\beta_3^*=1.75$.

CONCLUSION

In this paper, we gave an interpretation for mixed strategies to n-person games in normal form via the notion of resource allocation. In these games, a mixed strategy is an allocation strategy, as distinguished from other interpretations of a mixed strategy. Each player chooses a pure strategy with a probability that equals to the fraction of the maximum available resource allocated to that pure strategy over the total fraction of the resource the player allocates to all his pure strategies. We proved the existence of an NE in these games. Furthermore, we showed that an MBE may not exist in a resource allocation game unless there exists a strategy yielding zero for the associated nonlinear program.

ACKNOWLEDGEMENTS

This research is funded by the US Department of Education GAANN Fellowship, reward No. P200A130164.

REFERENCES

1. Maschler, M., Solan, E. and Zamir, S. (2013) Game Theory (Translated from the Hebrew by Ziv Hellman and edited by Mike Borns).
2. Chakravarty, S.R., Mitra, M. and Sarkar, P. (2015) A Course on Cooperative Game Theory. Cambridge University Press, Cambridge.
3. Nash, J. (1950) Equilibrium Points in n-Person Games. Proceedings of the National Academy of Sciences of the United States of America, 36, 48-49. https://doi.org/10.1073/pnas.36.1.48
4. Nash, J. (1951) Non-Cooperative Games. Annals of Mathematics, 54, 286-295. https://doi.org/10.2307/1969529
5. Border, K.C. (1989) Fixed Point Theorems with Applications to Economics and Game Theory. Cambridge University Press, Cambridge.
6. Berge, C. (1957) TheorieGenerale des jeuxanPersonnes. Gauthier-Villars.
7. Zhukovskiy, V. (1985) Some Problems of Nonantagonistic Differential Games. In: Kenderov, P., Ed., MatematiceskieMetody v IssledovaniiOperacij, 103-195.
8. Corley, H.W. (2015) A Mixed Cooperative Dual to the Nash Equilibrium. Game Theory, 2015, Article ID: 647246.
9. Batbileg, S. and Enkhbat, R. (2010) Global Optimization Approach to Game Theory. Mongolian Mathematical Society, 14, 2-11.
10. Batbileg, S. and Enkhbat, R. (2011) Global Optimization Approach to Nonzero Sum N Person Game. International Journal: Advanced Modeling and Optimization, 13, 59-66.
11. Nahhas, A. and Corley, H.W. (2017) A Nonlinear Programming Approach to Determine a Generalized Equilibrium for N-Person Normal Form Games. International Game Theory Review, 19, 1750011.
12. Samuelson, L. (2016) Game Theory in Economics and Beyond. Journal of Economic Perspectives, 30, 107-130. https://doi.org/10.1257/jep.30.4.107
13. Li, S.W. (2017) Obviously Strategy-Proof Mechanisms. American Economic Review, 107, 3257-3287. https://doi.org/10.1257/aer.20160425
14. Siniscalchi, M. (2016) Epistemic Game Theory: Beliefs and Types. The New Pal-grave Dictionary of Economics, 19, 1-7.

15. Perea, A. (2014) From Classical to Epistemic Game Theory. International Game Theory Review, 16, 1440001. https://doi.org/10.1142/S0219198914400015
16. Corley, H.W. (2017) Normative Utility Models for Pareto Scalar Equilibria in n-Person, Semi-Cooperative Games in Strategic Form. Theoretical Economics Letters, 7, 1667-1686. https://doi.org/10.4236/tel.2017.76113
17. Von Neumann, J. and Morgenstern, O. (1944) Theory of Games and Economic Behavior. Princeton University Press, Princeton.
18. Nash, J. (1953) Two-Person Cooperative Games. Econometrica: Journal of the Econometric Society, 21, 128-140. https://doi.org/10.2307/1906951
19. Aumann, R.J. (1987) What Is Game Theory Trying To Accomplish? In: Arrow, K. and Honkapohja, S., Eds., Frontiers of Economics, Basil Blackwell, Oxford, 5-46.
20. Rubinstein, A. (1991) Comments on the Interpretation of Game Theory. Econometrica: Journal of the Econometric Society, 59, 909-924. https://doi.org/10.2307/2938166
21. Harsanyi, J.C. (1973) Games with Randomly Disturbed Payoffs: A New Rationale for Mixed-Strategy Equilibrium Points. International Journal of Game Theory, 2, 1-23. https://doi.org/10.1007/BF01737554
22. Aumann, R. and Brandenburger, A. (1995) Epistemic Conditions for Nash Equilibrium. Econometrica: Journal of the Econometric Society, 63, 1161-1180. https://doi.org/10.2307/2171725
23. Joosten, R., Brenner, T. and Witt, U. (2003) Games with Frequency-Dependent stage Payoffs. International Journal of Game Theory, 31, 609-620. https://doi.org/10.1007/s001820300143

CHAPTER 18

On a Dynamic Optimization Technique for Resource Allocation Problems in a Production Company

Chuma R. Nwozo[1] and Charles I. Nkeki[2]

[1]Department of Mathematics, University of Ibadan, Ibadan, Nigeria

[2]Department of Mathematics, University of Benin, Benin City, Nigeria

ABSTRACT

This paper examines the allocation of resource to different tasks in a production company. The company produces the same kinds of goods and want to allocate m number of tasks to 50 number of machines. These machines are subject to breakdown. It is expected that the breakdown machines will be repaired and put into operation. From past records, the company estimated the profit the machines will generate from the various tasks at the first stage of the operation. Also, the company estimated the probability of breakdown of the machines for performing each of the tasks.

Citation: C. Nwozo and C. Nkeki, On a Dynamic Optimization Technique for Resource Allocation Problems in a Production Company, American Journal of Operations Research, Vol. 2 - No. 3, 2012, pp. 357-363. doi: 10.4236/ajor.2012.23043

The aim of this paper is to determine the expected maximize profit that will accrue to the company over T horizon. The profit that will accrued to the company was obtained as ₦4,571,100,000 **after 48 weeks of operation. At the** infinty horizon, the profit was obtained to be ₦20,491,000,000. It was found that adequate planning, prompt and effective maintainance can enhance the profitability of the company.

Keywords: Dynamic Optimization; Resource Allocation; Company; Machines; Tasks

INTRODUCTION

We consider the allocation of tasks to different machines in a production company. A certain number of machines is proposed to be purchased at the beginning of a planning horizon. From statistics, the company has an estimate of the profit each tasks is to yield at the first stage of operation. Also, the company estimates the probability of breakdown of the machines allocated to each tasks. When a machine breaks down, it goes in for repairs after which it returns to the factory for re-allocation at the beginning of the next period.

In this paper, we formulate the problem as a dynamic optimization (DO). Our approach builds on previous research. [1] used the stochastic programming technique of dynamic Programming in financial asset allocation problems for designing low-risk portfolios. [2] proposed the idea of using a parsimonious sufficient static in an application of approximate dynamic programming to inventtory management. [3] described an algorithm for computing parameter values to fit linear and separable concave approximations to the value function for large-scale problems in transportation and logistics. [4] described a more complicated variation of the algorithm that implores execution time and memory requirements. The improvement is critical for practical applications to realistic large-scale problems. [5] used DO for large-scale asset management problems for both single and multiple assets. [6] extended an approximate DO method to optimize the distribution operations of a company manufacturing certain products at multiple production plants and shipping to different customer locations for sales. [7] considered the allocation of buses from a single station to different routes in a transportation company in Nigeria.

In this section, we consider the methodology adopted in this paper. We start with the problem formulation.

PROBLEM FORMULATION

In this section, we consider the methodology adopted in this paper. We start with the problem formulation. Given a certain number of tasks that are to be allocated to different machines at the beginning of each time period, we expect some machine(s) breakdown at the end of each period. Due to the uncertainty in the number of breakdown machine(s), we assume that the states of the machines are random. The company must know the number of machines available for the next period before decision will be made on how to allocate the tasks to the remaining machines. The number of machines to be put into operation in the next period depends on the number of breakdown at the end of the previous period. Our aim is to maximize the total expected profit over a time horizon. We define the following notations which presented in Table 1.

In the next subsection, we define the one-period expected profit function and formulate the problem as a dynamic program.

THE OBJECTIVE AND ONE-PERIOD EXPECTED PROFIT FUNCTION

If the profit for allocating the k task to the machines at period t is Ψ_t^k, the state of the machines is s_t, number of machines allocated to operate on task k at period t under policy π, is $x_t^{\pi k}$ and the number of break down machines is b_t, then the profit that will accrue to the company over T-horizon is given by

$$\sum_{t=0}^{T}\sum_{k=1}^{m}\Psi_t^k\left(x_t^{\pi k}\left(s_{t-1}(b)\right)\right).$$

The expected maximum profit that will accrue to the company under policy π is given by

$$\mathrm{Y}_t^{\pi}(s_t)=E\left[\max_{x_t\in X(s_t)}\sum_{t=0}^{T}\sum_{k=1}^{m}\beta^t\Psi_t^k\left(x_t^{\pi k}\left(s_{t-1}(b)\right)\right)\right] \qquad (1)$$

subject to

$$\sum_{K=1}^{m}x_t^{\pi k}\left(s_{t-1}(b)\right)\le s_t,\quad t=0,1,\cdots,T \qquad (2)$$

$$x_t^{\pi k}\ge 0, t=1,\cdots,T;\ k=1,\cdots,m$$

Note: $X(s_t)$ is the set of possible solution of problem (1). Conditioning (1) on $s_t \in S$. We now have the following optimization problem (3)

$$\Upsilon_t^{\pi}(s_t) = E\left[\max_{x^{\pi} \in X(s)} \left\{\sum_{t'=t}^{T}\sum_{k=1}^{m} \beta^{t'} \Psi_{t'}^{k}\left(x_{t'}^{\pi^k}\left(s_{t'-1}(b)\right)\right)\right\} \Big/ s_t \in S_t\right] \tag{3}$$

Problem (1.3) maximizes the expected profit over $X(s_t)$ subject to

$$\sum_{k=1}^{m} x_t^{\pi^k}\left(s_{t-1}(b)\right) \le s_t, t = 0,1,\cdots,T$$

For the profit function $\Psi : S \to \mathbb{R}$, if we accumulate the profit of the first T-stage and add to it the terminal profit

$\Psi_T(s_T) = \sum_{k=1}^{m} \Psi_T^k(s_T)$ then (3) becomes

$$\Upsilon_t^{\pi}(s_t) = E\left[\max_{x^{\pi} \in X(s)} \left\{\sum_{t'=t}^{T-1}\sum_{k=1}^{m} \beta^{t'} \Psi_{t'}^{k}\left(x_{t'}^{\pi^k}\left(s_{t'-1}(b)\right)\right)\right\} + \beta^T \Psi_T(s_T) / s_t \in S\right] \tag{4}$$

subject to $\sum_{k=1}^{m} x_{t'}^{\pi^k}\left(s_{t'-1}(b)\right) \le s_t, t = 0,1,\cdots,T$.

$x_t^{\pi^k} \ge 0, t = 1,\cdots,T; k = 1,\cdots,m$.

DYNAMIC PROGRAMMING FORMULATION AND OPTIMALITY

Using s_t as the state variable at period t and S as the state space, we can formulate the problem as a dynamic program. The number of breakdown machines for task k at period t is given by $P_k x_t^k$, where P_k is the probability of break down machines for task k. Hence, total number of break down machines for all the tasks is given by

$$\sum_{k=1}^{m} P_k x_t^k, t = 1,2,\cdots,T$$

We therefore have that

Table 1. Notations and their Definitions.

Notation	Definition
β	discount factor, $0 < \beta < 1$.
S	the state space *i.e.* the set of all machines
T	set of time periods in the planning horizon.
π	is a rule which chooses an action based on current state of the system (policy).
$x_t^{\pi^k}$	number of machines allocated to task k at period t under the policy, π, $t \in [0, T]$.
s_t	number of functional machines at period t: $s_t \in S$, $t \in [0, T]$
b_t	number of breakdown machines at period t, $t \in [0, T]$
$x_t^k(s_{t-1}(b))$	readily available machines to be allocated in the next period.
Ψ_t^k	expected return from task k at period t., $t \in [0, T]$
Π	set of all admissible policy: $\pi \in \Pi$
s_0	number of machines at the beginning of the planning horizon.
$Y_t(s_t)$	objective function of our system, $t \in [0, T]$

$$b_t = \sum_{k=1}^{m} P_k x_t^k$$

Let s_t be the number of machines to be allocated in period t and let α be the percentage of break down (but repaired) machines that are expected to join the functional ones in period t, then the transformation equation for the system is given by

$$s_t = s_{t-1} - (1-\alpha)\sum_{k=1}^{m} P_k x_t^k, t = 1, \cdots, T \tag{5}$$

Observe that the transformation equation is a random variable.

We now set $1 - \alpha = \beta$ in (5) to have,

$$s_t = s_{t-1} - \beta\sum_{k=1}^{m} P_k x_t^k, t = 1, \cdots, T \tag{6}$$

In this case, the optimal policy can be found by computing the value functions through the optimization problem

$$F_t^{\pi}(s_t) = \max_{x^{\pi} \in X(s_t)} \sum_{k=1}^{m} \Psi_t^k \left(x_t^{\pi^k}\left(s_{t-1}(b)\right)\right) + E\{F_t(s_t)\}/S_t = s_t \tag{7}$$

subject to

$$\sum_{k=1}^{m} x_t^{\pi^k}\left(s_{t-1}(b)\right) \le s_t,\ t = 1,2,\cdots,T.$$

Equivalently,

$$\xi_t + \sum_{k=1}^{m} x_t^{\pi^k}\left(s_{t-1}(b)\right) = s_t, t = 1,\cdots,T; \xi_t \ge 0 \tag{8}$$

$$x_t^{\pi^k} \ge 0, t = 1,\cdots,T; k = 1,\cdots,m$$

Since all the available functional machines must be allocated in the next period, we have that $\xi_t = 0$, for all $t \in T$, which is the slack variable.

We show that (4) is equivalent to (7), and then use (4) and (7) interchangeably. The theorem below establish this claim.

Lemma 1.1: Let s_t be a state variable that captures the relevant history up to time t, and let $\Upsilon_t(s_{t+1})$ be some function measured at $t' \ge t+1$ conditionalon the random variable s_t. Then,

$$E\left[E\left\{\Upsilon_{t'} \middle| s_{t+1}\right\} \middle| s_t\right] = E\left[\Upsilon_{t'} \middle| s_t\right].$$

For the proof, see [3].

Theorem 1.1: Suppose that $\Upsilon_t^\pi(s_t)$ satisfies (4) and $F_t^\pi(s_t)$ satisfies (7), then $\Upsilon_t^\pi(s_t) = F_t^\pi(s_t)$.

Proof: We are to show that $\Upsilon_t^\pi(s_t) = F_t^\pi(s_t)$. We first use a standard method of DP. Obviously

$$\Upsilon_T^\pi(s_T) = F_T^\pi(s_T) = \Psi_T(s_T).$$

Suppose that it hold for t + 1, t + 2, ···, T, then we show that it is true for t.

We now write

$$F_t^\pi(s_t) = \max_{x^\pi \in X(s_t)} \sum_{k=1}^{m} \Psi_t^k\left(x_t^{\pi^k}\left(s_{t-1}(b)\right)\right)$$
$$+ E\left[E \max_{x_t} \sum_{t'=t+1}^{T-1} \sum_{k=1}^{m} \Psi_{t'}^k\left(x_{t'}^{\pi^k}\left(s_{t'}(b)\right)\right)\right.$$
$$\left. + \beta^T \Psi_T(s_T) \middle| S_t = s_t\right]$$

Applying Lemma 1.1, we have

$$F_t^{\pi}(s_t) = \max_{x^{\pi} \in X(s_t)} \sum_{k=1}^{m} \Psi_t^k \left(x_t^{\pi k} \left(s_{t-1}(b) \right) \right)$$
$$+ E\left[E \max_{x_t} \sum_{t'=t+1}^{T-1} \sum_{k=1}^{m} \Psi_{t'}^k \left(x_{t'}^{\pi k} \left(s_{t'}(b) \right) \right) \right.$$
$$\left. + \beta^T \Psi_T(s_T) \middle| S_t = s_t \right]$$

When we condition on $x_t^{\pi k}(s_t(b))$, we obtain

$$F_t^{\pi}(s_t) = E\left[\max_{x^{\pi} \in X(s_t)} \sum_{t'=t}^{T-1} \sum_{k=1}^{m} \Psi_t^k \left(x_t^{\pi k} \left(s_{t-1}(b) \right) \right) + \Psi_T(s_T) \middle| s_t \right]$$
$$= \Upsilon_t^{\pi}(s_t).$$

For any given objective function, we desire to find the best possible policy, π, that optimizes it, that is, we search for

$$\Upsilon_t^{*}(s_t) = \max_{\pi \in \Pi} \Upsilon_t^{\pi}(s_t).$$

This is obtained by solving the optimality equation

$$F_t(s_t) = \max_{x^{\pi k} \in X(s_t)} \left[\sum_{k=1}^{m} \Psi_t^k \left(x_t^{\pi k} \left(s_{t+1}(b) \right) \right) \right.$$
$$\left. + E\left\{ F_{t+1}(s_t) \right\} \right]$$
$$= \max_{x^{\pi k} \in X(s_t)} \left[\sum_{k=1}^{m} \Psi_t^k \left(x_t^{\pi k} \left(s_{t-1}(b) \right) \right) \right.$$
$$\left. + F_{t+1}\left(s_t - \beta \sum_{k=1}^{m} P_k x_{t+1}^k \right) \right] \tag{9}$$

If we find the set of F's that solves (9), then we have found the policy that optimizes $\Upsilon_t^{\pi}(s_t)$. The result below establishes this claim.

Theorem 1.2: The expression $F_t^{\pi}(s_t)$ is a solution to equation (1.9) if and only if

$$\Upsilon_t^{*}(s_t) = F_t(s_t) = \max_{\pi \in \Pi} Y_t^{\pi}(s_t).$$

The if part: This is shows by induction that $F_t(s_t) \geq \Upsilon_t^{*}(s_t)$, for all $s_t \in S$ and $t = 0, 1, \cdots, T-1$.

Since $F_T(s_T) = \Psi_T(s_T) = \Upsilon_T^{\pi}(s_T)$ for all s_T and $\pi \in \Pi$, we have that $F_T(s_T) = \Upsilon_T^{*}(s_T)$.

Suppose that $F_t(s_t) \geq Y_t^*(s_t) \geq Y_t^\pi(s_t)$, for t = n + 1, n + 2, ⋯, T, and let $\pi \in \Pi$ be an arbitrary policy. For t = n, we obtain the optimality equation as follows

$$F_n(s_n) = \max_{x^{\pi^k} \in X(s_n)} \left[\sum_{k=1}^{m} \Psi_n^k \left(x_n^{\pi^k} \left(s_{n-1}(b) \right) \right) + F_{n+1}\left(s_n - \beta \sum_{k=1}^{m} P_k x_{n+1}^k \right) \right].$$

By induction hypothesis, $F_{n+1}(s_{n+1}) \geq Y_{n+1}^*(s_{n+1})$. So we have

$$F_n(s_n) = \max_{x^{\pi^k} \in X(s_n)} \left[\sum_{k=1}^{m} \Psi_n^k \left(x_n^{\pi^k} \left(s_{n-1}(b) \right) \right) + Y_{n+1}^\pi \left(s_n - \beta \sum_{k=1}^{m} P_k x_{n+1}^k \right) \right], s' = s_{t+1}.$$

Also, we have that $Y_{n+1}(s_{n+1}) \leq Y_{n+1}^*(s_{n+1})$ for an arbitrary policy, $\pi \in \Pi$, hence

$$\begin{aligned} F_n(s_n) &\geq \max_{x^{\pi^k} \in X(s_n)} \left[\sum_{k=1}^{m} \Psi_n^k \left(x_n^{\pi^k} \left(s_{n-1}(b) \right) \right) + Y_{n+1}^\pi \left(s_n - \beta \sum_{k=1}^{m} P_k x_{n+1}^k \right) \right], s' = s_{t+1} \\ &\geq Y_n^\pi(s_n), \text{for all } \pi \in \Pi \\ &= \max_{\pi \in \Pi} Y_n^\pi(s_n). \end{aligned}$$

Only if part: We show that for any $\in > 0$, there exists $\pi \in \Pi$ that satisfies the following:

$$Y_n^\pi(s_n) + (T - n) \in \geq F_n(s_n) \tag{10}$$

We now define $F_n(s_n)$ as follows:

$$F_n(s_n) = \max_{x^{\pi^k} \in X(s_n)} \left[\sum_{k=1}^{m} \Psi_n^k \left(x_n^{\pi^k} \left(s_{n-1}(b) \right) \right) + F_{n+1}\left(s_n - \beta \sum_{k=1}^{m} P_k x_{n+1}^k \right) \right] \tag{11}$$

Let $x_n(s_n)$ be the decision rule that solves (1) and satisfies the following:

$$F_n(s_n) = \sum_{k=1}^{m} \Psi_n^k\left(x_n^{\pi^k}\left(s_{n-1}(b)\right)\right) + F_{n+1}\left(s_n - \beta\sum_{k=1}^{m} P_k x_{n+1}^k\right) + \in \tag{12}$$

We now prove (10) by induction. Assume that it is true for t = n + 1, n + 2, ⋯, T. But,

$$F_n(s_n) \geq \max_{x^{\pi^k} \in X(s_n)} \left[\sum_{k=1}^{m} \Psi_n^k\left(x_n^{\pi^k}\left(s_{n-1}(b)\right)\right) + \mathrm{Y}_{n+1}^{\pi}\left(s_n - \beta\sum_{k=1}^{m} P_k x_{n+1}^k\right)\right], s' = s_{t+1}$$

We now use the induction hypothesis which says

$\mathrm{Y}_n^{\pi}(s_n) \geq F_n(s_n) - (T-n)\in$ so that

$$F_n(s_n) \geq \sum_{k=1}^{m} \Psi_n^k\left(x_n^{\pi^k}\left(s_{n-1}(b)\right)\right) + F_{n+1}\left(s_n - \beta\sum_{k=1}^{m} P_k x_{n+1}^k\right) - (T-n-1)\in$$

$$\geq \sum_{k=1}^{m} \Psi_n^k\left(x_n^{\pi^k}\left(s_{n-1}(b)\right)\right) + F_{n+1}\left(s_n - \beta\sum_{k=1}^{m} P_k x_{n+1}^k\right) - (T-n-1)\in$$

$$\geq \sum_{k=1}^{m} \Psi_n^k\left(x_n^{\pi^k}\left(s_{n-1}(b)\right)\right) + F_{n+1}\left(s_n - \beta\sum_{k=1}^{m} P_k x_{n+1}^k\right) - (T-n)\in + \in$$

$$\geq \sum_{k=1}^{m} \Psi_n^k\left(x_n^{\pi^k}\left(s_{n-1}(b)\right)\right) + F_{n+1}\left(s_n - \beta\sum_{k=1}^{m} P_k x_{n+1}^k\right) + \in - (T-n)\in$$

$$\geq F_n(s_n) - (T-n)\in$$

Hence,

$$\mathrm{Y}_n^*(s_n) + (T-n)\in \geq \mathrm{Y}_n^{\pi}(s_n) + (T-n)\in \geq F_n(s_n) \geq \mathrm{Y}_n^*(s_n)$$

This result shows that solving the optimality equation also gives the optimal value function.

Theorem 1.3: 1) Let B(s) be the set of all bounded real-valued functions F: S → R. The mapping Γ: B(s) → B(s) is a contraction.

2) The operator Γ has a unique fixed point (given by F^*).

3) For any F, $\Gamma^{\infty} F = F^*$.

4) For any F, if ΓF ≤ F, then $F^* \leq \Gamma^t F$, $t \in \{0,1,\cdots\}$ ∀.

Note: Γ is called dynamic programming operator (See [5], for more detail).

Theorem 1.4: For any bounded return or scoring functions F_1: S → R and F_2: S → R, and for all t = 0, 1, 2, 3, ···, the inequality below holds

$$\max_{s\in S}\left|\left(\Gamma^t F_1\right)(s)-\left(\Gamma^t F_2\right)(s)\right| \leq \beta^t \max_{s\in S}\left|F_1(s)-F_2(s)\right|$$

See [8].

The next result shows that as T → ∞, $F^*(s) \to \Gamma^T F(s)$, ∀ $s \in S$. Thus, the profit per stage must be bounded i.e.

$\left|\sum_{k=1}^{m} \Psi^k\left(x^{\pi^k}(s(b))\right)\right| \leq \mu$ where μ is a positive constant.

We now state this claim formally as follows:

Theorem 1.5: For any bounded return or scoring function F: S → R, $F_T(s_0) \to F(s_0)$, T → ∞ that is,

$$F(s_0) = \lim_{T\to\infty} \Gamma^T(F)(s_0), \quad \forall\ s_0 \in S$$

Proof: Let H be a positive integer, $s_0 \in S$ and policy $\pi = \{\pi_0, \pi_1, \cdots\}$, we can decompose the return

$$F^{\pi}(s_0) = \lim_{T\to\infty} E\left\{\sum_{t=0}^{T-1}\sum_{k=1}^{m} \beta^t \Psi_t^k\left(x_t^{\pi^k}\left(s_{t-1}(b)\right)\right)\right\}$$

into the portion received over the first H stages and over the remaining stages.

$$F^{\pi}(s_0) = \lim_{T\to\infty} E\left\{\sum_{t=0}^{T-1}\sum_{k=1}^{m} \beta^t \Psi_t^k\left(x_t^{\pi^k}\left(s_t(b)\right)\right)\right\},$$

$$= E\left\{\sum_{t=0}^{H-1}\sum_{k=1}^{m} \beta^t \Psi_t^k\left(x_t^{\pi^k}\left(s_t(b)\right)\right)\right\}$$

$$+ \lim_{T\to\infty} E\left\{\sum_{t=H}^{T-1}\sum_{k=1}^{m} \beta^t \Psi_t^k\left(x_t^{\pi^k}\left(s_t(b)\right)\right)\right\}$$

But

$$\left|\lim_{T\to\infty} E\left\{\sum_{t=H}^{T-1}\sum_{k=1}^{m}\beta^t\Psi_t^k\left(x_t^{\pi^k}\left(s_t(b)\right)\right)\right\}\right| \le N\sum_{t=H}^{\infty}\beta^t = \frac{\beta^H N}{1-\beta}$$

Since $\sum_{t=H}^{\infty}\beta^t$

is a geometric progression and $0 < \beta < 1$.

Now

$$F^\pi(s_0) \le E\left[\sum_{t=0}^{H-1}\sum_{k=1}^{m}\beta^t\Psi_t^k\left(x_t^{\pi^k}\left(s_t(b)\right)\right) + \frac{\beta^H N}{1-\beta}\right]$$

using this relations, it follows that

$$F^\pi(s_0) - \frac{\beta^H N}{1-\beta} - \beta^H \max_{s_t\in S}\left|\Psi(s)\right|$$

$$\le E\left[\beta^H\Psi(s_H) + \sum_{t=0}^{H-1}\sum_{k=1}^{m}\beta^t\Psi_t^k\left(x_t^{\pi^k}\left(s_t(b)\right)\right)\right]$$

$$\le F^\pi(s_0) + \frac{\beta^H N}{1-\beta} + \beta^H \max_{s_t\in S}\left|\Psi(s)\right|.$$

By taking the maximum over π, we obtain for all S_0 and H.

$$F^*(s_0) - \frac{\beta^H N}{1-\beta} - \beta^H \max_{s_t\in S}\left|\Psi(s)\right|$$

$$\le \left(\Gamma^H F\right)(s_0) \le F^*(s_0) + \frac{\beta^H N}{1-\beta} + \beta^H \max_{s_t\in S}\left|\Psi(s)\right|,$$

and by taking the unit as $H \to \infty$, we have

$$F^*(s_0) \le \lim_{H\to\infty}\left(\Gamma^H F\right)(s_0) \le F^*(s_0)$$

Hence, $F^*(s_0) = \lim_{H\to\infty}\left(\Gamma^H F\right)(s_0), \forall\, s_0 \in S.$

This result shows that our optimization problem converges to a fixed point F^* in an infinite horizon.

We use value iteration algorithm for finite and infinite stage to solve our problem. The algorithm converges to an optimal policy.

Step 1: Initialization Set $F_0(s_0) = 0 \ \forall\, s_0 \in S.$

Set n = 0 Set $0 < \beta < 1$ Fix a tolerance parameter, $\varepsilon > 0$.

Step 2: For each $s_n \in S$, calculate

$$F_{n+1}(s_n) = \max_{x_n \in X} \sum_{k=1}^{m} \Psi_n^k \left(x_n^k \left(s_n(b)\right)\right) + F_n\left(s_{n-1} - \beta \sum_{k=1}^{m} P_k x_n^k\right) \tag{13}$$

Let x^{n+1} be the decision vector that solve (13).

Step 3: For β = 1:

If $\|F_{n+1} - F_n\| \le \varepsilon$, set $x^{\varepsilon} = x^{n+1}, F^{\pi^{\epsilon}} = F_{n+1}$ and stop; else set n = n + 1and return to step 2.

Step 4: For 0 < β < 1;

If $\|F_{n+1} - F_n\| \le \frac{1-\beta}{2\beta}$, set $x^{\varepsilon} = x^{n+1}$, $F^{\pi^{\epsilon}} = F_{n+1}$

and result stop; else set n = n + 1 and return to step 2.

The theorem below guarantees the convergent of the algorithm.

Theorem 1.6: If the algorithm with stopping parameter ε, terminates at iteration n with value function F_{n+1}, then

$$\|F_{n+1} - F^*\| \le \varepsilon/2 \tag{14}$$

In addition, if x^{ε} is the optimal decision rule and F^{π^k} is the value of this policy, then

$$\|F^{\pi^{\epsilon}} - F^*\| \le \varepsilon \tag{15}$$

This theorem implies that F^{π^k} is the fixed point of the equation $F = \Gamma(x^{\varepsilon}) F$. Since x^{ε} is the decision that solves ΓF_{n+1}, it implies that $\Gamma(x^{\varepsilon})F_{n+1} = \Gamma F_{n+1}$. Since $\|F^* - F_{n+1}\| = \|\Gamma F^* - \Gamma F_{n+1} + \Gamma F_{n+1} - F_{n+1}\|$ $F_{n+1} = \Gamma F_n$ and Γ is contraction, we have that

$$\|F^{\pi^{\epsilon}} - F_{n+1}\| \le \frac{\beta}{1-\beta}\|F_{n+1} - F_n\|$$

and

$$\|F_{n+1} - F^*\| \le \frac{\beta}{1-\beta}\|F_{n+1} - F_n\| \tag{16}$$

But the value iteration algorithm stops when

$$\left\|F_{n+1} - F_n\right\| \leq \varepsilon \frac{(1-\beta)}{2\beta} \tag{17}$$

From (16) and (17), we have

$$\left\|F_{n+1} - F^*\right\| \leq \frac{\beta}{1-\beta} \times \frac{\varepsilon(1-\beta)}{2\beta} = \varepsilon/2 \tag{18}$$

Similarly, $\left\|F_{n+1} - F^{\pi_\varepsilon}\right\| \leq \frac{\varepsilon}{2}$

Therefore,

$$\left\|F^* - F^{\pi^\epsilon}\right\| \leq \left\|F_{n+1} - F^{\pi\varepsilon}\right\| + \left\|F^* - F_{n+1}\right\| < \frac{\epsilon}{2} + \frac{\epsilon}{2} = \epsilon .$$

COMPUTATIONAL RESULT

A production company in Nigeria proposed to purchase 50 machines that can perform nine different tasks. These machines are subject to breakdown. The Table 2 gives information of their decisions.

The company further estimated that out of the number of breakdown machines per week, 95% will join the functional ones for the next period.

The aim of the company is to maximize profit over T horizon Let s_t represents the number of machines to be allocated in the next period, so that $s_t = x_t^k(s_{t-1}(b))$. Since s_0 is the number of machines at the beginning of the planning horizon,

$$\sum_{k=1}^{m} P_k x_t^k$$

is the total number of breakdown machines, and

$$\frac{19}{20}\sum_{k=1}^{m} P_k x_t^k$$

is expected to join the functional machines in the next period of operation, we have that

$s_t = s_{t-1} - \frac{1}{20}\sum_{k=1}^{m} P_k x_t^k, t = T, T-1, \cdots, 1$ which is our transformation equation and is a random variable.

We can now express our one-period expected return function as follows:

$$\Upsilon_t^{\pi}(s_t) = E\left[\max_{x_t^k \in X} \sum_{t=1}^{T} \sum_{k=1}^{m} \beta^t \Psi_t^k \left(x_t^k \left(s_{t-1}(b)\right)\right)\right] \tag{19}$$

subject to

$$\sum_{k=1}^{m} x_t^k \left(s_{t-1}(b)\right) = s_{t-1}, t = T, T-1, \cdots, 1$$

.

(The feasible region for stage t).

Since the company cannot allocate negative resources to any one task, we write $x_t^k \left(s_{t-1}(b_t)\right) \geq 0$, t = T, T – 1, ···, 1; k = 1, ···, m.

Of course, (19) is the same as

$$F_T(s_{t-1}) = \max_{x_t^k \in X} \left\{ \sum_{k=1}^{m} \Psi_t^k x_t^{\pi^k} \left(s_{t-1}(b_t)\right) + EF_{T-1}(s_t) \right\},$$

$$t = T, T-1, \cdots, 1,$$

$$= \max_{x_t^k \in X} \left[\sum_{k=1}^{m} \Psi_t^k \left(x_t^{\pi^k} \left(s_{t-1}(b_t)\right)\right) + EF_{T-1}\left(s_{t-1} - \frac{1}{20}\sum_{k=1}^{m} P_k x_t^k\right)\right],$$

t = T, T – 1, ···, 1, and $x_t^k \geq 0$, t = T, T – 1, ···, 1; k = 1, ···,m.

Note: Our problem has 9 tasks and 48 periods.

Hence, m = 9 and T = 48. Set $\varepsilon = \frac{1}{9}$.

Therefore,

$$\begin{aligned} &F_T(s_{t-1}) \\ &= \max_{x^k \in X} \{180000x_t^1 + 150000x_t^2 + 192000x_t^3 + 240000x_t^4 \\ &\quad + 228000x_t^5 + 120000x_t^6 + 168000x_t^7 + 222000x_t^8 \\ &\quad + 300000x_t^9 \\ &\quad + F_{T-1}\left(s_{t-1} - \frac{1}{20}\left[\frac{2}{11}x_t^1 + \frac{1}{10}x_t^2 + \frac{2}{13}x_t^3 + \frac{5}{21}x_t^4 \right.\right. \\ &\qquad + \frac{3}{13}x_t^5 + \frac{1}{12}x_t^6 + \frac{1}{9}x_t^7 \\ &\qquad \left.\left. + \frac{4}{15}x_t^8 + \frac{5}{18}x_t^9\right]\right)\} \\ &T = 1, 2, \cdots, 48; t = 48, 47, \cdots, 1. \end{aligned} \tag{20}$$

subject to:

$$\sum_{k=1}^{9} x_t^k = S_{t-1}, t = 48, 47, \cdots, 1,$$

$x_t^k \geq 0$, t = 48, 47, ⋯, 1; k = 1, ⋯, 9 which is a parametric linear programming problem with 9 variables.

A program using MatLab was used for (20). At the end, the following results were obtained.

The profit over 48 weeks is given by

Table 2. The Expected Initial Profit and Probability of Breakdown Machines.

Machines	Task x^1	Task x^2	Task x^3	Task x^4	Task x^5	Task x^6	Task x^7	Task x^8	Task x^9
Initial Expected Profit (in Naira)	180,000	150,000	192,000	240,000	228,000	120,000	168,000	222,000	300,000
Probability of Breakdown (P_k)	2/11	1/10	2/13	5/21	3/19	1/12	1/9	4/15	5/18

$$\begin{aligned} F_{48}(s_0) &= ₦91422000 s_o \\ &= ₦91422000 \times 50 \\ &= ₦4{,}571{,}100{,}000. \end{aligned}$$

By theorem 1.5, as T approach infinity, $F_T(s_0)$ approaches $F(s_0)$,

$$\begin{aligned} F(s_0) &= \lim_{T\to\infty}\left(\Gamma^T F\right)(s_0) \\ &= \lim_{T\to\infty} F_T(s_0) = ₦409820000 s_0 \\ &= ₦20{,}491{,}000{,}000. \end{aligned}$$

and the optimal policy $x_t^9(s_0)$, (t = 1, 2, ⋯), approaches the limit zero.

Discussion: Figure 1 shows the expected profit that will accrued to the company over a period of 48 weeks. We found that at 48 weeks, the maximum profit that accrued to the company to be ₦4,571,100,000. Figure 2 shows the expected profit that will accrued to the

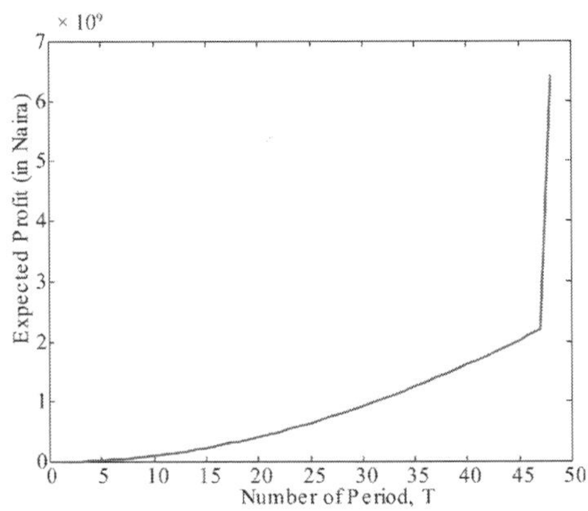

Figure 1. The Expected Profit that will accrued to the Company over a Period of 48 weeks.

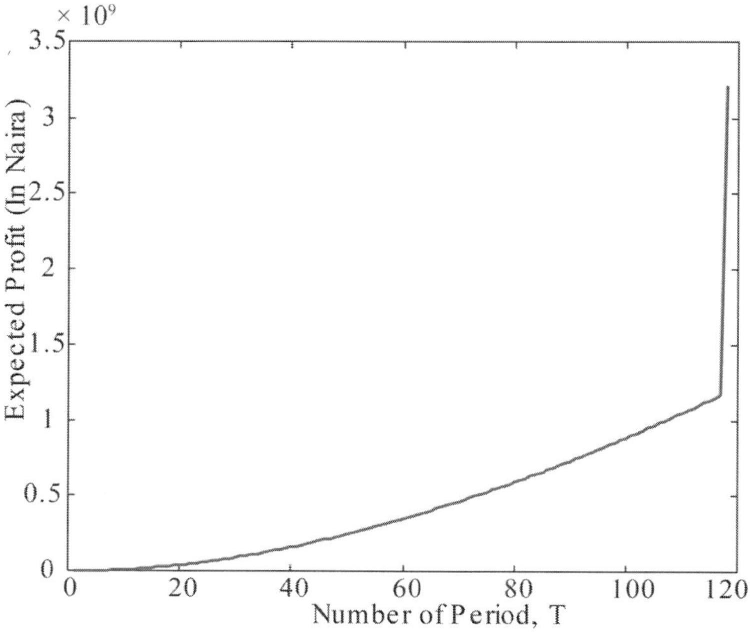

Figure 2. The Expected Profit that will accrued to the Company over an Infinite Period.

company over an infinite weeks. It was found that at infinity, the maximum profit that will accrued to the company to be ₦20,491,000,000.

CONCLUSION

Chuma R. Nwozo, Charles I. Nkeki Many production companies have for long been allocating resources to different tasks without putting into consideration certain factors that may hinder the realization of their objectives. This paper dealt with allocation of machines to tasks in order to maximize profit over finite and infinite horizon. Careful analysis of the situation reveals that adequate planning, prompt and effective maintainance can enhances the profitability of the company.

REFERENCES

1. J. M. Mulvey and H. Vladimirou, "Stochastic Network Programming for Financial Planning Problems," Manage00,000. ment Science, Vol. 38, No. 11, 1992, pp. 1642-1664. doi:10.1287/mnsc.38.11.1642
2. B. Van Roy, D. P. Bertsekas, Y. Lee and J. N. Tsitsiklis, "A Neuro-Dynamic Programming Approach to Retailer Inventory M the 36th IEEE Conference anagement," Proceedings of on Decision and Control, San Diego, 10-12 , pp. 4052-4057. C.1997.652501 December 1997 doi:10.1109/CD gramming, Elsevier, Amsterdam, 2003, pp. 555-635.
3. 5. W. B. Powell, "Approximate Dynamic Programming for Asset Manag versity, Princeton, 2004. 6. rogramming Algorithm for
4. W. B. Powell, "A Comparative Review of Alternative Algorithms for the Dynamic Vehicle Allocation Problem," In: B. Golden and A. Assad, Eds., Vehicle Routing: Methods and Studies, North Holland, Amsterdam, 1988, pp. 249-292.
5. W. B. Powell and H. Topaloglu, "Stochastic Programming in Transportation and Logistics," In: A. Ruszczynski and A. Shapiro, Eds., Handbook in Operation Research and Management Science, Volume on Stochastic Pro ement," Princeton Uni
6. 6. H. Topaloglu and S. Kunnumkal, "Approximate Dynamic Programming Methods for an Inventory Allocation Problem under Uncertainty," Cornell University, Ithaca, 200
7. 7. C. I. Nkeki, "On A Dynamic P Resource Allocation Problems," Unpublished M.Sc. Thesis, University of Ibadan, Ibadan, 2006.
8. 8. D. P. Bertsekas, "Dynamic Programming and Optimal Control," 2nd Edition, Vol. 2, Athena Scientific, Belmont, 2001.

Index

U

V

W